**Evolutionary Game Theory, Natural Selection, and Darwinian Dynamics**

All of life is a game and evolution by natural selection is no exception. Games have players, strategies, payoffs, and rules. In the game of life, organisms are the players, their heritable traits provide strategies, their births and deaths are the payoffs, and the environment sets the rules. The evolutionary game theory developed in this book provides the tools necessary for understanding many of Nature's mysteries. These include coevolution, speciation, and extinction as well as the major biological questions regarding fit of form and function, diversity of life, procession of life, and the distribution and abundance of life. Mathematics for the evolutionary game are developed based on Darwin's postulates leading to the concept of a fitness generating function ($G$-function). The $G$-function is a tool that simplifies notation and plays an important role in the development of the Darwinian dynamics that drive natural selection. Natural selection may result in special outcomes such as the evolutionarily stable strategy or ESS. An ESS maximum principle is formulated and its graphical representation as an adaptive landscape illuminates concepts such as adaptation, Fisher's Fundamental Theorem of Natural Selection, and the nature of life's evolutionary game.

THOMAS L. VINCENT is Professor Emeritus of Aerospace and Mechanical Engineering at the University of Arizona. His main research interests are in the areas of nonlinear control system design, optimal control and game theory, and evolution and adaptation of biological systems. He has 153 publications including 79 journal articles and 8 books.

JOEL S. BROWN is a Professor of Biology at the University of Illinois at Chicago. His main research interests lie in applying concepts from natural selection to behavioral, population, and community ecology with applications to conservation biology. Specific interests include the ecology of fear that studies the ecological and evolutionary implications of the non-lethal effects of predators on prey. He has 102 publications, including 88 journal articles.

# Evolutionary Game Theory, Natural Selection, and Darwinian Dynamics

**THOMAS L. VINCENT**
*Aerospace and Mechanical Engineering*
*University of Arizona*

**JOEL S. BROWN**
*Biological Sciences*
*University of Illinois at Chicago*

# CAMBRIDGE
## UNIVERSITY PRESS

University Printing House, Cambridge CB2 8BS, United Kingdom

Cambridge University Press is part of the University of Cambridge.

It furthers the University's mission by disseminating knowledge in the pursuit of education, learning and research at the highest international levels of excellence.

www.cambridge.org
Information on this title: www.cambridge.org/9780521841702

© T. L. Vincent and J. S. Brown 2005

This publication is in copyright. Subject to statutory exception and to the provisions of relevant collective licensing agreements, no reproduction of any part may take place without the written permission of Cambridge University Press.

First published 2005
Reprinted 2007
First paperback edition 2012

*A catalogue record for this publication is available from the British Library*

ISBN 978-0-521-84170-2 Hardback

Cambridge University Press has no responsibility for the persistence or accuracy of URLs for external or third-party internet websites referred to in this publication, and does not guarantee that any content on such websites is, or will remain, accurate or appropriate.

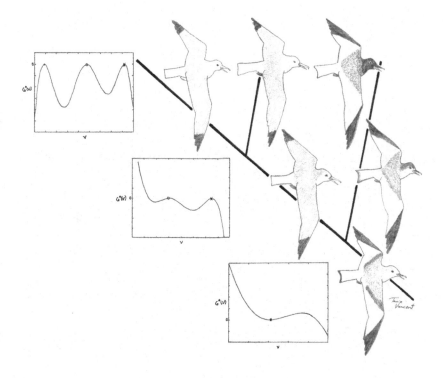

*What limit can be put to this power, acting during long ages and rigidly scrutinising the whole constitution, structure, and habits of each creature, – favouring the good and rejecting the bad? I can see no limit to this power, in slowly and beautifully adapting each form to the most complex relations of life.*
                                    Charles Darwin, *Origin of Species*, 1859

# Contents

|  | List of figures | page | x |
|---|---|---|---|
|  | Preface |  | xv |

| **1** | **Understanding natural selection** | **1** |
|---|---|---|
| 1.1 | Natural selection | 2 |
| 1.2 | Genetical approaches to natural selection | 7 |
| 1.3 | Natural selection as an evolutionary game | 10 |
| 1.4 | Road map | 21 |

| **2** | **Underlying mathematics and philosophy** | **26** |
|---|---|---|
| 2.1 | Scalars, vectors, and matrices | 28 |
| 2.2 | Dynamical systems | 33 |
| 2.3 | Biological population models | 39 |
| 2.4 | Examples of population models | 42 |
| 2.5 | Classical stability concepts | 49 |

| **3** | **The Darwinian game** | **61** |
|---|---|---|
| 3.1 | Classical games | 62 |
| 3.2 | Evolutionary games | 72 |
| 3.3 | Evolution by natural selection | 83 |

| **4** | **$G$-functions for the Darwinian game** | **88** |
|---|---|---|
| 4.1 | How to create a $G$-function | 89 |
| 4.2 | Types of $G$-functions | 91 |
| 4.3 | $G$-functions with scalar strategies | 92 |
| 4.4 | $G$-functions with vector strategies | 93 |
| 4.5 | $G$-functions with resources | 96 |
| 4.6 | Multiple $G$-functions | 99 |
| 4.7 | $G$-functions in terms of population frequency | 103 |

| | | |
|---|---|---|
| 4.8 | Multistage $G$-functions | 106 |
| 4.9 | Non-equilibrium dynamics | 110 |

## 5  Darwinian dynamics — 112

| | | |
|---|---|---|
| 5.1 | Strategy dynamics and the adaptive landscape | 113 |
| 5.2 | The source of new strategies: heritable variation and mutation | 116 |
| 5.3 | Ecological time and evolutionary time | 119 |
| 5.4 | $G$-functions with scalar strategies | 120 |
| 5.5 | $G$-functions with vector strategies | 131 |
| 5.6 | $G$-functions with resources | 140 |
| 5.7 | Multiple $G$-functions | 141 |
| 5.8 | $G$-functions in terms of population frequency | 143 |
| 5.9 | Multistage $G$-functions | 144 |
| 5.10 | Non-equilibrium Darwinian dynamics | 145 |
| 5.11 | Stability conditions for Darwinian dynamics | 147 |
| 5.12 | Variance dynamics | 149 |

## 6  Evolutionarily stable strategies — 151

| | | |
|---|---|---|
| 6.1 | Evolution of evolutionary stability | 153 |
| 6.2 | $G$-functions with scalar strategies | 160 |
| 6.3 | $G$-functions with vector strategies | 168 |
| 6.4 | $G$-functions with resources | 170 |
| 6.5 | Multiple $G$-functions | 174 |
| 6.6 | $G$-functions in terms of population frequency | 180 |
| 6.7 | Multistage $G$-functions | 183 |
| 6.8 | Non-equilibrium Darwinian dynamics | 188 |

## 7  The ESS maximum principle — 197

| | | |
|---|---|---|
| 7.1 | Maximum principle for $G$-functions with scalar strategies | 198 |
| 7.2 | Maximum principle for $G$-functions with vector strategies | 205 |
| 7.3 | Maximum principle for $G$-functions with resources | 211 |
| 7.4 | Maximum principle for multiple $G$-functions | 213 |
| 7.5 | Maximum principle for $G$-functions in terms of population frequency | 219 |
| 7.6 | Maximum principle for multistage $G$-functions | 222 |
| 7.7 | Maximum principle for non-equilibrium dynamics | 225 |

## 8  Speciation and extinction — 231

| | | |
|---|---|---|
| 8.1 | Species concepts | 234 |
| 8.2 | Strategy species concept | 236 |
| 8.3 | Variance dynamics | 243 |
| 8.4 | Mechanisms of speciation | 251 |

| | | |
|---|---|---|
| 8.5 | Predator–prey coevolution and community evolution | 264 |
| 8.6 | Wright's shifting balance theory and frequency-dependent selection | 266 |
| 8.7 | Microevolution and macroevolution | 268 |
| 8.8 | Incumbent replacement | 272 |
| 8.9 | Procession of life | 273 |
| **9** | **Matrix games** | **275** |
| 9.1 | A maximum principle for the matrix game | 277 |
| 9.2 | The 2 × 2 bi-linear game | 284 |
| 9.3 | Non-linear matrix games | 295 |
| **10** | **Evolutionary ecology** | **304** |
| 10.1 | Habitat selection | 304 |
| 10.2 | Consumer-resource games | 309 |
| 10.3 | Plant ecology | 324 |
| 10.4 | Foraging games | 333 |
| **11** | **Managing evolving systems** | **343** |
| 11.1 | Evolutionary response to harvesting | 344 |
| 11.2 | Resource management and conservation | 350 |
| 11.3 | Chemotherapy-driven evolution | 359 |
| | *References* | 364 |
| | *Index* | 377 |

# Figures

| | | |
|---|---|---|
| 2.1 | Dynamics of the logistic map | 36 |
| 2.2 | Alternative logistic map | 37 |
| 2.3 | Continuous logistic model | 38 |
| 2.4 | The carrying capacity and intraspecific competition as distribution functions | 46 |
| 3.1 | The prisoner's dilemma | 67 |
| 3.2 | The game of chicken | 68 |
| 3.3 | The ecological theater and evolutionary play | 75 |
| 5.1 | The species with highest carrying capacity survives | 114 |
| 5.2 | The "star" locates the strategy at the equilibrium value for $x^*$ | 115 |
| 5.3 | Strategy dynamics on the adaptive landscape for the Lotka–Volterra model with $\sigma_k^2 = 4$ | 128 |
| 5.4 | With $\sigma_k^2 = 12.5$ and $n = 1$, strategy dynamics produces an equilibrium point that is a local minimum | 129 |
| 5.5 | With $n = 2$, strategy dynamics allows for speciation | 130 |
| 5.6 | Low-speed strategy dynamics results in an equilibrium solution | 131 |
| 5.7 | High-speed strategy dynamics results in unstable Darwinian dynamics | 132 |
| 5.8 | With $r = 2.5$, strategy dynamics results in an equilibrium solution for $u_1$ and a four cycle solution for density | 146 |
| 5.9 | Increasing $r = 2.8$ results in a chaotic solution for the population density | 146 |
| 6.1 | The fitness set, depicted here by the interior of the top shaped region, represents the fitness in habitats $A$ and $B$ as a result of using every possible stratgy | 155 |
| 6.2 | The rational reaction set is given by the solid line | 158 |
| 6.3 | An ESS coalition of two under chaotic density dynamics | 192 |

## List of figures

| | | |
|---|---|---|
| 6.4 | When the ESS is strongly dependent on **x**, the strategy dynamics will also cycle | 194 |
| 6.5 | At a slower rate of evolution, the strategy dynamics becomes smoother | 194 |
| 6.6 | An ESS under non-equilibrium dynamics | 195 |
| 7.1 | At an ESS, $G^*(v)$ must take on a global maximum when $v = u_1$ | 201 |
| 7.2 | A convergent stable system will return to $\mathbf{x}^*$ when $\mathbf{u} = \mathbf{u}_c$ | 202 |
| 7.3 | The solution obtained does not satisfy the ESS maximum principle | 203 |
| 7.4 | An ESS coalition of two strategies as indicated by the open box and asterix | 204 |
| 7.5 | An ESS coalition of one strategy. Regardless of the number of starting species or their initial strategy values, Darwinian dynamics results in the single-strategy ESS | 207 |
| 7.6 | Decreasing the prey's niche breadth from that of Figure 7.5 changes the outcome. When the system is constrained to have a single species, then, regardless of initial conditions, it evolves to a local maximum. This single-species strategy is not an ESS | 209 |
| 7.7 | Darwinian dynamics results in ESS when the system starts with two or more species with sufficiently distinct initial strategy values. However, not all starting conditions need produce this result. For some starting conditions (with two or more species) the system will converge on the single, local, non-ESS peak of Figure 7.6 | 210 |
| 7.8 | Adaptive landscape for Bergmann's rule $G$-function. Because only a positive body size is allowed $G(v, \mathbf{u}^*, \mathbf{x}^*, y^*)$ has a unique maximum | 213 |
| 7.9 | Using $\sigma_b^2 = 10$ results in an ESS coalition with one prey and one predator. There is an illusion that the landscape for the prey dips. It is actually a true maximum as is the predator | 216 |
| 7.10 | Using $\sigma_b^2 = 4$ results in an ESS coalition with one prey and two predators | 217 |
| 7.11 | Using $\sigma_b^2 = 1$ results in an ESS coalition with two prey and two predators | 217 |
| 7.12 | Using $\sigma_b^2 = 0.75$ results in an ESS coalition with three prey and two predators | 218 |
| 7.13 | An ESS coalition of two strategies as indicated by the circle and asterisk | 221 |

| | | |
|---|---|---|
| 7.14 | A case where Darwinian dynamics does not result in an ESS solution | 221 |
| 7.15 | A multistage ESS coalition of one strategy | 225 |
| 7.16 | The ecological cycle in this case is a 4-cycle | 228 |
| 7.17 | The adaptive landscape at each of the four points of an ecological cycle. The time step and the value of the $G$-function at each peak are noted at the top of each graph | 228 |
| 7.18 | A plot of the adaptive landscape at each point of the 4-cycle | 229 |
| 8.1 | Adaptive dynamics can result in a stable minimum that is not an ESS | 238 |
| 8.2 | Using a narrow distribution of strategies about the archetype results in the clumping of strategies at the ends of the distribution | 238 |
| 8.3 | A wider distribution of strategies results in a clumping in the vicinity of the archetype as well as at the left end of the distribution | 239 |
| 8.4 | The two species are at evolutionarily stable maxima but they do not compose an ESS | 240 |
| 8.5 | In this case the strategies clump about a two species archetype denoted by the diamonds | 241 |
| 8.6 | The three-species ESS | 241 |
| 8.7 | By choosing a proper interval for the distribution of strategies, clumping is obtained around a three species archetype that together form an ESS | 242 |
| 8.8 | As the mean strategy approaches the ESS, the variance narrows | 244 |
| 8.9 | Shortly after the simulation starts, only those strategies in the neighborhood of the ESS have a positive fitness | 245 |
| 8.10 | The mean strategy changes with time in a fashion similar to that obtained using strategy dynamics with $\sigma^2 = 0.001$ | 246 |
| 8.11 | A clump of strategies evolves to the ESS | 248 |
| 8.12 | As time goes on, the clump of strategies straddles the ESS as given by $u_1 = 0.6065$ | 248 |
| 8.13 | After reaching the species archetype, the clump of strategies becomes a bimodal distribution | 250 |
| 8.14 | How the clump of strategies approaches the species archetype | 250 |
| 8.15 | With $m = 0.1$ the ESS is a coalition of one strategy | 255 |
| 8.16 | With $m = 0.005$ a single strategy evolves to a local minimum on the adaptive landscape | 255 |
| 8.17 | With $m = 0.005$ the ESS is a coalition of two strategies | 256 |

# List of figures

| | | |
|---|---|---|
| 8.18 | Decreasing $\sigma_k^2$ can also result in an ESS coalition of two strategies | 257 |
| 8.19 | The four species resulting from the two environmental conditions ($E_1$ to the left and $E_2$ to the right). Each figure shows the two co-existing species that have evolved to evolutionarily stable maxima | 259 |
| 8.20 | An adaptive radiation towards a five-species ESS. Sympatric speciation carries the system from a single species up to four species that converge on non-ESS evolutionarily stable maxima | 262 |
| 8.21 | An ESS coalition of five strategies for the Lotka–Volterra competition model | 263 |
| 8.22 | The adaptive radiation of the predator–prey model from a single prey and a single predator species to a non-ESS community of two prey and two predator species | 267 |
| 9.1 | The strategy $\mathbf{u}_1$ is a matrix-ESS | 289 |
| 9.2 | The adaptive landscape is linear for the bi-linear game | 289 |
| 9.3 | A coalition of two pure strategies exists for the game of chicken | 291 |
| 9.4 | The function $G(\mathbf{v}, \mathbf{u}, \mathbf{p}^*)$ takes on a proper maximum at $v = 0.5469$ | 297 |
| 9.5 | The matrix-ESS solution produces the maximum number of children | 301 |
| 10.1 | The solid line represents the fitness in habitat 1 and the curved dashed line the fitness in habitat 2. When the density reaches a level such that the two fitnesses are equal (designated by the square), any further increase in density is divided between the two habitats | 307 |
| 10.2 | The solution obtained in the previous example is found to be convergent stable | 314 |
| 10.3 | The solution obtained satisfies the ESS maximum principle | 314 |
| 10.4 | Strategy dynamics results in an ESS coalition of two strategies | 316 |
| 10.5 | The solution obtained satisfies the ESS maximum principle | 317 |
| 10.6 | When $R < mK_m$ equilibrium C is evolutionarily stable (left panel). When $R > mK_m$ equilibrium B is evolutionarily unstable (right panel) | 320 |
| 10.7 | After two years the cancer cells have evolved to a maximum on the adaptive landscape | 323 |
| 10.8 | After evolutionary constraints have been removed, cancer develops rapidly in the first year | 324 |

| | | |
|---|---|---|
| 11.1 | The first panel is the adaptive landscape for the Schaeffer model with no harvest ($E = 0$). The second and third panels illustrate how the adaptive landscape changes with size-restricted harvesting both before and after speciation | 357 |
| 11.2 | Before treatment, the cancer cells are at a local maximum on the adaptive landscape | 361 |
| 11.3 | During treatment the cancer cells evolve to a new, more deadly strategy | 362 |
| 11.4 | After treatment, the cancer cells are again at a local maximum on the adaptive landscape | 362 |

# Preface

Bernstein *et al.* (1983) coined the term "the Darwinian dynamic" to describe the dynamical process underlying natural selection. Michod (1999) adds "Darwinian dynamics are systems of equations that satisfy Darwin's conditions of variability, heritability, and the struggle to survive and reproduce." We take this same view. In fact, for several years, the authors have been collaborating on a particular unifying approach to Darwinian dynamics that puts the study of evolution in a sound mathematical framework by recognizing that natural selection is an evolutionary game. The objective of this book is to explain how the evolutionary game approach along with the concept of a fitness generating function (called a $G$-function) is used to formulate the equations for Darwinian dynamics. We then show how to use these equations to predict and/or simulate the outcome of evolution. The $G$-function also produces an adaptive landscape that is useful in analyzing results and drawing conclusions.

After 20 years of development, with our work spread over numerous publications, it was difficult, even for us, to see the whole picture. This book allowed us to draw together and unify our work within one cover. It should be a good reference for anyone interested in the mathematics of evolution. It can also function as a textbook. Working out the details of the examples provides ample homework problems.

This is a book quite unlike any other publication intended for the study of evolution. It might be thought of as mathematical Darwinism. Darwin used logical verbal arguments to understand evolution. Today, we think of evolution in terms of genetics, which involves the study of inheritance of genes from one generation to the next. Genetics seems to provide the ultimate tool for studying evolution, yet Darwin presented his theory without a proper appreciation of the work of Mendel (1866). It was not until the 1930s that Fisher (1930), Wright (1931), Dobzhansky (1937), and others combined evolution and genetics into what is known as the Modern Synthesis (Mayer and Provine, 1980). Although genetics has provided a framework for understanding evolution, it is not a necessary framework because Darwin's postulates *do not require any specific mechanism of inheritance*. Rather than taking a gene-focused view of evolution, we view natural selection with a focus on heritable phenotypes. Genes are critical as the recipe for inheritance, but it is the heritable phenotype that forms the interface between the organism and its environment.

Evolution by natural selection is an evolutionary game in the sense that it has players, strategies, strategy sets, and payoffs. The players are the individual organisms. Strategies are heritable phenotypes. A player's strategy set is the set of all evolutionarily feasible strategies. Payoffs in the evolutionary game are expressed in terms of fitness, where fitness is defined as the expected per capita growth rate of a given strategy within an ecological circumstance. The fitness of an individual directly influences changes in strategy frequency as that strategy passes from generation to generation. Evolution by natural selection has to do with the survival of a given strategy within a population of individuals using potentially many different strategies.

In the development of our approach, we work from Darwin's three simple postulates:

1. Like tends to beget like and there is heritable variation in traits associated with each type of organism.
2. Among organisms there is a struggle for existence.
3. Heritable traits influence the struggle for existence.

These postulates may be used to formulate fitness functions. The fitness functions are used to model both population dynamics and strategy dynamics for species within a community. Because fitness is influenced by all the strategies used in the community evolution by natural selection emerges naturally as an evolutionary game.

Generally, fitness functions have a symmetry property that allows for the identification of groupings of individuals. For example, in a prey–predator system the dynamics of each prey species is distinctly different from the dynamics of each predator species, and we would say that this system is composed of two different groups of individuals. However, each group may be made up of individuals of many different species. When considering only one group of individuals (e.g., all prey), every species within that group may possess a similar dynamic and we are able to group individuals on the basis that they have the same evolutionary potential. To capture this symmetry and to simplify notation we use the concept of a fitness generating function or $G$-function. There is a different $G$-function for every group of individuals that have the same evolutionary potential. For example, a prey–predator system will have one $G$-function for the prey and a different $G$-function for the predators.

We use $G$-functions to provide a mathematical interpretation of Darwin's postulates. The $G$-function is used to express both population dynamics and strategy dynamics. Together, strategy dynamics and population dynamics are the Darwinian dynamics.

In Chapter 1 we present an overview of natural selection as an evolutionary game and contrast this approach with one based on genetics. The bulk of the mathematical development occurs in Chapters 2, and 4–7. In each of these chapters we present the theory in terms of the "simplest problem" first before moving on to more complex problems. The reader may choose to move through these chapters focusing on the simplest problem. Chapter 3 defines the evolutionary game and introduces the $G$-function. Chapters 8–11 use the theory developed in the first seven chapters to examine speciation, extinction, matrix games, selected topics in evolutionary ecology, and some applications to conservation management. Some specific topics include community evolution, micro- and macroevolution, evolution of cooperation, habitat selection, carcinogenesis, plant ecology, resource management, and conservation.

The bibliography contains the names of many individuals who have co-authored papers with us. Their collaboration in the development of the $G$-function approach to evolutionary games has been vital and welcome. We are indebted to all of them. In particular we are grateful to Yosef Cohen for the time he spent in helping us get this book started and for sharing material with us. We also owe a great deal of thanks to Chris Whelan for his careful reading of the entire manuscript and his invaluable suggestions. Finally, we thank Tania Vincent for her artwork.

# 1
# Understanding natural selection

The following observations about patterns in nature have captured the imagination of humans for millennia.

1. **Fit of form and function (*FF&F*)**: different organisms appear remarkably well suited and engineered for their particular environments. The high-crowned molars of zebras and white rhinoceros act as mulching mowers for grinding grass, and protect against the inevitable wear imposed by the silica content of grass. Black rhinos, on the other hand, have lower crowned molars favoring efficient mastication of leaves and foliage. None of these animals has the sharp and stabbing canines like those of lions. Distinct **species**[1] of organisms apply themselves to different ecological tasks using their appropriate sets of tools. For example, zebras and white rhinoceros feed on grass, black rhinos browse leaves from shrubs, and lions kill and eat zebras.
2. **Diversity of life**: we share this planet with a phenomenal array of different life forms. These forms range from delicate mosses and annual flowering plants to awesome whales and fearsome sharks. While many of these forms differ in subtle ways, most can be readily recognized and categorized as types or species quite distinct from others. This is possible because the extant denizens of our planet do not exhibit a continuum of morphological variation from bacteria to redwood tree. Rather, the morphologies and characteristics of living organisms cluster like conspicuous and discrete galaxies in morpho-space.
3. **Procession of life**: despite the variety and discreteness of life, organisms seem connected by design rules of increasing levels of complexity. Notions such as the tree of life identify a regular, yet increasing, sophistication of organisms in terms of size, behavior, and the number and specialization of

---

[1] A formal definition of species is given in Subsection 8.2.2.

traits. The early idea of a **bauplan** recognized the fixity of certain design rules among definable groups of species. Linnaeus in his binomial nomenclature used design rules to place organisms in the tree of life. Modern systematics and taxonomy, now more than ever, rely on the hierarchical structuring of traits among collections of species to assign names and position within life's tree.

4. **Distribution and abundance of organisms**: this is the central question of ecology. Paleolithic peoples probably pondered this as the central question of survival. Organisms are not spread randomly in space and time. Furthermore, some organisms seem ubiquitous and excessive in numbers (various species of crow, for their size, are particularly abundant around the globe) while others puzzle us with their rarity (the introduced Eurasian tree sparrow has a toe-hold in the city of St. Louis while its congener, the European house sparrow, occupies the rest of North America).

These observations must predate recorded history. Yet a satisfactory and unified answer to why the above four patterns exist has been available for only about 150 years with the development of Darwin's theory of evolution by natural selection. More recently, game theory (the mathematics used to study conflicts of interest among two or more players) – is being successfully applied to modeling natural selection. The classical game theory of economics, sociology, and engineering has existed as a formal discipline since the 1940s and 1950s, while game theory as a formalism for natural selection has existed since the 1970s.

The objective of this book is to show that the synthesis of Darwin's ideas within the context of an evolutionary game provides a most useful tool for understanding the four patterns of nature. Because the use of evolutionary game theory to model natural selection requires a moderate amount of mathematics, we provide all of the concepts and mathematical tools needed in the chapters that follow.

In this chapter, we start by discussing Darwin's marvelous idea of natural selection, introduce life as an evolutionary game, and explain why we favor a game theoretic approach as a complement to the more familiar and orthodox genetical approaches to natural selection.

## 1.1 Natural selection

### 1.1.1 Historical perspective

It is appropriate that well into the Age of Enlightenment the field of evolutionary ecology resided within the intellectual pursuit of Natural Philosophy. Natural

Philosophy encompassed all aspects of the sciences. Then, as today, philosophy (literally the love of wisdom) pursues the facts and principles of reality. Ecology falls into this quest for understanding Nature's reality, and natural philosophers recognized a wisdom to nature. All organisms exhibit in their characteristics excellent engineering in fit of form and function (*FF&F*) and the engineering shows a commonality and connectedness of design across all life from simple to complex (*procession of life*). It is remarkable that, over the ages, the diverse natural philosophies have all recognized a design and engineering component to nature. And, until the mid 1850s, all of these philosophies drew a very logical connection between human tools and organisms as tools designed by nature.

The connection between the tools in the human household and organisms in nature's house is compelling. Hence, essentially all pre-Darwinian natural philosophies took the next logical step. Tools exist because humans design and fashion them with purpose and forethought of intent, a watch is proof of a watchmaker.[2] Commonality of features among watches reflects the watchmaker's trademark and level of technology. It then follows that biodiversity is a reflection of a Creator, of gods, of Mother Earth, or of some other personified force that shows intent and purpose in the conscious design of its organisms. For most cultures over most of history this logical construction held sway. Just as humans make tools so something greater (singular or plural, masculine or feminine) made life. This philosophical view of life provided a seamless blend for people's ecological knowledge and spiritual beliefs. In the nineteenth century (Darwin's Century as Eiseley (1958), aptly calls it) in Western Europe, and in England in particular, this viewpoint began to lose favor as applicable to biology.

Lyell's geology showed how ongoing forces and non-personified natural processes could explain the forms, types, and layering of rocks (Lyell, 1830). And within many of these distinctly non-living rocks were the distinct remains of previous life. Erosion, sedimentation, compression, and volcanism provided for geological changes with time. Could the fates of rocks and life be tied together? Could similarly non-personified natural forces explain the origins and changes of life with time? The essentialists (linked to Greek ideas of life mirroring or manifesting some deeper fixed reality and truth) and biblical creationists (Genesis as scientific treatise) scrambled to make sense of

---

[2] Apparently William Paley was the first to use the analogy. "... suppose that I had found a watch upon the ground ... this mechanism being observed the inference we think is inevitable, that the watch must have had a maker, ... or shall it, all at once turn us around to an opposite conclusion, namely that no art or skill whatever has been concerned in the business. Can this be maintained without absurdity?" Evidence of evolution reveals a universe without design, hence Dawkins's (1986) useful metaphor of the blind watchmaker.

these new findings and ideas. In its more complex forms, scientific creationism stretched biblical days into millennia, and recognized multiple creations and destructions of life, of which Noah's Flood was but one particularly noteworthy example (Schroeder, 1997). But those seeking a "uniformitarian" explanation for life also had major conceptual and logical hurdles. Yes, geology and life seemed to share a common fate, but erosion, sedimentation, and volcanism do not form the characteristics of organisms. Empirically, life might change its characteristics with time, but what were life's natural processes?

Evolution built around heritable change with time was a potentially attractive force. Most natural philosophers accepted the presence of this force within animal and plant breeding, and many social philosophies emphasized the connections between human bloodlines and human hierarchies. But, as a force for change, it was presumed to be rather limited and in most cases useful only for protecting good blood from bad. Few saw breeding as providing the force or opportunity for truly novel evolutionary change. Early attempts at linking evolution to *FF&F* and *procession of life* still clung to the notion of foreordained or consciously driven improvement. Some espoused a kind of creationist–evolutionist blend: a view that saw God creating life at all levels followed by the evolution of these forms up a chain of being towards humans, angels, and beyond. Lamarck advanced a tenable theory of evolution via "self improvement." Just as an individual can be conditioned physically for a task, perhaps a species can condition their heritable characteristics towards needs and particular tasks, leading to the inheritance of conditioned or acquired traits. Two aspects of this theory of evolution are interesting. First, Darwin did not see Lamarck as incompatible with natural selection and in fact viewed the inheritance of acquired traits as one of several likely ways for introducing heritable variation. Second, Lamarckism could have been correct as a scientific perspective. If **pangenesis** (the equal contributions of all units of the body to the heritable blueprint for the organism's offspring) had been correct, then acquired (or discarded) traits could manifest as heritable change, and natural selection could work within this context. And indeed, in prokaryotes, and some plants where there are fewer clear boundaries between the somatic cell line and the gametic cell line, manifestations of Lamarckian evolution do occur comfortably within the framework of natural selection. But, the raindrops that eroded and formed Lyellian geology still eluded evolutionist thinking.

Darwin found the raindrops in deceptively simple ecological processes – surplus births and subsequent famine. The Struggle for Existence (loosely associated with Malthus (1796)) recognizes a reality of ecology. Organisms are capable of having many more offspring than the environment can possibly support. Darwin's genius was in making the link between heritable variation (however it came about!) with the Struggle for Existence in which less satisfactory individuals die. Just as raindrops sculpt landscapes by eroding softer and harder stones at different rates, the ecological raindrops of births and deaths striking the softer and harder rocks of heritable characteristics sculpt life. It is not hard to see how many a natural philosopher would find repugnant the deep social irony of natural selection as beautifully described in *"Darwin's Dangerous Idea,"* (Dennett, 1995). This repugnance resonates today in the writings of intellectuals such as Gould (1998). The "noble" excellence exhibited by *FF&F* and *procession of life* is engineered by the scourges put upon it as manifested by "poverty" and "famine."

### 1.1.2 As Darwin saw it

Evolution is the physical, genetic, or behavioral change in populations of biological organisms over time. Evolution's more interesting and significant manifestations result from **natural selection**, a process that engineers biological systems. Natural selection works within genetic, developmental, and environmental constraints to shape biological organisms in ways that make them appear adapted to their environments. Understanding an evolutionary design has its roots in **Darwin's postulates** (Darwin, 1859). As Sober (1984, p. 21) notes, Darwin's postulates are really two drawn out, discursive propositions. Darwin saw heritable variation leading to evolution, and evolution leading to new species and to new distributions of characteristics within species. Drawing from Lewontin (1974), we will separate Darwin's argument into three postulates:

1. Like tends to beget like and there is heritable variation in traits associated with each type of organism.
2. Among organisms there is a struggle for existence.
3. Heritable traits influence the struggle for existence.

The first postulate was generally well known at the time and had been used by plant and animal breeders for centuries to improve native strains. The second postulate was influenced by Malthus's *Essay on Population* (1976) with the thesis that resources can only increase arithmetically while human populations grow geometrically. Darwin extended this idea into the general phenomenon of competition among individuals of the same or different species for limiting resources. Darwin's last postulate provided the key for understanding the consequences of evolution. For a particular environment, this postulate results in an increase in phenotypically well endowed individuals who are better able to survive and reproduce than less well endowed individuals.

Darwin used logical verbal arguments to model evolution. His views on inheritance were both orthodox for the day and flawed. Today, we think of evolution in terms of genetics, which involves the study of inheritance of genes from one generation to the next. Genetics seems to provide the ultimate tool for studying evolution, yet it is a curious fact that Darwin presented his theory in the absence of any understanding of genes as presented by Mendel (1866). It was not until the 1930s that Fisher (1930), Wright (1931), Haldane (1932), Dobzhansky (1937), and others combined evolution and genetics into what is known as the Modern Synthesis (Mayer and Provine, 1980). Genetics has provided a framework for understanding evolution, yet it need not be the essential core for modeling or understanding evolution by natural selection. Darwin's postulates do not require any specific mechanism of inheritance. This observation is in accordance with the development presented in this book. Since Darwin's three postulates constitute a fundamental principle that can be used to explain and predict evolution, we use these principles in developing a non-genetical mathematical framework for natural selection. The framework is *not* non-genetical in the sense of not having some mechanism for inheritance, and an understanding of the recipe of inheritance, as in the case of modern genetics, is paramount to Darwin's first postulate (as well as to bioengineering, medical genetics, animal and plant breeding programs, taxonomy, DNA fingerprinting, etc.). The framework is non-genetical in the sense that an actual genetic system for allowing natural selection is an auxiliary hypothesis. In the same manner natural selection is merely an auxiliary hypothesis (among several evolutionary forces) for changes to a genetic system. We propose that evolution by natural selection is a dynamic game. Our objective is to develop an evolutionary game theory that can be used as a fundamental modeling tool for understanding natural selection.

### 1.1.3 The Modern Synthesis

The Modern Synthesis that began in the 1900s and was completed by the 1930s is often viewed as a critical step in formalizing natural selection (Sober, 1984). The lack of a mechanism for inheritance hampered development of rigorous mathematical models of natural selection, which in turn hampered application and advancement. The "rediscovery" of Mendel's Laws in the 1900s (Pearson, 1904; Hardy, 1908) energized work on breeding and inheritance, and drew into question the compatibility of Mendel's particulate inheritance with natural selection. Fisher (1930), Wright (1931), Haldane (1932), and others ushered in a golden age of population genetics by placing the study of evolution on a firm mathematical foundation. In creating this foundation, they showed the compatibility of Mendelian genes, loci, and alleles with natural selection, the evolution of quantitative traits, and systematics. In addition, the recipe of inheritance provided insights into other forces of evolution (mutation and genetic drift) and into interactions that might occur genetically within and between organisms. Genetic interactions within an organism could be **epistatic** (many genes at different loci may contribute non-additively to a particular trait) and **pleiotropic** (a single gene may contribute to the phenotype of several traits). Among individuals, natural selection could be **density dependent** and/or **frequency dependent** depending on whether the population's size and/or gene frequencies influence the success of individuals with particular phenotypes, respectively.

The Modern Synthesis led to the primacy of genes over heritable phenotypes as the objects of evolution. This primacy seems self evident. In the Modern Synthesis, evolution is defined as a change in gene frequency. However, natural selection in terms of $FF\&F$ must involve the ecological consequences of heritable phenotypes. Can a strictly genetical approach be sufficient for modeling natural selection? Models of gene-frequency dynamics determine what has been selected but cannot necessarily determine what survival or fecundity aptitudes of the organism have been selected for. The $FF\&F$ requires understanding both what has been selected and why. The "why" requires a focus on heritable phenotypes, particularly when natural selection is frequency dependent. So, while the Modern Synthesis provided a huge advance in our understanding of evolution, taxonomy, and gene dynamics, it may have unwittingly hampered a fuller appreciation of natural selection by subordinating heritable phenotypes to their genetic recipes.

## 1.2 Genetical approaches to natural selection

**Population genetics** (modeling changes in the frequency of particular alleles within a population) and **quantitative genetics** (modeling the change with time of quantitative traits under the assumption that many alleles and loci contribute more or less additively to the trait value within an interbreeding population) are the concepts currently used for thinking about and modeling evolution where evolution is defined as a change in gene frequency. This outlook guided research to examine how genetic variability and genetic constraints direct and restrict evolutionary change (Crow and Kimura, 1970).

Viewing evolution as change in gene frequency can produce reasonable results in terms of producing an *FF&F*. For example, consider the case where the fitness conferred by a gene on an individual is **density independent** (independent of the population size) and **frequency independent** (independent of gene frequencies). In this case, the gene dynamics favor the genes that confer the highest per capita rate of growth on the population. In the situation where the fitness conferred by a gene is density dependent and frequency independent, then gene dynamics favors genes that maximize the population's size. In both of these cases the gene dynamics favors survival of the fittest if fitness is defined either as population growth rate or population size. However, as soon as evolution is frequency dependent, that is the fitness conferred by a gene on an individual is influenced by the frequencies of other genes in the population, then the linkage between the consequence of natural selection operating on genes and some corresponding measure of fitness at the population level disappears. The endpoint of the gene dynamics no longer optimizes any obvious measure of ecological success. This will be the most common situation as plausible genetic interactions such as epistasis, pleiotropy, and heterozygote superiority all introduce frequency dependence. The decoupling of change in gene frequency from some measure of ecological success for the individual organism or the population has unintended and unfortunate consequences for the question of *FF&F*. When evolution by natural selection becomes simply the endpoint of genetic dynamics, evolution by natural selection becomes potentially tautological. The fittest genes are those that survive and so survival of the fittest becomes a truism. Or it encourages a view of a life in which genes are the engineers of blindly programmed robots that serve only to reproduce more genes (paraphrased from *The Selfish Gene* (Dawkins, 1976)). The wings of a bird are no longer for flying; rather they are a part of the machinery for proliferating genes. The *FF&F* concept is lost in favor of the dynamical system of gene frequencies.

In this book, the focus will be on the wing rather than the genes coding for the wing. Characters such as wings will be modeled as evolutionary **strategies** (heritable phenotypes). Even under frequency-dependent selection, the resulting game theory analysis will reveal both what has been selected and why. The *FF&F* requires us to study strategies as the outcome of an evolutionary process (accessible using gene-frequency dynamic models), and to study strategies by their function (tricky when using strictly genetical models of frequency-dependent selection). A game theoretic approach is needed because frequency-dependent selection is ubiquitous in natural selection and plays the key role in the *diversity of life* and the *distribution and abundance of organisms*.

## 1.2 Genetical approaches to natural selection

A consequence of the strictly genetical approach used in current textbooks on evolution[3] is a narrow perspective on genetic variability and a decoupling of the concepts of **microevolution** (small evolutionary changes) from concepts of **macroevolution** (large evolutionary changes and the stuff of the *procession of life*). In evolution courses, such traits as tongue rolling and blood type serve to emphasize the idea of genetic variability. Once the genetic variability has been identified, the loci and alleles specified, and the consequences of genes for survival and fecundity defined, then population genetics brings mathematical rigor to subsequent changes in gene frequencies brought about by natural selection. Unwittingly though, the focus on extant genetic variability greatly reduces our appreciation of the complete set of heritable variation on which natural selection operates. Subsequent analyses give the impression that natural selection is a finishing school for microevolution but is inapplicable to macroevolution. Natural selection becomes subordinated to the known and accepted machinery of population and quantitative genetics which then gets subordinated to explaining readily observable evolutionary changes within populations. By not being able to apply the genetical approach to the big interesting evolutionary changes that separate species, families, orders, and classes from each other, evolutionists have proposed macroevolutionary forces such as genetic revolutions, species selection, and phylogenetic constraints and inertia that have little grounding in natural selection (Eldridge and Gould, 1972; Stanely, 1979; Vermeij, 1994). Current evolutionary teaching reflects this split in intellectual thinking. The rigors of population and quantitative genetics are used to show how natural selection can shape characteristics of populations, and then this machinery is discarded and replaced when the course moves on to the really interesting questions of speciation, biogeographic patterns, and the evolution of characters that define and separate the higher taxa of life. Because macroevolution does not fit comfortably within population genetics, natural selection becomes separated from the question of the *diversity of life* and the *procession of life* by virtue of its association with genetical models.

Genetical views of natural selection often ignore the most appealing applications of natural selection to *FF&F*, *diversity of life*, and *procession of life*. This happens because a genetical basis for natural selection cannot comfortably account for the seemingly limitless, though constrained, set of heritable variability available to natural selection, and it subordinates the organism's ecology to the genetic mechanism. But in Darwin's original formulation it is the ecological interactions operating on the set of evolutionarily feasible phenotypes

---

[3] Frequency-dependent selection often gets short shrift in these textbooks. Usually the most interesting examples of natural selection cited involve frequency dependence (at least implicitly) even as the formalisms for conceptualizing frequency dependence receive minimal attention.

that sculpt and refine species towards an *FF&F*. This aspect of Darwin's perspective on natural selection represents an **adaptationist research program** which studies the advantages that particular characters might confer on the individual. Fields such as physiological ecology, functional morphology, and behavioral ecology (particularly in the guise of foraging theory and sociobiology) produce more or less plausible hypotheses for the adaptation of an organism's heritable traits.

The adaptationist approach to natural selection is appealing in that it seems to contain the spirit of Darwin's original idea. However, it is built on a poor foundation. As scathingly noted in "The Spandrels of San Marco" (Gould and Lewontin, 1979), the intuitively appealing explanations for the value of traits to an organism rested on non-rigorous and often indefensible notions of what is valuable to an organism and what is heritably feasible. The adaptationist paradigm in the 1970s lacked formal fitness functions, formal statements of what was feasibly heritable, and formal evolutionary dynamics. Here's the dilemma. Genetical approaches have been successful at modeling what is selected but lack insights into why a character has been selected. Adaptationist approaches have been successful at proposing why a character has been selected for, but often lack a modeling framework.

Here we take another look at the adaptationist approach as embodying the spirit of Darwin's theory of natural selection. While we applaud the formalism and rigor of population genetic and quantitative genetic approaches to evolution, we regard life as a game, and that a game theoretic approach provides the right tools and a sufficient level of rigor for an adaptationist approach to evolution by natural selection. In this book, we present life as a game and develop the formalism necessary to model evolution by natural selection as a game. To make the transition from a strictly genetical perspective to a game theoretic one, we view evolution as a change in heritable phenotypes rather than as a change in gene frequency. From this viewpoint, we recover the sense of natural selection as an optimization process leading to adaptations, and support the engineer's perspective that organisms are designed for a function.

## 1.3 Natural selection as an evolutionary game

The long loop of Henle within a kangaroo rat's kidney allows it to produce exceedingly concentrated urine. Because of this and other physiological adaptations (Schmidt-Nielson, 1979), the kangaroo rat can inhabit deserts, eat little more than seeds, and never drink a drop of water in its lifetime. Mussels inhabiting inter-tidal habitats have strong abyssal threads that lash them to the

## 1.3 Natural selection as an evolutionary game

rocks and prevent them from being swept away by crashing waves. At low tide they clam up tight to prevent desiccation and at high tide they open up to filter food particles from the swirling water. These physiological, morphological, and behavioral traits seem sensible in the organisms' ecology. They meet challenges and exploit opportunities. These traits must be heritable, and natural selection has produced this *FF&F*. The kangaroo rat's kidney and the mussel's threads represent evolutionary design features that allow these animals to survive in a seemingly optimal way under the circumstances. Explanations for these adaptations fall squarely within Darwin's theory.

For some species of frogs, choruses of males sing at ponds and waterways in order to attract females. In Costa Rica, for instance, the calls also attract the attention of frog-eating bats (Tuttle and Ryan, 1981). The males, as in many species of the animal world, are literally dying for love. A chorusing male calls for a mate, but may call in its predator. Yet it still choruses. Why? And, more fundamentally, is this adaptation consistent with *FF&F*? Such chorusing is not the only way for females to find mates. There are other frog species that achieve match-making more quietly. There are costs and benefits to chorusing, such calling seems to make it easier for females to find mates, allows males to advertise their presence, reduces male survivorship, and feeds bats. A collective reduction in calling volume by the males would probably hamper bats' effectiveness as predators with negligible effect on females' access to mates. This apparent paradox in behavior is resolved when mating is viewed as a game played with many other frogs. The male frog's best chorusing strategy depends upon the strategy choice of females and the calling strategy of other males. In the frog's mating game, chorusing functions primarily to attract mates away from other males. An economist would quickly see this mating system as an **advertising game**, where advertising expenditures serve primarily to divert customers from one's competitors rather than to increase the overall pool of customers.

Given the strategies used by the females and other males, the male's chorus strategy is optimal in the sense of maximizing his reproductive success that depends on the product of survivorship and mating success. Suppose that mating success increases and survivorship declines with the volume of an individual's chorusing; and that the reverse happens in response to the chorusing volume of other males. If most males chorus quietly, it behooves a male to call louder – it gains more from mating success than it loses in survivorship. If most males are exceptionally loud, it may behoove the male to call more quietly – it gains more from survivorship than it loses from fewer mating opportunities. Between these advertising extremes, a best strategy exists where the male calls at the same volume level as all of the other males. The whole business of calling loudly may seem maladaptive but, given the circumstances, the calling behavior of the individual male frog maximizes his reproductive success.

There are similarities in other situations that, at first glance, seem radically different from the perspective of the organisms' natural histories. For instance, what does the evolution of height in trees have to do with chorusing in male frogs? They are fundamentally the same evolutionary game! Woody plants invest resources (photosynthate) into non-productive wood that presumably could have been used for reproduction or the production of seemingly more useful structures such as leaves and roots. The trunks of trees precipitate all sorts of additional challenges for the plant: susceptibility to tipping over in high winds, greater surface area for the invasion of pathogens and herbivores, hydrostatic problems associated with nutrient exchange between roots and leaves, and weight and balance issues associated with supporting tall, narrow structures. Trees behave as trees for the sake of sunlight. Since the pool of available sunlight is fixed, a tree grows to escape the shade of other trees. At the optimum height, being taller than the others around you incurs greater costs in height than benefits in light. Being shorter than others incurs greater losses in light than benefits in less height. Tree trunks function primarily to gather light away from other plants. In the presence of canopy trees, all striving to be shade free, there exist opportunities for a diversity of light-gathering strategies. Light gaps provide opportunities for fast-growing, light-loving shrubs and herbs. The seasonality in temperate zones provides spring windows of light for early flowering herbs that grow and flower before the canopy trees have had time to leaf out. The flecks of light that pass through the sieve of canopy leaves provide opportunities for shade-tolerant plant species. The game of light competition may not only select for tree trunks (*FF&F*), but also select for a co-existing diversity of strategies (*diversity of life*).

Two important properties are essential elements of any game. First, each player's success depends not only on its own strategy but on the strategies of the other players. Second, the players' objectives are generally at cross purposes; that is, the combination of strategies among players that is preferable to one player need not be the arrangement most preferable to any of the other players.

Nature abounds with games. The frogs and trees engage in different forms of the **tragedy of the commons** (Hardin, 1968) in which gains to individuals at the expense of the group encourage organisms (and humans) invariably to overuse their common pool of resources. Pursuit–evasion games occur within most, if not all, predator–prey systems. Arms races occur in the competition for mates, food, space, and safety. Games of chicken occur in many example of interference competition. The prisoner's dilemma characterizes many games involving the evolution of cooperative behaviors.

All of life is a game. Organisms are not masters of their own fate. In evolving favorable characteristics, no organism is free from the evolutionary meddling of others. Indubitably, the advantage or functioning of many heritable traits can only be understood in the context of the traits of others, that include the same and/or different species. From the perspective of a game, the strategies used by trees and frogs can be not only understood, but actually predicted.

Darwin recognized life as a game. In describing sexual selection, he saw that many behavioral and morphological traits of sexually reproducing animals only made sense in terms of mate competition. The peacock's feathers, male leks, and elaborate courtship rituals, rather than functioning directly for survivorship or food, serve the suitor in terms of quality and quantity of mates. Altruism, traits that seem to provide "public" goods to others at a "private" cost to the individual, was troubling for Darwin and early proponents of natural selection. Enlightened self-interest provided clues. Costly but nice behaviors may yield the individual indirect or delayed rewards from within its social context. While recognizing the challenge of some of nature's games for natural selection, Darwin and his contemporaries did not possess the formal tools of game theory to logically and deductively assess the traits whose adaptive significance lies in the traits of others.

Wright (1931) (see also Wright, 1932) sought the optimality of traits supported by natural selection. His construct of the **fitness landscape**, where a measure of fitness is plotted against gene frequency, provided a visual representation of adaptations as peaks of these landscapes. **Fisher's Fundamental Theorem of Natural Selection** (Fisher, 1930) provided a dynamic for how natural selection would drive gene frequencies to these peaks. He showed how the rate of evolution in response to natural selection proceeds up gradients of higher fitness and at a rate proportional to the product of **heritable variability** (additive genetic variability) and the magnitude of the fitness gradient. This aspect of natural selection as an optimization process that climbs the slopes of fitness landscapes worked well for density-independent and frequency-independent selection. In these cases, the landscape is rigid in response to gene frequencies within the population. Gene frequencies change, but the shape of the landscape does not. Under both frequency dependence and density dependence, the landscape is no longer rigid in response to changes in gene frequencies, and optimization of fitness is no longer applicable. Because Wright's analysis came before the development of game theory, formal analysis of a **flexible landscape** was not possible.

Fisher[4] in addressing the evolution of sex ratios recognized gender frequencies as a game and flirted with tools of game theory. He saw how, in diploid, sexually reproducing organisms, the ultimate fitness payoffs to each gender must be equal. Furthermore, a mother producing offspring achieves fitness in her grandchildren through both her sons and her daughters. Sons enter a kind of lottery for half of the genetic payoffs to the next generation, while daughters join a lottery for the other half. This occurs whether the pool of individuals in the next generation is fixed, or is determined solely by the number of females, or some combination of males and females. If the male side of the lottery is oversubscribed it is best to produce daughters, and vice versa in producing sons. Hence, it is best to devote equal reproductive resources towards the production of male and female offspring. In formalizing this centerpiece of sex-ratio theory, Fisher (1930) anticipated Maynard Smith's (1976) ESS concept and Bishop and Cannings's (1976) general result that, in a matrix game with a mixed-strategy solution, each pure strategy of the ESS will yield equal payoffs to the player.

Hamilton (1963) in his contribution to the evolution of cooperation by kin selection touched upon optimality as a game. He formulated the idea that the advantage of cooperative behavior may lie in its inclusive fitness effect on relatives. Haldane (1932) anticipated the idea of **inclusive fitness** when he observed that a person's self-sacrificing strategy would actually benefit if the person were willing to sacrifice his life to save three of his brothers who are presumed to have a fifty–fifty chance of sharing this strategy. The likelihood of like interacting with like (interactions among relatives providing such a context) can render altruistic behaviors adaptive. **Altruism** can function to maximize an individual's genetic prospects given circumstances of non-random interactions.

Levins (1968), with his concept of fitness sets and evolution in heterogeneous environments, provided a theory of evolution by natural selection based on strategies (i.e., heritable phenotypes). He also advanced the approach of using heritable phenotypes to see how natural selection may promote diversity. He first proposed a strategy set for a quantitative trait. The strategy set represents all of the evolutionarily feasible values for the strategy. He then assumed that the environment by virtue of heterogeneity offers different opportunities to the organism. Next, he reasoned that natural selection will always favor strategy values that simultaneously improve an organism's prospects in both habitats of the heterogeneous environment. Those remaining strategies that trade-off performance in the two habitats become the **active edge** of the

---

[4] Among Fisher's contributions to genetics and statistics is his measure of indeterminacy, now called Fisher information. This concept now plays an important role in theoretical physics (Frieden, 1998).

## 1.3 Natural selection as an evolutionary game

fitness set (Levins, 1968). The active edge has similar properties to the **Pareto-optimal solution** from game theory. A Pareto-optimal solution has the property that it is not possible to change strategies among individuals so as to maintain or increase everyone's payoff (Vincent and Grantham, 1981). Levins then fixes the environmental context with respect to the scale (fine- versus coarse-grained for small vs. large patches of habitat, respectively) and frequency of the two habitats. One can then solve for an evolutionarily stable strategy by pitting the active edge of the fitness set against the environmental circumstances. This produces an optimum that may favor either a single generalist strategy or two specialist strategies. These strategies produce the highest per capita growth rate (fitness) given the circumstances.

Ever since Darwin, natural selection has been viewed as an optimizing process. One that promotes heritable phenotypes (strategies) that are optimal in the sense of being the best (maximizing fitness) given the circumstances. However, when the circumstances include the strategies used by others, evolution can no longer be viewed in terms of a simple optimization process, rather it is a game. As a game, natural selection combines evolutionary principles of inheritance with ecological principles of population interactions to produce what Hutchinson (1965) called the **ecological theater** and **evolutionary play**.

### 1.3.1 Game theory and evolution

The genesis of a formal theory of games can be traced to the publication of *Theory of Games and Economic Behavior* by von Neumann and Morgenstern (1944). Game theory had its beginnings with the two-player matrix game, which introduced the concepts of conflict, strategy, and payoff. There continues to be a large and growing literature on the theory and application of matrix games in economics, particularly from an evolutionary perspective (Samuelson, 1997; Weibull, 1997).

Games are generally divided into three major classes: matrix games, continuous static games, and differential games. Matrix games have a finite number of strategy choices. After each player makes a choice, the payoff to each player is determined by an element of a matrix. One player's strategy corresponds to the selection of a row and the other player's strategy corresponds to the selection of a column. All possible payoffs are given by the elements of the matrix. The particular payoff a player receives is determined by this combination of strategies. In continuous static games, the strategies and payoffs are related in a continuous rather than discrete manner (Vincent and Grantham, 1981). The game is static in the sense that an individual's strategy is constant. Differential games (Issacs, 1965) are characterized by continuously time-varying strategies and payoffs with a dynamical system governed by ordinary differential equations.

In a conventional game, a rational player's objective is to choose a strategy that maximizes his or her payoff. When an individual's payoff is a function not only of his or her own strategy, but the strategies of other players, we have the conditions that define and separate a game problem from an optimization problem. A game consists of players, strategies and strategy sets, payoffs, and rules for determining how the strategies employed by the players result in their respective payoffs. Unlike optimization theory, which is dominated by the concept of a maximum, game theory has a variety of solution concepts for predicting the game's outcomes and the players' optimal choice of strategies (e.g., min-max) von Neumann and Morgenstern 1944), Nash (Nash, 1951), Pareto-optimal (Pareto, 1896), and Stackelberg (von Stackelberg, 1952)).

Maynard Smith, an aeronautical-engineer-turned-biologist, was one of the first to examine evolution as a mathematical game in his book *Evolution and the Theory of Games* (Maynard Smith, 1982). The hawk–dove, sex ratio, Prisoners Dilemma, and other interesting matrix games are discussed and analyzed by Maynard Smith in the context of his concept of an evolutionarily stable strategy (Maynard Smith and Price, 1973). Our approach to evolutionary game theory is from a perspective that has its roots in Maynard Smith's pioneering work (Vincent, 1985; Brown and Vincent, 1987a,c; Vincent et al., 1993, 1996).

Evolution by natural selection is an **evolutionary game** in the sense that it has players, strategies, strategy sets, and payoffs. The players are the individual organisms. Strategies are heritable phenotypes. A player's strategy set is the set of all evolutionarily feasible strategies. Payoffs in the evolutionary game are expressed in terms of fitness, where fitness is defined as the expected per capita growth rate for a given strategy and ecological circumstance. The fitness of an individual directly influences changes in the strategy's frequency within the population as that strategy is passed from generation to generation. Evolution, then, has to do with the survival of a given strategy within a population of individuals using potentially many different strategies.

Several features distinguish evolutionary games from classical games. First, the evolutionary game does not fall into one of the three major classes of games. It is a hybrid with some similarity to both continuous static games and differential games. Furthermore, in classical game theory, the focus is on the players who strive to choose strategies that optimize their payoffs; whereas, in the evolutionary game, the focus is on strategies that will persist through time. Through births and deaths, the players come and go, but their strategies

## 1.3 Natural selection as an evolutionary game

pass on from generation to generation. In classical game theory, the players choose their strategies from a well-defined strategy set given as part of the game definition. In the evolutionary game, players generally inherit their strategies and occasionally acquire a novel strategy as a mutation. The strategy set is determined by genetic, physical, and environmental constraints that may change with time. In classical game theory, each player may have a separate strategy set and separate payoffs associated with its strategies. In the evolutionary game, there will be groups of **evolutionarily identical individuals** who have the same strategy set and experience the same expected payoffs from using the same strategies. In classical game theory, rationality or self-interest provides the optimizing agent that encourages players to select sensible strategies. In the evolutionary game, natural selection serves as the agent of optimization. We know we are dealing with a classical game when admiration is reserved for the winners (e.g., members of Congress), and we are dealing with an evolutionary game when admiration is reserved for the survivors (e.g., cockroaches). This is not to say that winners could not also be survivors, but, as we shall see, survivors need not be winners in the usual sense.

The evolutionary game has an inner and an outer game (Vincent and Brown, 1984b). The **inner game** involves only ecological processes and can be considered as a classical game. For the inner game, players interact with others and receive payoffs in accordance with their own and others' strategies. Evolution takes place in the **outer game**. It is the dynamical link, via inheritance and fitness, whereby the players' payoffs become translated into changes in strategy frequencies.

### 1.3.2 Games Nature plays

Darwin's second postulate states that there is a struggle for existence among organisms. This struggle may be simulated using population dynamics models. Such models contain many parameters, such as growth rates, resources uptake rates, predation rates, and carrying capacities. These parameters, in turn, depend on the strategies (i.e., heritable traits) used by the various species in the population. The influence of strategies on the struggle for existence is obtained by varying model parameters. If we can identify strategies and embed Darwin's first postulate into the system, the strategies in the model will have the capacity to evolve. This gives us all the elements needed for the formulation of an evolutionary game.

Some elements of the evolutionary game are readily apparent in population dynamic models. Start by choosing initial population sizes and strategies for each species in the community. Solve these time-dependent population dynamic models until the system asymptotically approaches an equilibrium solution (or until the system develops a persistent state of oscillations or non-equilibrium dynamics). Those species at a non-zero equilibrium number represent the survivors. One discovers that, in general, starting with many different strategies results in relatively few surviving strategies. This represents the struggle for existence in these models. We seek a special type of stability for the evolutionary game that will focus on the main feature: the ability of systems to evolve. For a given strategy to persist through time, it must be able to maintain a viable population in the face of the introduction of new strategies and through the subsequent evolution of the strategies.

### 1.3.3 ESS concept

It is one thing to be able to characterize natural selection as a game, it is quite another to determine an appropriate solution concept for that game. Every evolving system produces trajectories of changing strategy values. But do these trajectories have a stable endpoint, and do these endpoints have anything in common? In particular, do these endpoints of evolution by natural selection have the common property of producing the best strategy given the circumstances? One might think that the best strategy is one that maximizes some measure of collective reward. If this were the case, then the conventional game concept of a Pareto-optimal solution (Vincent and Grantham, 1981; Pareto, 1896) would provide an interesting starting point. However, due to the nature of the evolutionary game, this solution concept is not appropriate.

Maynard Smith and Price (1973) provided a solution concept for evolutionary games with their definition of an **evolutionarily stable strategy** (ESS). They reasoned that for a strategy to be evolutionarily stable it must be able to resist invasion from alternative strategies. With their definition, this resistance to invasion applied only when almost everyone in the population is using the ESS. In other words an ESS must be better than all alternative strategies when the ESS is common throughout the population. Soon it was recognized that the ESS definition had similarities to the **Nash equilibrium** (Nash, 1951) of classical game theory (Auslander *et al.*, 1978). A Nash equilibrium and an ESS are "no-regret" strategies in the sense that if everyone in a population is playing a Nash strategy (in a conventional game) or an ESS (in an evolutionary game), then no one individual can benefit from unilaterally changing their strategy. However, there are important differences between the Nash definition used in continuous static games and an ESS. The former focuses on payoffs with no recourse to any dynamic on the population sizes of individuals possessing particular strategies;

whereas, an ESS must also address population dynamics as influenced by the strategies present within and among the interacting populations.

The origins of a formal evolutionary game theory can be traced to the introduction of the ESS concept and its application to matrix games. It had its initial intellectual beachhead on the shores of simple matrix games (Riechert and Hammerstein, 1983). Matrix games, involving pairwise animal behaviors, form portions of the conceptual underpinning of almost all current investigations into animal social behaviors, including behaviors such as territoriality, dominance hierarchies, cooperation, group foraging, vigilance, group size, female choice, mate competition, and breeding strategies. Evolutionary game theory has been expanded to include continuous games where strategy sets are continuous rather than discrete (Vincent and Brown, 1984$b$). Continuous games can model quantitative traits such as body size, flowering date, and niche specialization. Because the ESS may contain several coexisting strategies, evolutionary game theory can be applied to the maintenance of polymorphisms within populations (Bishop and Cannings, 1976) or to the question of maintaining species diversity (Brown and Vincent, 1987$a$). Evolutionary game theory can apply to questions of coevolution (Lawlor and Maynard Smith, 1976), speciation (Dieckmann and Doebeli, 1999), and the evolution of community organization (Brown and Vincent, 1992). In particular, the last 30 years has seen advances in the types of ecological and evolutionary dynamics inherent in evolutionary game theory, and in the array of stability properties associated with these dynamics (Metz *et al.*, 1996; Cohen *et al.*, 1999).

Early on it was appreciated that the ESS definition of Maynard Smith captured one notion of evolutionary stability but missed a second (Taylor and Jonker, 1978; Zeeman, 1981). The original definition requires only that the ESS be **resistant to invasion** by a rare alternative strategy (a necessary condition for evolutionary stability), but this condition does not ensure that a population will actually evolve to the ESS. This requires a second notion of evolutionary stability, that of **convergence stability**. Convergence stability implies that a population will evolve to an ESS when its strategy composition is near but not at the ESS (Eshel, 1983). The presence or absence of these two forms of evolutionary stability yield a variety of possible solutions to evolutionary games. Notable among non-ESS solutions are evolutionarily stable minima at which a population's strategy actually evolves to a fitness minimum (Brown and Pavlovic, 1992; Abrams *et al.*, 1993$b$). At such points, an individual by using its current strategy can do no worse! Remarkably, such perverse situations can be convergent stable. They are not ESS and interestingly they open the door to new models of adaptive speciation or raise questions about whether natural selection actually leaves species with strategies that not only appear unfit, but seem to be the worst given the circumstances (Brown and Pavlovic, 1992; Cohen *et al.*, 1999).

### 1.3.4 Scope of evolutionary game theory

Evolutionary game theory began with the objectives of solving questions of natural selection involving the behavior of animals confronted with pairwise conflict. Such conflict is easily modeled by matrix games and the bulk of previous books on evolutionary games emphasize behaviors and this class of games (Maynard Smith, 1982; Hofbauer and Sigmund, 1988; Cressman, 2003). Evolutionary game theory can easily model frequency-dependent selection. But, from a genetics-based perspective on evolution, evolutionary game theory is often viewed as a short-cut method for solving problems with frequency dependence that applied only to organisms with asexual reproduction. This is a bit ironic, as sex, genders, and sex ratios are strategies that likely evolved in response to natural selection and which form a part of the evolutionary game. Many important issues regarding the application of evolutionary game theory in the absence of explicit genetics have been resolved. In general, conclusions obtained using evolutionary game models or their genetic counterparts are the same or similar. This is particularly the case for quantitative genetics models and evolutionary game models based upon the adaptive dynamics of quantitative traits. Evolutionary game theory is sufficiently developed to provide a framework for evolution of adaptations within species, the coevolution of traits among species, speciation, and micro- versus macroevolution.

**Adaptation** has been thought of in two contexts. The first of these describes the *FF&F*, the extent to which a heritable trait has been shaped via natural selection within a specified environmental context. The second context is similar to the first, but not directly tied to the endpoint of natural selection. Adaptation is often used to describe how a trait or characteristic of an individual serves it well. For instance, one might refer to how well squirrels have adapted to urban habitats. This statement conveys how squirrels seem ideally suited to backyards and how they have used their behavioral flexibility to adjust to peanut butter for food and attics as homes. However, they have not (at least not completely) adapted in the sense of evolving new strategies in response to a novel habitat. It is just that squirrels have a suite of existing traits that permit them to thrive with humans (this association is most evident with fox squirrels and gray squirrels in the United States, but this train of thought applies equally well to those specific mammals and birds that thrive in the backyards of people anywhere in the world). As a game theoretic concept we define an **adaptation** as the particular strategies which make up the ESS. Such a view of adaptation combines the appealing parts of the two contexts within which adaptation is commonly used. The **ESS** is the endpoint of evolution by natural selection. Furthermore, as a "no-regret" strategy, an individual's strategy serves it better than any alternative strategy. In this sense, adaptations resulting from natural selection are optimal given the circumstances, and represent the *FF&F* (Mitchell and Valone, 1990).

We use the term **adaptive dynamics** (Metz *et al.*, 1996) to describe the change in the frequency of strategies within the population and the term **strategy dynamics** to describe how a strategy associated with a particular species changes with time. Because the term adaptive dynamics has more than one related meaning in the literature,[5] we use it only in the restricted sense above. The strategy dynamics equations are an important part of evolution by natural selection that drives a population to an ESS, to other solutions, or to non-equilibrium evolutionary dynamics. Because of strategy dynamics, Fisher's Fundamental Theorem of Natural Selection also applies to evolutionary games.

**Co-adaptation** in genetics describes reciprocal evolutionary responses of genes at different loci to each other's effect on the individual's fitness. Via co-adaptation, the gene favored by natural selection at one locus may be influenced by the genes at other loci. In the context of adaptation, this makes sense when one considers how different traits interact to determine an organism's success. For instance, one often sees co-adaptation between behaviors and morphologies. Organisms with specialist morphologies often feed selectively, while those with generalist morphologies feed opportunistically. By allowing strategies to be a vector of traits, evolutionary game theory can model co-adaptations. In this case, the ESS for one trait is influenced by the values of other traits within the organism. For instance, in desert annuals seed-dispersal mechanisms, seed dormancy, and xeric (dry) adapted leaves can all assist the plant in bet-hedging against droughts and bad years. Hence, a plant with highly xeric leaves requires less seed dormancy than one with mesic (wet) leaves. Or a plant with seed-dispersal traits such as hooks for animal dispersal or awns for wind dispersal requires less seed dormancy or less xeric leaves. At the ESS we expect these traits to become co-adapted, and evolutionary game theory provides the modeling tools needed for such co-adaptations.

## 1.4 Road map

As with living systems, the evolutionary game theory that we present in this book evolved from a relatively straightforward theory developed to deal with

---

[5] For example, the term adaptive dynamics also serves to describe changes in gene frequency from population genetics models. Not surprisingly, the equations for change in strategy frequency have a parallel to those of Wright (1932) for changes in gene frequency.

biological problems with little structure to a more complex theory that is applicable to biological problems with lots of structure. Fortunately, the route to complexity is based on an underlying principle that does not change as we add layers of complexity. As a consequence, one can understand the basic ideas by focusing on the easy problem first and leaving the more complicated (and more interesting) problems for later. We structure this book accordingly.

The bulk of the mathematical development occurs in Chapters 2, 4, 5, 6, and 7. Chapter 2 provides an introduction to population modeling, population dynamics, equilibrium points, and the stability of these points. Chapter 3 provides an introduction to classical and evolutionary game theory. Life is viewed as a game and the stage is set for game theory to provide the tools for modeling the ecological and evolutionary dynamics associated with natural selection, with the introduction of the **fitness generating function** ($G$-function) concept. Individuals are said to be of the same **G-function** if they possess the same set of evolutionarily feasible strategies and experience the same fitness consequences of possessing a given strategy within a given environment. There is a close connection between a $G$-function and the German term **bauplan**, which is an old descriptor for classifying organisms by what appear to be common design features or design rules. The $G$-function may be thought of as describing both an organism's bauplan and the environmental conditions that the organisms must deal with.

Chapters 4, 5, 6, and 7 are structured in parallel to address the concepts needed to model and analyze evolutionary games. In Chapter 4 we present a recipe for making an evolutionary game, starting with a model of population ecology and showing how to construct a $G$-function from it. The $G$-function is used in the development of both the ecological and evolutionary dynamics (Chapter 5). We refer to the combination of population dynamics (ecological changes in the population sizes) and strategy dynamics (evolutionary changes in a species' strategy value) as **Darwinian dynamics**. The Darwinian dynamics may converge on population and strategy values that are both convergent stable and which cannot be invaded by rare alternative strategies. The strategy values obtained in this way are **evolutionarily stable strategies** (**ESS**). Chapter 6 expands on the ESS concept of Maynard Smith to provide a formal definition of an ESS. An ESS must be convergent stable and optimal in the sense of maximizing an individual's fitness given the circumstances and strategies of others. The ESS maximum principle of Chapter 7 provides necessary conditions For determining an ESS. In terms of the $G$-function the ESS is an evolutionary optimum that represents the $FF\&F$. Our approach in each of Chapters 4–7 is to present the theory in terms of the simplest problem first before moving to the more complex problems.

An ESS can possess a diversity of species or strategies (Chapter 8). These co-existing species can emerge from both within and between $G$-functions. When diversity is promoted within a $G$-function the evolutionary model can include speciation and the means for generating diversity. When diversity occurs between $G$-functions the evolutionary model can include coevolution, microevolution (evolutionary changes within $G$-functions), and macroevolution (evolutionary changes resulting in new $G$-functions).

Much of evolutionary game theory focuses on matrix games. In Chapter 9 we revisit matrix games from the perspective of a more general theory of evolutionary games. Chapter 10 provides examples for applying the modeling tools of this book to topics in evolutionary ecology including habitat selection, resource competition, plant competition, and foraging games between predators and prey. The theory also has direct applications to problems involving the management and conservation of evolving systems (Chapter 11).

In Chapters 4–7 the following classes of biological problems are examined in detail. The same basic ideas are applicable to each class; however, the features become more complex as the problems become more complex.

### 1.4.1 The simplest problem

The simplest problem is one that might mimic early life on Earth. We assume that the evolving populations are based on a single $G$-function. There is only one bauplan and a single (scalar) strategy that defines a phenotype. There is a corresponding constraint set from which strategies may evolve within upper and lower bounds (e.g., negative body size not allowed). The adaptive strategies influence the fitness of the organisms relative to one another. Moreover the abiotic environment is so stable that population dynamics results in stable ecological equilibria. Evolution in such a system proceeds from the fact that not all phenotypes can co-exist. Rather, only certain phenotypes or combinations of phenotypes can survive and persist through time when confronted with a new phenotype. The main objective for the simplest problem and all others is to be able to identify those phenotypes or species that can persist through time.

### 1.4.2 Vector strategies

Even the simplest of organisms possess a number of different heritable traits. Each functioning protein, each structural character, and each physiological or behavioral pathway can represent a different, and sometimes independent, heritable trait. Many adaptive characteristics of an organism likely influence fitness.

The first generalization of the simplest problem is to introduce vector strategies with a corresponding constraint set. Many constraints are obvious, while others are not obvious such as those that result from physical or genetic limitations, or non-independencies among different heritable traits. Each element of a vector strategy describes a different heritable trait of interest.

### 1.4.3 Evolving systems with resources

This class of systems is useful when dealing with plants or animals feeding on an explicit resource that is not itself evolving. However, each affects the dynamics of the other. In some situations, competition between the evolving organisms occurs solely through non-evolving resources.

### 1.4.4 Multiple $G$-functions

An important generalization of the simplest problem is the introduction of additional $G$-functions. Multiple $G$-functions are required when a population contains several bauplans[6] or there is a single bauplan with organisms in more than one environmental setting. This allows the modeling of biological systems at different tropic levels as well as the introduction of novel types within a given tropic level. For example, a simple one-tropic-level system could involve just prey and predators. Two $G$-functions would be required, one for the prey and one for the predators. Evolution in this case could result in several prey species (all of the same bauplan) and several predators (all of the same bauplan). Two $G$-functions could also be used to define a simple tropic system with two levels, for example plants and herbivores. More-complicated systems are studied by introducing additional $G$-functions.

### 1.4.5 Frequency dynamics

The majority of the book takes a population dynamics point of view that deals with changes in population sizes with time. However, there is an alternative point of view that is important in the study of genetics and matrix games. Instead of thinking of a biological system in terms of the density a population has as a result of its corresponding strategy, one can think in terms of the frequency at which a given strategy is found in the population. The two points of view are totally interchangeable; however, there are advantages to both points of view.

---

[6] The original German spelling for the plural of bauplan is Bauplaene. Since we are using this term in a modern and somewhat different context, we choose to use the English plural.

The frequency dynamics point of view is particularly useful if there is no density dependence in the model, often the case in matrix games.

### 1.4.6 Multistage systems

A given bauplan may be more complicated than a bauplan based on lumping all life stages together into one variable. While this approach is desirable and useful for a lot of systems it will not be valid for all. Hence the theory is extended to include multistage $G$-functions (e.g., pupa, larva, adult).

### 1.4.7 Non-equilibrium dynamics

It is generally recognized that equilibrium dynamics represents an idealization, useful for study but not realistic in the real world. We agree. Having an abiotic environment that is fixed is not sufficient for stable equilibrium dynamics, but it certainly helps. Not only does assuming the existence of stable equilibrium dynamics simplify the theory, but valid conclusions and insights about evolution can still be made. For those situations where non-equilibrium dynamics is generic to the system, the theory is extended to include this condition.

# 2
# Underlying mathematics and philosophy

Darwin used lengthy, sometimes discursive, yet convincing, verbal arguments in his *Origin of Species*. Darwin's postulates, as discussed in Chapter 1 and upon which his theory is built, apply broadly to the explanation or understanding of evolution. However, as verbal concepts they are limited to persuasion with few formal predictive capabilities. For example, one can understand why Darwin's finches have particular beak characteristics (Weiner, 1994). One can even predict that natural selection will tend to increase beak size during periods of drought. However, one cannot use verbal Darwinian arguments to predict the exact beak size appropriate to a particular species of finch. In fact, the ability to make such a prediction based on pure Darwinian principles is impossible unless these principles can be translated into a mathematical language. Only then can his theory be used not only to explain, but also to make predictions. Furthermore, without a mathematical framework, it is difficult or impossible to understand how a trait such as cooperation evolves.

Making Darwin's theory rigorous and predictive has been an achievement of population genetics and quantitative genetics approaches to evolution. These approaches often get bogged down in the genetic details and, consequently, lose a sense of the ecological interactions that take place to determine evolution by natural selection. Furthermore, while the genetic approach may determine those "selfish" genes that are propagated through time, it is the trait that the genes code for that actually is selected. The genes are selected but the heritable phenotypes constitute what are selected for. The heritable phenotypes, not the genes, are the adaptations. Until one focuses on the function that traits serve rather than just their heritable recipe, one cannot answer why a trait has evolved or how it is maintained within a population. This provides the motivation to focus on heritable traits rather than the specific genetical recipe as the unit of adaptive evolution. We aim for a rigorous and predictive approach to Darwinian evolution that considers differences in the population growth rates of individuals possessing different strategies, as in Hutchinson's "ecological theater" and "evolutionary play" (Hutchinson, 1965).

In this book, we develop a theory for **Darwinian dynamics**. Dynamics, in the physical context, is the study of the relationship between motion and the forces affecting motion. We think of Darwinian dynamics as the study of the relationship between the evolution of heritable traits and natural selection as the force affecting evolution. (Genetic drift and mutation are other important forces of evolution, but outside of the central focus of this book.) Motion is change in position of a physical entity. Evolution is the change in character of a biological entity. The key here is the observation that both physical and biological systems change with time. If we want to understand and predict these changes, a mathematics is required that describes that change. Putting Darwinian dynamics into a mathematical framework similar to that used in ordinary population dynamics from ecology results in a theory for understanding and predicting evolution by natural selection – a theory that is testable and accessible through experiments.

The ecological process is modeled using standard population ecology approaches. However, we identify in these population dynamic models certain adaptive parameters called **strategies**. The strategies in the ecological models change with time and hence evolve. We show how a strategy dynamic can be derived from standard population dynamic models when a distribution of strategies is present among individuals of a population. The combination of population dynamics and strategy dynamics defines the Darwinian dynamics for an evolving system. Generally speaking, the population dynamics determines changes in population density with time whereas the strategy dynamics determines changes in the distribution of strategies within the population.

It is common in evolutionary models to work in terms of **frequency** of genes or strategies in the population. Because strategy frequency is easily determined by knowing the density of species within a population, we prefer, instead, to focus on the densities of species. The population density approach permits a close tie-in with population ecology, and it clarifies the density-dependent processes of most models. However, we also develop a frequency approach that is useful when discussing matrix games, or games without explicit consideration of population density.

As in any treatment of evolutionary ecology, we require a species concept. Any student of biology knows that species is a rather loaded concept, one that is often euphemistically left "constructively ambiguous." We use a definition that identifies individuals in a population as being of the same **species** if they all share the same $G$-function and have strategies that are closely clumped together. In this context, the mean of this clump is used to define a species. This

**strategy-species definition** is intended to be useful and applicable to defining different types of organisms at the taxonomic subdivision below genus. This definition will not remove all of the important issues and controversies over species definitions. Furthermore, the definition does not preclude ambiguities over whether two closely related types of individuals belong to the same species or to two different species. This fuzziness should exist in any evolutionary model that deals with the processes and outcomes of speciation resulting in the formation of new species and higher taxa. We advocate the strategy-species concept as useful for modeling Darwinian dynamics and the propagation and persistence of species diversity. The species concept is formalized and discussed in much more detail in Chapter 8.

Before building game theory models of natural selection, we need a notational system that can serve usefully throughout the book. From the perspective of the strategy-species definition, the remainder of this chapter is devoted to developing the notation, mathematics, and characterization of the concepts needed to describe Darwinian dynamics.

## 2.1 Scalars, vectors, and matrices

What level of mathematics do we need to describe the ecological and evolutionary changes associated with life as a game? Algebra is not enough. It is a static theory useful for calculating the outcome of a single event such as finding the roots of quadratic equations. Calculus with its concept of a derivative comes closer. Differential equations or difference equations (a form of iterated algebra) are needed to describe Darwinian dynamics. In this book we use both. A mathematical dynamical model for the evolutionary game may involve many variables, and requires vector and matrix notation to facilitate the efficient modeling of evolution.

A **scalar**, $u$, is a quantity with only one dimension. It can be represented as a point on a line. Height and temperature as well as the beak length of Darwin's finches are examples of scalars. A **vector, u**, is a quantity of dimension greater than 1. It can be represented as a point in a higher-dimensional space such as a plane (two dimensions), cube (three dimensions), or hypercube (greater than three dimensions). The length, height, and width of a bird's beak jointly represent a vector. We will use boldface to indicate a vector **u** and italic to indicate one of its scalar components $u_i$. In order to put together a model for an evolving biological system we need to deal with potentially many different vector quantities. For example, we need a vector **x** to describe the population sizes of all of the different species within a population and a vector **u** to describe

## 2.1 Scalars, vectors, and matrices

all the different strategies. The strategy of a given species may itself be a vector of traits (length, height, and width of bird beaks), and so the vector **u** may in fact be a vector of vectors.

Consider a population of $n_s$ different species.[1] We use the scalar $x_i$ to represent the number, density or biomass[2] of individuals of species $i$ and the vector **x** to represent the population densities of the $n_s$ different species. While we have assumed that there are $n_s$ different species, this number need not be fixed. In fact, one of the features of Darwinian dynamics is that the evolutionary process may determine the number of species $n_s$ as the product of evolution. The vector **x** is composed of scalar components as described by the vector

$$\mathbf{x} = \begin{bmatrix} x_1 & \cdots & x_{n_s} \end{bmatrix}.$$

All individuals in the population are identified by the heritable phenotypes or strategies which characterize that individual as belonging to a particular species. The notation

$$\mathbf{u}_i = \begin{bmatrix} u_{i1} & \cdots & u_{in_{u_i}} \end{bmatrix}$$

is used to denote the strategy vector of individuals of species $i$. The first subscript in the vector of traits refers to the species and the second subscript denotes the particular trait (e.g., bill length, bill depth, or bill width) of the strategy vector used by that species. Since the number of traits may vary from species to species, the size of a strategy vector may vary with species. We use the notation $n_{u_i}$ to denote the number of components in the strategy vector of species $i$. When we need to refer to all of the strategies in the population, the notation

$$\mathbf{u} = \begin{bmatrix} \mathbf{u}_1 & | & \cdots & | & \mathbf{u}_{n_s} \end{bmatrix}$$

is used. In general, **u** is a vector formed from the catenation of all of the species strategy vectors. The vertical bars are used to emphasize how this catenation leads to a natural partitioning of **u**. The notation is simplified considerably when all strategies are scalars. In this case we drop the double subscripts and refer to the strategy for species $i$ by $u_i$. In this case the vector of all strategies used by all of the species in the population is given by

$$\mathbf{u} = \begin{bmatrix} u_1 & \cdots & u_{n_s} \end{bmatrix}.$$

---

[1] We are using what may be an unfamilier subscript notation (e.g., $n_s$) in order to avoid a proliferation of symbols. The advantage of this notation is that it is mnemonic and hierarchical (e.g., $n_s$ refers to the number of species).

[2] Since populations may be measured in terms of number, density, or biomass, all of these terms are used interchangeably.

In many ecological circumstances the growth rate of a population depends upon the abundance of some **resource**. For plants, this resource may be nitrogen, phosphorus, etc. This introduces an environmental feedback where the population size of plants may influence the availability of resources, and the availability of these resources influences the growth of plants. Such environmental feedbacks are common as animals and plants modify features and resources in their environment. The significance of including such feedbacks in ecological models is recognized by studies and models of consumer–resource dynamics (Tilman, 1982), ecological engineering (Jones *et al.*, 1994), niche construction (Odling-Smee *et al.*, 2003), and feedback environments (Getz, 1999). In consumer-resource models, the primary interaction among individuals is through their use, protection, or reduction of a common limiting factor. There is often an indirect effect of individuals on each other, where the strategy of one individual influences the availability of the resource to another. In ecological engineering, the denning behavior, life style, or rooting strategy of individual organisms influences some aspect of the biotic or abiotic environment. For instance, in models of succession, a pioneer plant species may be able to colonize a recently disturbed system. The presence of the species may stabilize soil, promote nutrient build-up, alter moisture regimes, and perhaps facilitate the invasion of successive species as physical conditions change in response to the presence of these successive species. **Niche construction** considers ecological engineering from the perspective of how organisms may possess strategies specifically designed to make the environment more hospitable to the individual. For instance, prairie dogs require extensive burrow systems for protection and open sightlines aboveground for the detection of predators. Consequently, prairie dogs live in colonies, constantly create and remodel burrows and dens, and they will collectively chew on and decimate woody plants and shrubs that obstruct their sightlines. All of these interactions of organisms create a feedback environment in which the strategies and population sizes of organisms influence some property of the environment which in turn influences the fitness of the organisms.

Resources and environmental features are explicitly modeled by including resource dynamics in the evolutionary model. Let $n_y$ be the number of such resources. The vector of all resources is written as

$$\mathbf{y} = \begin{bmatrix} y_1 & \cdots & y_{n_y} \end{bmatrix}.$$

**Example 2.1.1 (three-species, two-strategy, three-resource system)**
*Suppose that a biotic community has three different plant species with densities given by*

$$\mathbf{x} = \begin{bmatrix} x_1 & x_2 & x_3 \end{bmatrix}$$

in which each species has a set of two strategies: stem height and root biomass,

$$\mathbf{u} = [\, \mathbf{u}_1 \mid \mathbf{u}_2 \mid \mathbf{u}_3 \,]$$
$$= [\, u_{11} \quad u_{12} \mid u_{21} \quad u_{22} \mid u_{31} \quad u_{32} \,]$$

and relies on three resources: light, available nitrogen, and water.

$$\mathbf{y} = [\, y_1 \quad y_2 \quad y_3 \,].$$

*In this case the state of the community, at any point in time, is specified by the three vectors* $\mathbf{u}$, $\mathbf{x}$, *and* $\mathbf{y}$. *In this example* $n_s = 3$, $n_{u_1} = n_{u_2} = n_{u_3} = 2$ *and* $n_y = 3$.

### 2.1.1 Elementary operations

As with scalars, elementary operations apply to vectors and matrices. A **matrix** is an array composed of several vectors. For example, a two-row, three-column ($2 \times 3$) matrix

$$\mathbf{M} = \begin{bmatrix} m_{11} & m_{12} & m_{13} \\ m_{21} & m_{22} & m_{23} \end{bmatrix}$$

may be thought of as being composed of either three column vectors or two row vectors. While a matrix cannot be thought of as a point in space, it does represent a convenient generalization of a scalar and vector. That is, a scalar is a $1 \times 1$ matrix, a column vector is an $n_s \times 1$ matrix and a row vector is a $1 \times n_s$ matrix. However, the usefulness of matrix notation goes beyond this as will be evident in later chapters. Both bold capital letters and calligraphic fonts are used to designate matrices. Since scalars and vectors are included as special classes of matrices we need only define the elementary operations in terms of matrices.

#### 2.1.1.1 Addition

Two matrices are added (or subtracted) component by component:

$$\begin{bmatrix} a_{11} & a_{12} & a_{13} \\ a_{21} & a_{22} & a_{23} \end{bmatrix} + \begin{bmatrix} b_{11} & b_{12} & b_{13} \\ b_{21} & b_{22} & b_{23} \end{bmatrix} = \begin{bmatrix} a_{11}+b_{11} & a_{12}+b_{12} & a_{13}+b_{13} \\ a_{21}+b_{21} & a_{22}+b_{22} & a_{23}+b_{23} \end{bmatrix}$$

or in matrix notation

$$\mathbf{A} + \mathbf{B} = \mathbf{C}.$$

For addition to make sense, the matrices must have the same dimensions.

### 2.1.1.2 Multiplication

Two matrices are multiplied in such a way that one takes the inner product of each of the row vectors of the first matrix with each of the column vectors of the second matrix. An inner product results in a scalar, by summing the product of each element of the row vector with its corresponding element in the column matrix. When all combinations of inner products of row and column vectors have been calculated the result is a new matrix whose dimension has the same number of rows as the first matrix and the same number of columns as the second matrix

$$\begin{bmatrix} a_{11} & a_{12} & a_{13} \\ a_{21} & a_{22} & a_{23} \end{bmatrix} \begin{bmatrix} b_{11} & b_{12} \\ b_{21} & b_{22} \\ b_{31} & b_{32} \end{bmatrix}$$
$$= \begin{bmatrix} a_{11}b_{11}+a_{12}b_{21}+a_{13}b_{31} & a_{11}b_{12}+a_{12}b_{22}+a_{13}b_{32} \\ a_{21}b_{11}+a_{22}b_{21}+a_{23}b_{31} & a_{21}b_{12}+a_{22}b_{22}+a_{23}b_{32} \end{bmatrix}$$

or in matrix notation

$$\mathbf{AB} = \mathbf{C}.$$

For multiplication to make sense, the number of columns in the first matrix must equal the number of rows in the second matrix. When the number of rows in the first matrix is equal to the number of columns in the second matrix, as is the case in the above example, the product produces a **square matrix** (same numbers of rows and columns).

### 2.1.1.3 Division

Division is defined for square matrixes in terms of an inverse. Let $\mathbf{M}$ be a square matrix of dimension $n \times n$ then $\mathbf{M}^{-1}$ is its inverse, provided that

$$\mathbf{MM}^{-1} = \mathcal{I} \tag{2.1}$$

where $\mathcal{I}$ is the $n \times n$ **identity matrix** with ones along its diagonal and zeros everywhere else

$$\mathcal{I} = \begin{bmatrix} 1 & 0 & \cdots & 0 & 0 \\ 0 & 1 & \cdots & 0 & 0 \\ \vdots & \vdots & \ddots & \vdots & \vdots \\ 0 & 0 & \cdots & 1 & 0 \\ 0 & 0 & \cdots & 0 & 1 \end{bmatrix}. \tag{2.2}$$

If an inverse exists, then the solution to the system of equations

$$\mathbf{Ax} = \mathbf{b}$$

where $\mathbf{A}$ is dimension $n_s \times n_s$, $\mathbf{x}$ is dimension $n_s \times 1$, and $\mathbf{b}$ is dimension $n_s \times 1$ is given simply by

$$\mathbf{x} = \mathbf{A}^{-1}\mathbf{b}.$$

When the inverse of a matrix exists, it can be found using the definition (2.1). When the size of a matrix is large, all of the elementary operations are cumbersome to perform by hand. Fortunately several software packages are available to render such calculations easy.

## 2.2 Dynamical systems

Biological systems are dynamical systems since the **state** of the system, defined by $\mathbf{x}$, $\mathbf{u}$, and $\mathbf{y}$, can (and usually does) change with time. Knowledge of the state, at any point in time, represents all the information needed to predict the future states of the biological system by means of dynamical equations. In modeling vernacular, $\mathbf{x}$, $\mathbf{u}$, and $\mathbf{y}$ are called **state variables**. Dynamical systems are usually modeled using difference equations[3] or differential equations.[4] We will consider both types of equations here; however, we will restrict the class of differential equations to ordinary differential equations.[5] Both the difference equations and the ordinary differential equations will be written in what is called **state-space notation**. For difference equations this means that the equations involve only the current state and the state one time period in the future. For differential equations this means that the equations involve only the current state and first-order derivatives of the state. In state-space form, the order of the system of equations will always be the same as the number of equations.

**Example 2.2.1 (state-space notation for difference equations)** *The following second-order difference equation*

$$z(t+2) + z(t+1) + z = 0$$

*may be written in state-space form as two first-order equations*

$$x_1(t+1) = x_2$$
$$x_2(t+1) = -x_1 - x_2$$

---

[3] A difference equation is a relationship between consecutive elements of a sequence in terms of current and future (and/or past) states. The current state is designated without an argument (e.g., $x$) and future states are designated with an argument [e.g., $x(t+1)$ is the state one time unit in the future from current time $t$].
[4] Equations that involve dependent variables (e.g., states) and their derivatives with respect to one or more independent variables are called differential equations.
[5] Differential equations that involve only one independent variable (e.g., time) are called ordinary differential equations.

where $z = x_1$. This result is easily verified by starting with $z = x_1$, stepping forward in time and making appropriate substitutions as follows

$$z(t+1) = x_1(t+1) = x_2$$
$$z(t+2) = x_2(t+1) = -x_1 - x_2 = -z - z(t+1).$$

**Example 2.2.2 (state-space notation for differential equations)** The following second-order ordinary differential equation

$$\ddot{z} + \dot{z} + z = 0$$

may be written in state-space form as two first-order equations

$$\dot{x}_1 = x_2$$
$$\dot{x}_2 = -x_1 - x_2$$

where $z = x_1$. This result is easily verified by starting with $z = x_1$, taking two derivatives and making appropriate substitutions as follows:

$$\dot{z} = \dot{x}_1 = x_2$$
$$\ddot{z} = \dot{x}_2 = -x_1 - x_2 = -z - \dot{z}.$$

Since we wish to model the interactions of systems that may have many different species utilizing many different resources, there may be many state variables involved with the order of the system quite large. It is for this reason that we use the state-space notation in modeling the dynamics. Under this notation, the next time step or the first derivative of each state variable is given on the left side of the equation and the function producing this change is given on the right side.

### 2.2.1 Difference equations

Changes in a species' population density are often modeled by means of **difference equations** using only the state variable **x**. For example, the dynamics for three different species $\mathbf{x} = \begin{bmatrix} x_1 & x_2 & x_3 \end{bmatrix}$ is modeled by

$$x_1(t+1) = f_1(x_1, x_2, x_3)$$
$$x_2(t+1) = f_2(x_1, x_2, x_3)$$
$$x_3(t+1) = f_3(x_1, x_2, x_3).$$

In general, the number of state variables can be large, so the equivalent notation

$$x_i(t+1) = f_i(\mathbf{x}) \quad i = 1, \cdots, 3 \tag{2.3}$$

## 2.2 Dynamical systems

is used. Difference equations produce a dynamic for **x** by means of **iteration**. That is, given an initial point in state space **x**(0), the state of the system in the future is calculated by first substituting **x**(0) into the right-hand side of (2.3) to determine **x**(1), then substituting **x**(1) into the right-hand side of (2.3) to determine **x**(2), and so on. In this way the state of the system for any future generation is determined. Note that $t$ actually plays the role of a counter. However, for biological models, $t$ generally is time scaled into empirically relevant units such as years[6] so that $t$ is the current year and $t+1$ is one year later.

A difference equation is also referred to as a **map** since it charts the current state to a future state. One unique feature of a difference equation is that even a one-dimensional equation with a non-linear right-hand side can produce dynamics that is far from simple. May (1976) was one of the first to point out that very simple biological models could produce very complicated population dynamics. Examples 2.2.3–2.2.5 examine three different versions of the **logistic equation**. The first two examples are discrete forms that produce stable asymptotic motion, periodic motion, or chaotic motion by simply changing the value of the parameter $r$ in the model.

**Example 2.2.3 (discrete logistic equation)** *The discrete one-dimensional logistic equation is given by*

$$x(t+1) = x\left[1 + \frac{r}{K}(K - x)\right]$$

*where $x$ is the population density at time $t$, $r$ is the intrinsic growth rate determined by the physiology of the individual species and $K$ is a constant known as the carrying capacity determined by the species characteristic and/or environmental factors. The dynamics of this system varies considerably depending on the value for the constant $r$. Consider iterating this equation starting from $x(0) = 0.1$, using $K = 1$ and $r = 1.5$, 2.5, and 3. Figure 2.1 illustrates the results obtained by iterating this equation 20 times. Note that when $r = 1.5$, stable asymptotic motion to the equilibrium solution[7] $x = K$ is obtained. However, with $r = 2.5$ a two-point cycle about the equilibrium solution is obtained and finally with $r = 3$, chaotic motion about the equilibrium solution occurs.*

---

[6] Or generations, in which case the units of generations must remain fixed even if actual generation time itself changes.

[7] The values of **x** at which the system dynamics produces no change in **x** with time is traditionally called a fixed point when using discrete equations and an equilibrium point when using differential equations. We choose to use "equilibrium" to refer to both. See Subsection 2.5.1 for more details.

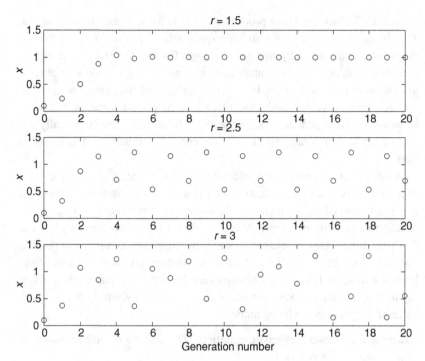

**Figure 2.1** Dynamics of the logistic map.

**Example 2.2.4 (discrete exponential logistic equation)** *An alternative discrete representation of the logistic map is given by*

$$x(t+1) = x \exp\left[\frac{r}{K}(K-x)\right].$$

*Figure 2.2 illustrates the dynamics of this map using the same initial condition, and K and r values as in the previous example. Similar, but not exactly the same population dynamics results.*

### 2.2.2 Differential equations

A continuous dynamical system is one in which the state of the system changes in a continuous fashion, such as the flight of a bird. Such systems can be described by a system of ordinary differential equations, one equation for each state variable. A system with two species and four resources would have six equations. Applications in biological systems include biomass models for plants and bacterial systems. However, the differential equation approach is often used

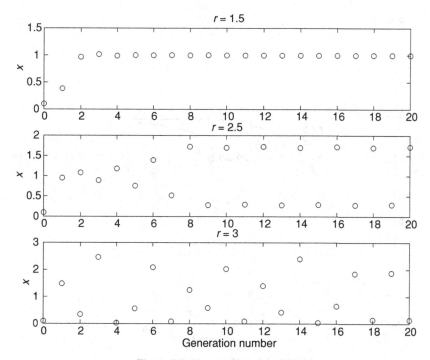

**Figure 2.2** Alternate logistic map.

to model animal populations as well when the number of individuals or units is sufficiently large to make the continuous approximation valid.

Again, suppose that the population dynamics of a system is represented by using only the state variable $\mathbf{x} = \begin{bmatrix} x_1 & x_2 & x_3 \end{bmatrix}$ of the form

$$\dot{x}_1 = f_1(x_1, x_2, x_3)$$
$$\dot{x}_2 = f_2(x_1, x_2, x_3)$$
$$\dot{x}_3 = f_3(x_1, x_2, x_3)$$

or equivalently

$$\dot{x}_i = f_i(\mathbf{x}), \quad i = 1, \cdots, 3. \tag{2.4}$$

where the dot denotes differentiation with respect to time $t$ (i.e., $\dot{x}_1 = dx_1/dt$).

Differential equations produce a temporal sequence of values for **x** by means of **integration**. Given an initial point in state space $\mathbf{x}(0)$, the state of the system for all future time is determined by integrating the system of equations (2.4)

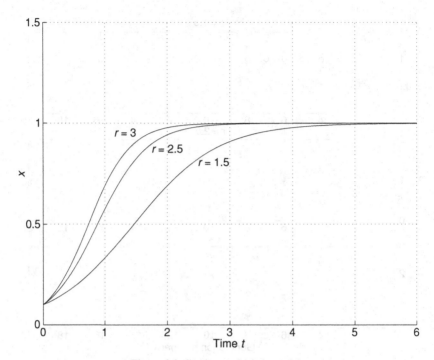

**Figure 2.3** Continuous logistic model.

starting from the initial state $\mathbf{x}(0)$. Generally an analytical solution is not available and numerical methods must be used. All integration results presented in this book were produced using integration routines available with Matlab.

**Example 2.2.5 (continuous logistic equation)** *The continuous version of the logistic model is given by*

$$\dot{x} = x\left[\frac{r}{K}(K-x)\right].$$

*Figure 2.3 illustrates the dynamics using the same initial condition, and K and r values as in the previous two examples. Note that, in each case, the continuous model approaches the equilibrium solution $x = K$ and is similar to the dynamics of the other two models only when $r = 1.5$. In fact, differential equations of the form of (2.4) cannot produce cyclic motion unless the dimension of $\mathbf{x}$ is 2 or greater, nor can chaotic motion be produced unless the dimension of $\mathbf{x}$ is 3 or greater.*

## 2.3 Biological population models

We will be using either difference equations or differential equations for modeling biological systems. The above three examples illustrate some possibilities for a system composed of one species. In each of the models, $x$ appears as a multiplicative factor on the right-hand side of each equation.

$$x(t+1) = x\left[1 + \frac{r}{K}(K-x)\right]$$

$$x(t+1) = x\exp\left[\frac{r}{K}(K-x)\right]$$

$$\dot{x} = x\left[\frac{r}{K}(K-x)\right].$$

This is a characteristic feature of biological models. Population growth, like money in the bank, compounds with time at a growth rate multiplied by the amount present. The logistic equation includes a density-dependent term (in brackets) that models a decrease in per capita growth rate as population size increases. At the non-zero equilibrium, $x = K$, the growth rate equals 1 in the difference equation models and zero in the differential equation model. Under unregulated growth, $K \to \infty$, the above equations simply become

$$x(t+1) = x(1+r)$$

$$x(t+1) = x\exp r$$

$$\dot{x} = xr$$

that produces exponential growth for any $r > 0$.

### 2.3.1 A special class of dynamical systems

The multiplicative nature of unregulated growth puts population models into a special class of dynamical systems in which the right-hand sides, representing the change or rate of change in $x_i$, always include a term for per capita growth rate, $F_i$, multiplied by $x_i$. We write the equations used to model population dynamics in the form

$$\begin{aligned} x_i(t+1) &= x_i F_i(\mathbf{u}, \mathbf{x}, \mathbf{y}) \\ \dot{x}_i &= x_i F_i(\mathbf{u}, \mathbf{x}, \mathbf{y}). \end{aligned} \quad (2.5)$$

This notation allows for the fact that, in general, the term $F_i$ depends on strategies $\mathbf{u}$, population density $\mathbf{x}$, and resources $\mathbf{y}$, all of the state variables that can change with time. Parameters that do not change with time are generally not included in the arguments of the functions $F_i$.

A unique feature of the differential equation system is that as long as $F_i$ is continuous in its arguments and as long as **u** and **y** cannot change discontinuously, if positive values are chosen for $x_i$ then for all future time $x_i \geq 0$. Indeed, this result is what we would expect for any population dynamic model. However, this result does not automatically follow for the difference equation model, but must be built in by using realistic functions for $F_i$.

### 2.3.2 The fitness concept with scalar $F_i$

The same **fitness function** applies to all individuals of the same species. For example, the fitness function for individuals of species $i$ is a scalar function of the vectors **u**, **x**, and **y** and is designated[8] by $F_i(\mathbf{u}, \mathbf{x}, \mathbf{y})$. For the discrete difference equation models, **fitness** is defined as the per capita change in population density (the finite growth rate) from one time period to the next; whereas, for differential equation models, fitness is defined as the current per capita rate of change in population density (the instantaneous growth rate). Under these definitions, we see that the $F_i$ values in (2.5) are indeed fitness functions.

For the first type of discrete model we assume that the functions $F_i$ are of the form

$$F_i(\mathbf{u}, \mathbf{x}, \mathbf{y}) = 1 + H_i(\mathbf{u}, \mathbf{x}, \mathbf{y}).$$

For the second type of discrete model we assume that the functions $F_i$ are of the form

$$F_i(\mathbf{u}, \mathbf{x}, \mathbf{y}) = \exp H_i(\mathbf{u}, \mathbf{x}, \mathbf{y})$$

and for the differential equation models we have

$$F_i(\mathbf{u}, \mathbf{x}, \mathbf{y}) = H_i(\mathbf{u}, \mathbf{x}, \mathbf{y}).$$

That is, the classes of population model that we will study are of the form

$$x_i(t+1) = x_i[1 + H_i(\mathbf{u}, \mathbf{x}, \mathbf{y})] \tag{2.6}$$

$$x_i(t+1) = x_i \exp H_i(\mathbf{u}, \mathbf{x}, \mathbf{y}) \tag{2.7}$$

$$\dot{x}_i = x_i H_i(\mathbf{u}, \mathbf{x}, \mathbf{y}). \tag{2.8}$$

By using the $H$ notation rather than the $F$ notation, we are able to express results for all three models simultaneously and use the $H$ notation throughout the book. Because $F$ and $H$ are simply related, it is more convenient to refer to $H$ as the **fitness function** and call $F$ the **population projection function**. The

---

[8] In those situations in which it is desirable to distinguish between different life stages of an individual, the fitness function becomes a matrix designated by $\mathbf{F}_i(\mathbf{u}, \mathbf{x}, \mathbf{y})$.

terminology for $F$ is borrowed from matrix population model theory (Caswell, 1989; Stearns, 1992). When we discuss multistage models, we cannot define fitness as simply a finite growth rate or an instantaneous growth rate (as was done here for the scalar case) but we can define a population projection matrix $F$ and a fitness matrix $\mathbf{H}$. In other words the definitions given here are special cases of a more general situation.

### 2.3.3 Continuous versus discrete modeling with scalar fitness

The two discrete equations (2.6) and (2.7) are "derivable" from the continuous differential equation (2.8). In order to obtain the first discrete equation, the differential equation is approximated as

$$\frac{x_i(t + \Delta t) - x_i}{\Delta t} = x_i H_i\,(\mathbf{u}, \mathbf{x}, \mathbf{y})$$

$$x_i(t + \Delta t) = x_i + x_i H_i\,(\mathbf{u}, \mathbf{x}, \mathbf{y})\,\Delta t.$$

Setting $\Delta t = 1$ yields (2.6).

In order to obtain (2.7), the continuous model is first written as

$$\frac{dx_i}{x_i} = H_i\,(\mathbf{u}, \mathbf{x}, \mathbf{y})\,dt.$$

Integrating both sides over one time interval, we get

$$\int_{x_i}^{x_i(t+1)} \frac{dx_i}{x_i} = \int_t^{t+1} H_i\,(\mathbf{u}, \mathbf{x}, \mathbf{y})\,d\tau$$

or

$$\ln \frac{x_i\,(t+1)}{x_i} = \int_t^{t+1} H_i\,(\mathbf{u}, \mathbf{x}, \mathbf{y})\,d\tau. \qquad (2.9)$$

By definition, the state variables in a discrete process remain constant between the time intervals. This in turn implies that $H_i\,(\mathbf{u}, \mathbf{x}, \mathbf{y})$ remains constant over the unit time interval, allowing integration of (2.9) to yield

$$\ln \frac{x_i\,(t+1)}{x_i\,(t)} = H_i\,(\mathbf{u}, \mathbf{x}, \mathbf{y})$$

or

$$x_i\,(t+1) = x_i \exp H_i\,(\mathbf{u}, \mathbf{x}, \mathbf{y}).$$

The two discrete versions of the logistic equation given in Examples 2.2.3 and 2.2.4 are related to the continuous model in Example 2.2.5 in exactly this way.

## 2.4 Examples of population models

We model population dynamics by selecting from among the forms in (2.6)–(2.8). A given scalar fitness function $H_i(\mathbf{u}, \mathbf{x}, \mathbf{y})$ can be used with any one of the dynamical models and we do not need to specify *a priori* which of the three models we intend to use. We need develop only one theory that applies to all three models. In general, $H_i$ will explicitly depend on the population densities $\mathbf{x}$, existing strategies $\mathbf{u}$, and resources $\mathbf{y}$. However, the first few models we look at do not include the dependence on $\mathbf{u}$ and $\mathbf{y}$.

### 2.4.1 Single-species logistic model

One of the earliest models used to describe the population dynamics of a single species is given by the Verhulst–Pearl equation (Verhulst, 1844; Pearl, 1924)

$$\dot{x} = x\left[\alpha\left(\beta - x\right)\right] \tag{2.10}$$

where $\alpha$ and $\beta$ are constant parameters. The significant feature of this equation is that it has an equilibrium solution other than zero. In its modern form with $\alpha = r/K$, and $\beta = K$, it is known as the logistic equation that we represent in terms of the scalar fitness function

$$H(x) = \frac{r}{K}(K - x). \tag{2.11}$$

The parameters $r$ and $K$, as noted in Example 2.2.3, are the intrinsic rate of growth and the carrying capacity respectively. All three population models have a non-zero equilibrium at $x = K$. When $x$ is small, growth will be exponential for the continuous model and exponential-like for the difference equation models.

Logistic population growth is often a poor approximation to growth rates of actual populations (Slobodkin, 2001). Furthermore the model may not have a clear mechanistic interpretation in terms of intrinsic growth rates and carrying capacities. However, conceptually it represents the simplest first-order approximation of any population in which fitness declines with population size (Turchin, 2001). It is also a valuable conceptual starting point for similar models involving interspecific interactions such as competition and predation.

### 2.4.2 Lotka–Volterra models for many species of individuals

The classical model for a one-predator, one-prey system is due to Lotka (Lotka, 1932) and Volterra (Volterra, 1926).

$$\begin{aligned} \dot{x}_1 &= x_1\left(\alpha - \beta x_2\right) \\ \dot{x}_2 &= x_2\left(-\gamma + \delta x_1\right) \end{aligned} \tag{2.12}$$

where $x_1$ is the prey population density and $x_2$ is the predator population density. All of the parameters are assumed to be positive. The parameter $\alpha$ is related to the birth rate of the prey, $\gamma$ to the death rate of the predator, $\beta$ and $\delta$ to the interactions between prey and predators. This model is the simplest form for a two species interaction. It has the feature of producing equilibrium points that have neutral stability. It is also extremely useful for developing more-sophisticated models of competition and predation. Equations (2.12) can also be used to model competition between two species by changing the sign of the parameters $\gamma$ and $\delta$.

The Lotka–Volterra (L–V) competition model is a generalization of (2.12) for $n_s$ species. All species have a positive growth term with all interspecific interactions negative, including a negative intraspecific interaction. The following is the generalization of the L–V competition equations for $n_s$ separate species

$$H_1(\mathbf{x}) = \frac{r_1}{K_1}\left(K_1 - \sum_{j=1}^{n_s} a_{1j} x_j\right)$$

$$\vdots = \vdots$$

$$H_{n_s}(\mathbf{x}) = \frac{r_{n_s}}{K_{n_s}}\left(K_{n_s} - \sum_{j=1}^{n_s} a_{n_s j} x_j\right).$$

Note that, when $n_s = 1$ and $a_{11} = 1$, this system reduces to the logistic model and when $n_s = 2$, $a_{11} = a_{22} = 0$, $a_{12} = \beta K_1/\alpha$, $a_{21} = \delta K_2/\gamma$, $r_1 = \alpha$, and $r_2 = -\gamma$, this system reduces to the original Lotka–Volterra prey–predator model. Like logistic population growth, this model can be a poor predictor of actual population dynamics in multi-species competitive interactions. But, it does represent the simplest first-order approximation of any multi-species system under inter- and intraspecific competition. The generalized Lotka–Volterra model is frequently used to model co-existence of similar competitors (Goh, 1980) while more comprehensive models are used for prey–predator systems.

The various constants in the Lotka–Volterra competition model have their own notation. The **intrinsic growth rate** $r_i$ is the (exponential) rate of growth a species would have when $\mathbf{x}$ is near-zero density. The **carrying capacity** $K_i$ is the non-zero equilibrium density for any species when all other species are at zero density. The **competition coefficient** $a_{ij}$ determines the competitive effect of species $j$ on species $i$.

### 2.4.3 Leslie model of one prey and one predator

The following model (Brown and Vincent, 1992) extends the original Lotka–Volterra prey–predator model to include a density-dependent death-rate term for the prey, a birth-rate term for the predator and a more realistic death-rate term for the predator (Leslie, 1945).

$$H_1 = \frac{r_1}{K_1}(K_1 - x_1) - bx_2$$

$$H_2 = r_2\left(1 - \frac{x_2}{cbx_1}\right). \tag{2.13}$$

In this model, $H_1$ is the fitness function of the prey and $H_2$ is the fitness function of the predator. The death-rate term of the predator has two density-dependent terms. The first increases the death rate (or reduces the birth rate) with an increase in the density of predators and the second decreases the death rate with an increase in the density of prey. The constant $b$ determines the effectiveness of predators in killing prey. The constant $c$ relates to the nutritional value of the prey to the predators. Unlike some models of predator–prey interactions that assume no direct negative effects of predators on each other (Rosenzweig and MacArthur, 1963), this model assumes that predators directly interfere with each other. This direct negative effect of predators on themselves increases the stability of equilibrium points that have positive sizes for the prey and predator populations.

### 2.4.4 Many prey and many predators model

The above model may be generalized to a community with many prey species and many predator species. To do so, the prey model needs to include the competitive effects of each prey species on the others and the mortality induced by each of the predator species. Furthermore, the predator models must include the negative direct effects of the combined populations of predators and the benefits accrued from capturing each of the different prey species. Assume that $n_p$ is the number of prey species and $n_s$ is the number of prey plus predator species

$$H_1(\mathbf{x}) = \frac{r_1}{K_1}\left(K_1 - \sum_{j=1}^{n_p} a_{1j}x_j\right) - \sum_{j=n_p+1}^{n_s} b_{1j}x_j$$

$$\vdots = \vdots$$

$$H_{n_p}(\mathbf{x}) = \frac{r_{n_p}}{K_{n_p}}\left(K_{n_p} - \sum_{j=1}^{n_p} a_{n_p j}x_j\right) - \sum_{j=n_p+1}^{n_s} b_{n_p j}x_j$$

$$H_{n_p+1}(\mathbf{x}) = r_{n_p+1}\left(1 - \frac{\sum_{j=n_p+1}^{n_s} x_j}{c\sum_{j=1}^{n_p} b_{n_p+1,j}x_j}\right)$$

## 2.4 Examples of population models

$$\vdots = \vdots$$

$$H_{n_s}(\mathbf{x}) = r_{n_s} \left( 1 - \frac{\sum_{j=n_p+1}^{n_s} x_j}{c \sum_{j=1}^{n_p} b_{n_s j} x_j} \right).$$

### 2.4.5 Identifying strategies in the Lotka–Volterra model

Many of the parameters used in the above models could be either strategies or functions of strategies. Heritable phenotypes of organisms likely influence their ability to capture prey, evade predators, efficiently metabolize food, etc. All such phenotypes have effects on the corresponding parameters in the organism's fitness function. The strategies used by these phenotypes become the **u** used in the fitness functions. The methods presented in this book require that the strategies be explicitly identified. Let's see how this could be done with the Lotka–Volterra model. All of the parameters in this model could possibly be strategies themselves. However, more likely there are trade-offs between parameters. For example, $r_i$, $K_i$, and $a_{ij}$ may all depend on metabolic rates and conversion efficiencies and, if these more basic strategies were changed to increase, say, $r_i$, it is likely that $K_i$ would decrease with possible changes in $a_{ij}$ as well.

Consider the situation where $r_i$ is constant, $K_i$ is a function of strategy $u_i$, and $a_{ij}$ is a function of all of the strategies **u**. The Lotka–Volterra model is then expressed as

$$H_1(\mathbf{u}, \mathbf{x}) = \frac{r_1}{K_1(u_i)} \left( K_1(u_i) - \sum_{j=1}^{n_s} a_{1j}(\mathbf{u}) x_j \right)$$

$$\vdots = \vdots$$

$$H_{n_s}(\mathbf{u}, \mathbf{x}) = \frac{r_{n_s}}{K_{n_s}(u_{n_s})} \left( K_{n_s}(u_{n_s}) - \sum_{j=1}^{n_s} a_{n_s j}(\mathbf{u}) x_j \right).$$

To complete the model, specific relationships for $K_i$ and $a_{ij}$ must be given. For example, the following functional forms have been used in models of coevolution (Roughgarden, 1983; Vincent et al., 1993)

$$K_i(u_i) = K_m \exp\left(-\frac{u_i^2}{2\sigma_k^2}\right)$$

$$a(\mathbf{u}) = \exp\left[-\frac{(u_i - u_j)^2}{2\sigma_a^2}\right].$$

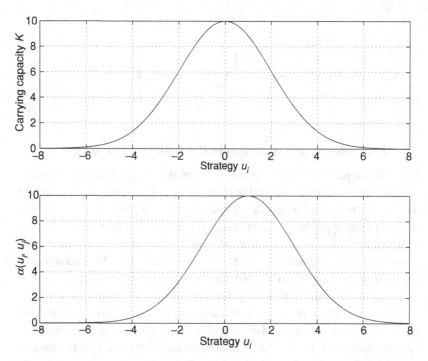

**Figure 2.4** The carrying capacity and intraspecific competition as distribution functions.

Both of these equations describe normal distribution functions with the mean value determined by the choices for $u_i$ and $u_j$ as illustrated in Figure 2.4 using the values $K_m = 10$, $\sigma_k^2 = \sigma_\alpha^2 = 4$, and $u_j = 1$. In this formulation, we have assumed that the variances and maximum values are fixed.

### 2.4.6 Consumer-resource models

The models presented so far do not explicitly include the utilization of resources. Rather, resource use is incorporated implicitly through terms such as carrying capacity. **Consumer-resource models** have an explicit dependence on resources and are used extensively in the modeling of plants (Tilman, 1982). As the consumer, the plant requires resources such as soil nutrients and sunlight. An example of this situation (Vincent and Vincent, 1996) uses the $H$ functions

$$H_1(\mathbf{u}, \mathbf{x}, \mathbf{y}) = \frac{r y_1 u_1}{y_1 + k_{y_1}} - R - d$$

$$H_2(\mathbf{u}, \mathbf{x}, \mathbf{y}) = \frac{r y_2 (1 - u_2)}{y_2 + k_{y_2}} - R - d$$

to model how the biomasses of two different plant species change with time. The $H_1$ function is for a plant species that is nutrient limited and $H_2$ is for a plant species that is light limited. The strategies $u_1$ and $u_2$ are the fractions of biomasses each plant species devotes to roots, $y_1$ is available soil nutrients, $y_2$ is the light availability, and the remaining parameters $r$, $R$, $d$, $k_{y_1}$, and $k_{y_2}$ are fixed parameters particular to this model. The nutrient availability in the soil is modeled by a differential equation of the form

$$\dot{y}_1 = N_1(\mathbf{u}, \mathbf{x}, y_1)$$

and the light availability is modeled by an algebraic equation of the form

$$N_2(\mathbf{u}, \mathbf{x}, y_2) = 0.$$

Note that the fitness functions do not depend on the plant biomasses $x_1$ and $x_2$ and that only the strategy of the first species is contained in the first fitness function and only the strategy of the second species is contained in the second fitness function. Competition between the plants takes place only through the resources $y_1$ and $y_2$. Since the resources available to each of the plants depend on the biomass of each species as well as the strategies used by each species these fitness functions along with the resource equations do, indeed, model competition.

### 2.4.7 Multistage models

In many biological models, we may not want to combine all individuals of a population into a single scalar variable $x_i$. This situation applies when one needs to consider distinct age classes, developmental states, or life-history stages. For instance, many insect species exhibit larval, pupal, and adult stages, all possessing quite distinct ecologies and needs. Other organisms may be modeled by considering how age explicitly influences fecundity and mortality. Classes of individuals within a multistage model may also simply represent those individuals experiencing different conditions within a spatially heterogeneous environment.

If a particular species has, say, three life stages then we can refer to this species by means of the vector

$$\mathbf{x}_i = \begin{bmatrix} x_{i1} & x_{i2} & x_{i3} \end{bmatrix}.$$

We consider only multistage population dynamic models of the form

$$\mathbf{x}_i(t+1) = \mathbf{x}_i \mathbf{F}_i^T(\mathbf{u}, \mathbf{x}, \mathbf{y})$$
$$\dot{\mathbf{x}}_i = \mathbf{x}_i \mathbf{F}_i^T(\mathbf{u}, \mathbf{x}, \mathbf{y})$$

where $\mathbf{F}_i$ are matrices and $\mathbf{F}_i^T$ is the **transpose** of $\mathbf{F}_i$ obtained by interchanging the rows and columns of $\mathbf{F}_i$. In this case, $\mathbf{F}_i$ is a **population projection matrix**. We do not include an exponential difference form since such a form cannot be derived from differential equations as described in Subsection 2.3.3.

**Example 2.4.1 (transpose notation)** *To see why the transpose notation is needed, suppose that we had the population projection matrix*

$$\mathbf{F} = \begin{bmatrix} F_{11} & F_{12} \\ F_{21} & F_{22} \end{bmatrix}$$

*with a single-species system described by the differential equation*

$$\dot{x}_1 = x_1 F_{11} + x_2 F_{12}$$
$$\dot{x}_2 = x_1 F_{21} + x_2 F_{22}.$$

*It follows that*

$$\begin{bmatrix} \dot{x}_1 & \dot{x}_2 \end{bmatrix} = \begin{bmatrix} x_1 & x_2 \end{bmatrix} \begin{bmatrix} F_{11} & F_{21} \\ F_{12} & F_{22} \end{bmatrix}.$$

*That is*

$$\dot{\mathbf{x}} = \mathbf{x} \mathbf{F}^T.$$

Similarly to the scalar case we express difference and differential equation models in terms of a **fitness matrix** $\mathbf{H}_i$

$$\mathbf{x}_i(t+1) = \mathbf{x}_i \left[ \mathbf{I} + \mathbf{H}_i^T(\mathbf{u}, \mathbf{x}, \mathbf{y}) \right]$$
$$\dot{\mathbf{x}}_i = \mathbf{x}_i \mathbf{H}_i^T(\mathbf{u}, \mathbf{x}, \mathbf{y}).$$

where $\mathbf{I}$ is the **identity matrix** (of appropriate dimension) as given by (2.2). The difference equation model can be "derived" from the differential equation model in the same way as the scaler case as in Subsection 2.3.3.

**Example 2.4.2 (two-stage system)** *Consider the following two-stage differential equation system*

$$\dot{x}_{i1} = x_{i1} \left( f(u_i) - \sum_{j=1}^{n_x} u_j x_{j1} \right) + x_{i2} u_i$$

$$\dot{x}_{i2} = x_{i1} u_i - x_{i2} \left( \sum_{j=1}^{n_x} x_{j1} \right).$$

*The fitness matrix for this case is given by*

$$\mathbf{H}_i(\mathbf{u}, \mathbf{x}) = \begin{bmatrix} f(u_i) - \sum_{j=1}^{n_x} u_j x_{j1} & u_i \\ u_i & -\left(\sum_{j=1}^{n_x} x_{j1}\right) \end{bmatrix}$$

*Note that the product*

$$\mathbf{x}_i \mathbf{H}_i^{\mathrm{T}}(\mathbf{u}, \mathbf{x}) = \begin{bmatrix} x_{i1} & x_{i2} \end{bmatrix} \begin{bmatrix} f(u_i) - \sum_{j=1}^{n_x} u_j x_{j1} & u_i \\ u_i & -\left(\sum_{j=1}^{n_x} x_{j1}\right) \end{bmatrix}$$

*results in the right-hand side of the original set of equations. In this case, since $\mathbf{H}_i$ is symmetric, $\mathbf{H}_i = \mathbf{H}_i^{\mathrm{T}}$.*

Multistage models will be discussed further in Chapters 4, 6, 7, and 10.

## 2.5 Classical stability concepts

The stability of dynamical systems, such as (2.6)–(2.8), is always with reference to some nominal operating condition such as a trajectory or equilibrium point. The primary focus in this book will be stability with reference to equilibrium points. In discussing the stability of evolutionary games, we must keep in mind that there may be up to three coupled dynamical systems running simultaneously. These determine how $\mathbf{x}$, $\mathbf{u}$, and $\mathbf{y}$ change with time. As a consequence, more than one definition of stability is needed in order to distinguish the **ecological stability** associated with $\mathbf{x}$ and $\mathbf{y}$ from the **evolutionary stability** associated with $\mathbf{u}$. Both of these stability concepts differ from classical stability concepts that are defined for systems of the form

$$\mathbf{x}(t+1) = \mathbf{f}(\mathbf{x})$$
$$\text{or} \tag{2.14}$$
$$\dot{\mathbf{x}} = \mathbf{f}(\mathbf{x}).$$

Nevertheless, classical stability represents a starting point for the study of stability of the evolutionary game. We review some of these concepts in this section. Ecological stability and evolutionary stability, the focal points of interest in the Darwinian game, will be formally introduced in Chapters 5 and 6.

Stability is so pervasive in our lives that it seems to be the natural order of things. We tend not to give it much thought unless something goes wrong. While we are apt to think of stability in terms of a system (for example, one might think that the Earth, Moon, and Sun represent a stable system) **stability** does not actually refer to the system itself, but to a specific trajectory or equilibrium point associated with that system. The actual orbits of the Earth about the Sun and the Moon about the Earth are stable orbits, but if we were to place the Earth, Moon, and Sun on some arbitrary orbits, it is likely that in a short time the Moon would be sent off to infinity! As a more down to earth example, consider a ball and a mixing bowl as our system. If we place the ball in the bowl, the bottom of the bowl represents a stable equilibrium point. The combination of gravity, friction, and the bowl's curvature guarantees that the ball will always return to the bottom of the bowl when displaced from this position. However, if we turn the bowl over and place the ball at the top point, this point is still an equilibrium point, but it is no longer stable. A **stable equilibrium point** or trajectory is one to which the system returns when perturbed from that equilibrium point or trajectory. A **nominal operating condition** refers to a specific point or trajectory, while stability refers to the properties of the dynamics of the system in the neighborhood of the nominal condition. While the notion of a stable equilibrium point is relatively clear, we need to be precise in our mathematical formulation.

### 2.5.1 Equilibrium solutions

Given the dynamical system (2.14), if $\mathbf{f}(\mathbf{x})$ is continuous in $\mathbf{x}$, then the dynamical systems will have a **fixed point** (difference equations) or **equilibrium point** (differential equations) at $\mathbf{x}^*$ if and only if

$$\mathbf{f}(\mathbf{x}^*) = \mathbf{x}^*$$

or

$$\mathbf{f}(\mathbf{x}^*) = \mathbf{0}.$$

Even though the requirements $f(\mathbf{x}^*) = x^*$ and $f(\mathbf{x}^*) = 0$ are different, they both impose the same condition: no change in population size occurs with time. It follows that the terms fixed point and equilibrium point both refer to the same condition. For brevity, we use the term equilibrium point to refer to this condition for both systems.

To simplify the discussion of stability we will assume that the function $\mathbf{f}(\cdot)$ is not only continuous but also has continuous partial derivatives of any order required by the ensuing analysis.

## 2.5.2 Asymptotic stability

An equilibrium point, $\mathbf{x}^*$, is **stable** (more precisely, Lyapunov stable (Vincent and Grantham, 1997)) if any trajectory $\mathbf{x}(t)$ that starts near $\mathbf{x}^*$ remains near $\mathbf{x}^*$ for all $t \geq 0$. If, in addition the Euclidean length of the difference

$$\|\mathbf{x}(t) - \mathbf{x}^*\| \to 0 \quad \text{as } t \to \infty \quad (2.15)$$

then $\mathbf{x}^*$ is **asymptotically stable**. The Euclidean length of a vector $\mathbf{x}$ is defined by

$$\|\mathbf{x}\| = \sqrt{\mathbf{x}^T\mathbf{x}} = \sqrt{x_1^2 + \cdots + x_{n_s}^2}.$$

If the solution $\mathbf{x}^*$ is asymptotically stable for every solution $\mathbf{x}(t)$ from all possible initial conditions then $\mathbf{x}^*$ is **globally asymptotically stable**.

Note that Lyapunov stability does not require that the system returns to the equilibrium point, only that it stays near by. Asymptotic stability does require the system to return to the equilibrium point, but in general asymptotic stability is only local for non-linear systems. For linear systems and some non-linear systems, asymptotic stability is global.

**Example 2.5.1 (Lotka–Volterra predator–prey model)** *From (2.12) this model is given by*

$$\begin{aligned} \dot{x}_1 &= \alpha x_1 - \beta x_1 x_2 \\ \dot{x}_2 &= -\gamma x_2 + \delta x_1 x_2, \end{aligned} \quad (2.16)$$

*and non-zero equilibrium populations are given by*

$$x_1^* = \frac{\gamma}{\delta}, \quad x_2^* = \frac{\alpha}{\beta}. \quad (2.17)$$

*Since the trajectories for this system are closed periodic solutions to orbits about the equilibrium solution given by $V(\mathbf{x}) = C$ where $C$ is a constant and where*

$$V(x_1, x_2) = \gamma \left[ \frac{x_1}{x_1^*} - \ln\left(\frac{x_1}{x_1^*}\right) \right] + \alpha \left[ \frac{x_2}{x_2^*} - \ln\left(\frac{x_2}{x_2^*}\right) \right] \quad (2.18)$$

*it follows that the equilibrium at $\mathbf{x}^*$ is stable, but not asymptotically stable.*

**Example 2.5.2 (continuous logistic equation)** *From (2.11) the differential equation model is given by*

$$\dot{x} = x\frac{r}{K}(K - x)$$

*with the solution*

$$x(t) = \frac{K}{1 + CKe^{-rt}}$$

where C is a constant determined from the initial condition x (0) by

$$C = \frac{1}{x(0)} - \frac{1}{K}.$$

*It follows that for any non-zero initial condition the solution will always asymptotically approach the equilibrium point*

$$x^* = K.$$

In non-linear systems, it often happens that an equilibrium solution $\mathbf{x}^*$ is asymptotically stable, but the equilibrium point is not globally asymptotically stable. The **domain of attraction** $\mathcal{D}$ for an asymptotically stable equilibrium point is the set of initial states $\mathbf{x}(0)$ from which solutions $\mathbf{x}(t)$ converge to $\mathbf{x}^*$ as $t \to \infty$. Another useful concept is that of an invariant set. A set $X$ is a **positively invariant set** if for every $\mathbf{x}(0)$ in $X$ the corresponding trajectory $\mathbf{x}(t)$ remains in $X$ for all future times $t \geq 0$. A set $X$ is an **invariant set** if $\mathbf{x}(0) \in X$ implies $\mathbf{x}(t) \in X$ for all time, $-\infty < t < \infty$.

### 2.5.3 Linearization

We can investigate the stability of an equilibrium point for a system of non-linear equations, at least locally, by examining the **linearized equations** of motion. This approach is known as **Lyapunov's First (or Indirect) Method** and it can provide local stability properties in many applications (Vincent and Grantham, 1997).

Let $\mathbf{X}^*$ be an equilibrium solution to the non-linear system

$$\mathbf{X}(t+1) = \mathbf{f}(\mathbf{X})$$
$$\text{or} \qquad (2.19)$$
$$\dot{\mathbf{X}} = \mathbf{f}(\mathbf{X})$$

that satisfies $\mathbf{f}(\mathbf{X}^*) = \mathbf{X}^*$ (discrete system) or $\mathbf{f}(\mathbf{X}^*) = 0$ (continuous system). In order to investigate the stability of $\mathbf{X}^*$ we need to examine the nature of solutions with initial conditions near $\mathbf{X}^*$. This is done by finding a linear approximation to (2.19) and then examining the stability properties of the linear system. We first define a **perturbation solution** as the difference between the actual solution to the non-linear system and the equilibrium solution $\mathbf{x}(t) = \mathbf{X}(t) - \mathbf{X}^*$. If we choose $\mathbf{x}(0)$ to be in a small neighborhood of $\mathbf{X}^*$, then from Taylor's Theorem, as long as the perturbation solution $\mathbf{x}(t)$ remains small, its motion is determined by a system of linear **state perturbation equations**

$$\mathbf{x}(t+1) = \mathbf{A}\mathbf{x}$$
$$\text{or} \qquad (2.20)$$
$$\dot{\mathbf{x}} = \mathbf{A}\mathbf{x}$$

## 2.5 Classical stability concepts

where **A** is the $n_s \times n_s$ matrix of partial derivatives

$$\mathbf{A} = \frac{\partial \mathbf{f}(\mathbf{X})}{\partial \mathbf{X}} = \begin{bmatrix} \frac{\partial f_1(\mathbf{X})}{\partial X_1} & \cdots & \frac{\partial f_1(\mathbf{X})}{\partial X_{n_s}} \\ \vdots & \ddots & \vdots \\ \frac{\partial f_{n_s}(\mathbf{X})}{\partial X_1} & \cdots & \frac{\partial f_{n_s}(\mathbf{X})}{\partial X_{n_s}} \end{bmatrix}_{\mathbf{X}=\mathbf{X}^*}$$

evaluated at the equilibrium point $\mathbf{X}^*$. As long as the perturbation solution remains small the solution to the non-linear system is closely approximated by

$$\mathbf{X}(t) = \mathbf{X}^* + \mathbf{x}(t)$$

where $\mathbf{x}(t)$ is the solution to the linear system (2.20). Furthermore the local stability properties of $\mathbf{X}^*$ will be exactly the same as the stability properties of the equilibrium solution for the linear system (i.e., $\mathbf{x}^* = 0$).

In summary, we determine the **local stability** of the non-linear equilibrium point $\mathbf{X}^*$ by examining the stability of the equilibrium solution $\mathbf{x}^* = 0$ for the linearized system (2.20). Note that for linear systems, local stability implies **global stability**, but this need not be the case for non-linear systems and results must be interpreted accordingly.

- If $\mathbf{x}^* = 0$ is asymptotically stable this implies that $\mathbf{X}^*$ is locally asymptotically stable.
- If $\mathbf{x}^* = 0$ is stable, this implies that $\mathbf{X}^*$ is locally stable.
- If $\mathbf{x}^* = 0$ is unstable, this implies that $\mathbf{X}^*$ may or may not be locally stable.

In the first case, if $\mathbf{X}(0)$ is outside the domain of attraction to $\mathbf{X}^*$ the non-linear system may move away from $\mathbf{X}^*$ (to another equilibrium point or to infinity). In the last case, the non-linear system will move away from $\mathbf{X}^*$ but may remain in the neighborhood of $\mathbf{X}^*$. In order to use linearization we need some additional information about linear systems.

### 2.5.4 Equilibrium point stability for linear difference equations

Consider the $n$th iteration of the difference equation (2.20) as given by

$$\mathbf{x}(2) = \mathbf{A}\mathbf{x}(1)$$
$$\mathbf{x}(3) = \mathbf{A}\mathbf{x}(2) = \mathbf{A}^2\mathbf{x}(1)$$
$$\vdots$$
$$\mathbf{x}(n) = \mathbf{A}^{n-1}\mathbf{x}(1).$$

We see that the initial point $\mathbf{x}(1)$ is propagated one iteration to the next by adding one more power to $\mathbf{A}$. Given a non-zero vector for $\mathbf{x}(1)$, we see the only way that $\mathbf{x}(n) \to \mathbf{x}^* = 0$ as $n \to \infty$ is for $\mathbf{A}^{n-1} \to \mathbf{0}$ as $n \to \infty$. To find conditions which will guarantee this, rewrite

$$\mathbf{x}(t+1) = \mathbf{A}\mathbf{x} \qquad (2.21)$$

as

$$\mathbf{z}(t+1) = \Lambda \mathbf{z}.$$

where $\mathbf{x}$ and $\mathbf{z}$ are related through the non-singular transformation

$$\mathbf{z} = \mathbf{M}^{-1}\mathbf{x} \iff \mathbf{x} = \mathbf{M}\mathbf{z} \qquad (2.22)$$

and

$$\Lambda = \mathbf{M}^{-1}\mathbf{A}\mathbf{M}.$$

If the eigenvalues of $\mathbf{A}$ are distinct and if $\mathbf{M}$ is chosen to be a matrix of the corresponding eigenvectors, then $\Lambda$ will be a diagonal matrix with the eigenvectors down the diagonal (Grantham and Vincent, 1993). This means that the set of $\mathbf{z}$ equations is completely decoupled and of the form

$$z_1(t+1) = \lambda_1 z_1$$
$$z_2(t+1) = \lambda_2 z_2$$
$$\vdots = \vdots$$

where $\lambda_i$ are the **eigenvalues**. It follows that

$$\mathbf{z}(n) = \Lambda^{n-1}\mathbf{z}(1)$$

and that $\Lambda^{n-1} \to \mathbf{0}$ provided that $|\lambda_i| < 1$. Because (2.22) is a non-singular transformation, this also provides the requirements on $\mathbf{A}$. In other words *for the constant-coefficient linear system (2.21) the origin is globally asymptotically stable if and only if the absolute values of all of the eigenvalues of the matrix $\mathbf{A}$ are less than 1.*

**Example 2.5.3 (Leslie predator–prey discrete model)** *From (2.13) we have the difference equation form*

$$x_1(t+1) = x_1\left[1 + \frac{r_1}{K_1}(K_1 - x_1) - bx_2\right]$$

$$x_2(t+1) = x_2\left[1 + r_2\left(1 - \frac{x_2}{cbx_1}\right)\right].$$

## 2.5 Classical stability concepts

In order to simplify the analysis assume that

$$r_1 = 0.25$$
$$r_2 = 0.1$$
$$K_1 = 100$$
$$b = 0.1$$
$$c = 1$$

the equilibrium solutions are obtained by solving

$$x_1 \left[ 1 + \frac{0.25}{100} (100 - x_1) - 0.1 x_2 \right] = x_1$$

$$x_2 \left[ 1 + 0.1 \left( 1 - \frac{x_2}{0.1 x_1} \right) \right] = x_2$$

yielding the solutions $\{x_2 = 0, x_1 = 100\}$, $\{x_1 = 20, x_2 = 2\}$. Calculating the partial derivatives

$$\frac{\partial \left\{ x_1 \left[ 1 + \frac{0.25}{100} (100 - x_1) - 0.1 x_2 \right] \right\}}{\partial x_1} = 1.25 - 0.005 x_1 - 0.1 x_2$$

$$\frac{\partial \left\{ x_1 \left[ 1 + \frac{0.25}{100} (100 - x_1) - 0.1 x_2 \right] \right\}}{\partial x_2} = -0.1 x_1$$

$$\frac{\partial \left\{ x_2 \left[ 1 + 0.1 \left( 1 - \frac{x_2}{0.1 x_1} \right) \right] \right\}}{\partial x_1} = \frac{x_2^2}{x_1^2}$$

$$\frac{\partial \left\{ x_2 \left[ 1 + 0.1 \left( 1 - \frac{x_2}{0.1 x_1} \right) \right] \right\}}{\partial x_2} = \frac{11 x_1 - 10 x_2}{10 x_1} - \frac{x_2}{x_1}$$

allows us to evaluate the matrix

$$\mathbf{A} = \begin{bmatrix} 1.25 - 0.005 x_1 - 0.1 x_2 & -0.1 x_1 \\ \frac{x_2^2}{x_1^2} & \frac{11 x_1 - 10 x_2}{10 x_1} - \frac{x_2}{x_1} \end{bmatrix}_{x_1=20, x_2=2} = \begin{bmatrix} 0.95 & -2.0 \\ \frac{1}{100} & 0.9 \end{bmatrix}.$$

The eigenvalues of this matrix are

$$0.925 \pm 0.139\,19 i.$$

Because the absolute value of this complex pair

$$|0.925 \pm 0.139\,19 i| = 0.935\,41$$

is less than 1, the non-zero equilibrium point for the non-linear system is locally asymptotically stable.

### 2.5.5 Equilibrium point stability for linear differential equations

A similar procedure may also be used with the differential equations of the form

$$\dot{\mathbf{x}} = \mathbf{A}\mathbf{x}. \tag{2.23}$$

If we decouple this system using (2.22) to obtain

$$\dot{\mathbf{z}} = \Lambda \mathbf{z}$$

we have a system of decoupled equations of the form

$$\dot{z}_1 = \lambda_1 z_1$$
$$\dot{z}_2 = \lambda_2 z_2$$
$$\vdots = \vdots$$

each one of which has a solution of the form

$$z_j(t) = z_j(0)\,e^{\lambda_j t}.$$

For $z_{ji}(t) \to 0$ as $t \to \infty$, it is required that

$$e^{\lambda_j t} = e^{(\sigma_j \pm i\omega_j)t} = e^{\sigma_j t}(\cos \omega_{ji} t \pm i \sin \omega_j t) \to 0 \tag{2.24}$$

as $t \to \infty$ where $i$ in (2.24) refers to the complex number $i = \sqrt{-1}$. Thus $e^{\lambda_j t} \to 0$ as $t \to \infty$ provided that $\sigma_j < 1$. Because (2.22) is a non-singular transformation, this also provides the requirements on $\mathbf{A}$. That is, in order for the origin to be stable for the constant-coefficient linear system, all of the eigenvalues of $\mathbf{A}$ must have $\text{Re}(\mu_j) \leq 0$, where $\text{Re}(\cdot)$ denotes real parts. If any of the eigenvalues has a positive real part then there is at least one solution $\mathbf{x}(t)$, starting arbitrarily near $\mathbf{x} = \mathbf{0}$, for which $\|\mathbf{x}(t)\| \to \infty$ as $t \to \infty$, which implies that the origin is unstable. Therefore, we conclude: *for the constant-coefficient linear system (2.23) the origin is* **globally asymptotically stable** *if and only if all of the eigenvalues of the matrix* $\mathbf{A}$ *have negative real parts.*

**Example 2.5.4 (Leslie predator–prey continuous model)** *From (2.13) we have the differential equation form*

$$\dot{x}_1 = x_1 \left[ \frac{r_1}{K_1}(K_1 - x_1) - bx_2 \right]$$

$$\dot{x}_2 = x_2 \left[ r_2 \left(1 - \frac{x_2}{cbx_1}\right) \right].$$

## 2.5 Classical stability concepts

Using the same parameter values as in the discrete case

$$r_1 = 0.25$$
$$r_2 = 0.1$$
$$K_1 = 100$$
$$b = 0.1$$
$$c = 1$$

the equilibrium solutions are obtained by solving

$$x_1 \left[ \frac{0.25}{100} (100 - x_1) - 0.1 x_2 \right] = 0$$

$$x_2 \left[ 0.1 \left( 1 - \frac{x_2}{0.1 x_1} \right) \right] = 0,$$

yielding the same solutions as in the discrete case, $\{x_2 = 0, x_1 = 100\}$, $\{x_1 = 20, x_2 = 2\}$. Calculating the partial derivatives

$$\frac{\partial \left\{ x_1 \left[ \frac{0.25}{100} (100 - x_1) - 0.1 x_2 \right] \right\}}{\partial x_1} = 0.25 - 0.005 x_1 - 0.1 x_2$$

$$\frac{\partial \left\{ x_1 \left[ \frac{0.25}{100} (100 - x_1) - 0.1 x_2 \right] \right\}}{\partial x_2} = -0.1 x_1$$

$$\frac{\partial \left\{ x_2 \left[ 0.1 \left( 1 - \frac{x_2}{0.1 x_1} \right) \right] \right\}}{\partial x_1} = \frac{x_2^2}{x_1^2}$$

$$\frac{\partial \left\{ x_2 \left[ 0.1 \left( 1 - \frac{x_2}{0.1 x_1} \right) \right] \right\}}{\partial x_2} = \frac{x_1 - 20 x_2}{10 x_1}$$

allows us to evaluate the matrix

$$\mathbf{A} = \begin{bmatrix} 0.25 - 0.005 x_1 - 0.1 x_2 & -0.1 x_1 \\ \frac{x_2^2}{x_1^2} & \frac{x_1 - 20 x_2}{10 x_1} \end{bmatrix}_{x_1 = 20, x_2 = 2} = \begin{bmatrix} -0.05 & -2.0 \\ \frac{1}{100} & -0.1 \end{bmatrix}.$$

The eigenvalues of this matrix are

$$-0.075 \pm 0.139\,19 i.$$

Because the eigenvalues are complex, the motion in the vicinity of the equilibrium point is oscillatory and because the real parts are negative the equilibrium point is locally asymptotically stable.

### 2.5.6 Other situations

Additional information on equilibrium point stability may be found in Vincent and Grantham (1997). Even for linear models there are complications associated with repeated eigenvalues or the borderline situation where the eigenvalue is purely imaginary in the continuous case, or the absolute value of the eigenvalues equals 1 in the discrete case. For continuous systems, if one of the eigenvalues has a zero real part and the others have negative real parts then the origin for the linear system is Lyapunov stable, but not asymptotically stable. This is because the eigenvalue with the zero real part corresponds to terms in the solution having a constant amplitude. It does not grow with time, nor does it decay to zero. If there are repeated eigenvalues with zero real parts and the corresponding eigenvectors are not linearly independent, then some of the coefficients in the solution will be polynomials in time with no counteracting exponential decay factor and the origin will be unstable for the linear system.

We draw conclusions about the **local asymptotic stability** or instability of an equilibrium point for a non-linear system, based on the stability or instability of the linearized system. We are guaranteed that, if the stability condition is satisfied, then the equilibrium point will provide asymptotic stability, at least in some neighborhood of the equilibrium point for the non-linear system. However, if the linearized system has an unstable equilibrium point, this does not imply that any solution in the neighborhood of the equilibrium point necessarily goes to infinity for the non-linear system.

### 2.5.7 Non-equilibrium dynamics

In general, the state vector **x** associated with a dynamical system will have one or more equilibrium points $\mathbf{x}^*$. If the system is initially placed at $\mathbf{x}^*$, then, by definition, it will remain there. However, the fact that a system will remain at $\mathbf{x}^*$ does not imply anything about the stability of such a point. In particular if the system is initially placed in the neighborhood of $\mathbf{x}^*$, will it remain in the neighborhood, ultimately returning to $\mathbf{x}^*$; not return, but stay in some bounded region of the equilibrium point; or, will the motion become unbounded with one or more of the components of the state vector **x** becoming infinite? In the first case the equilibrium point is said to be **asymptotically stable**. In the second, it is said to be **stable** and in the third case the equilibrium point is said to be **unstable**. There are many types of bounded motion corresponding to the second case that we lump together under the heading **non-equilibrium dynamics**.

### 2.5.7.1 Linear systems

A linear dynamical system has the form

$$\mathbf{x}(t+1) = \mathbf{A}\mathbf{x}$$

or

$$\dot{\mathbf{x}} = \mathbf{A}\mathbf{x}$$

where $\mathbf{x}$ is a state vector of dimension $n_s$ and $\mathbf{A}$ is a constant $n_s \times n_s$ matrix. If $\mathbf{A}^{-1}$ exists then

$$\mathbf{x}^* = 0$$

is the only equilibrium solution.[9] The stability of this equilibrium point is determined by the eigenvalues of the $\mathbf{A}$ matrix. If one or more eigenvalues have a positive real part then if the initial state is any point other than $\mathbf{x}^*$ the motion will be unbounded.[10] However, if all the eigenvalues have non-positive real parts then $\mathbf{x}^*$ is stable[11] and two types of motion are possible. These are asymptotic stability (all of the eigenvalues have negative real parts) and **periodic orbits** (with some of the eigenvalues having zero real parts). A simple pendulum with and without friction at the pivot point illustrates both types of motion. The pendulum with friction will always return to its downward equilibrium position when displaced. If we start a frictionless pendulum at an angle $+\theta$ from the vertical, it will swing forever between $-\theta$ and $+\theta$.

### 2.5.7.2 Non-linear systems

Non-linear systems of the form

$$\mathbf{x}(t+1) = \mathbf{F}(\mathbf{x})$$

or

$$\dot{\mathbf{x}} = \mathbf{F}(\mathbf{x})$$

may have many equilibrium points, $\mathbf{x}^*$, defined by

$$\mathbf{F}(\mathbf{x}^*) = \mathbf{x}^*$$

---

[9] If $\mathbf{A}^{-1}$ does not exist, then there is a set of equilibrium points equal to a subspace (line, plane, etc.) through $\mathbf{x} = 0$ of dimension $n_s - \text{Rank}[\mathbf{A}]$.

[10] Unstable motion is also possible with non-positive real parts if there are repeated eigenvalues with zero real parts and the corresponding eigenvectors are not linearly independent.

[11] Provided that any repeated eigenvalues with zero real parts have eigenvectors that are linearly dependent.

or
$$F(x^*) = 0.$$

In addition to asymptotic stability and periodic orbits, there are three additional types of stable motion possible: limit cycles, quasi-periodic orbits, and chaotic motion. A **stable limit cycle** is a periodic trajectory with no periodic neighbors. Rather, neighboring trajectories are attracted to the limit cycle. The trajectory making up the limit cycle is much like the track on a roller coaster. It may twist and turn in space, but it always ends up where it started. A **quasi-periodic orbit** is much like the periodic orbits found in linear systems except that each time it returns near its initial point it is displaced by a small amount. One may think of the trajectory winding around the surface of a torus, the trajectory never repeats, but after an infinite time it becomes arbitrarily close to being periodic. **Chaotic motion** produces trajectories that wander erratically on a bounded chaotic attractor without repeating themselves, with neighboring trajectories separating exponentially with time. As the name implies, a **chaotic attractor** attracts trajectories that are not initially part of it. Motion on the attractor is "quasi-periodic" in the sense that every point on the chaotic attractor gets a close visit time and again as the trajectory winds its way on the attractor, but the time intervals for such an event are random and it never happens in exactly the same way. A simple example is given by the discrete logistic equation (May, 1976). For further discussion of non-equilibrium motion in an evolutionary context see Rand *et al.* (1994).

# 3
# The Darwinian game

Because evolution occurs within an ecological setting, the concepts and models of population ecology are integral to evolutionary game theory. The organisms' environment and ecologies provide the "rules," the context to which evolution responds. The transition from an ecological model to an evolutionary model can be made seamless. Examples include the Logistic growth model, Lotka–Volterra competition equations, models of predator–prey interactions, and consumer-resource models. In fact, any model or characterization of population dynamics can be reformulated as an evolutionary game. One need only identify **evolutionary strategies** that determine fitness and population growth rates. Conjoining an ecological model of population growth with heritable strategies puts the model in an evolutionary game setting. Not surprisingly, then, evolutionary game theory is well suited for addressing *FF&F* (fit of form and function) under all of nature's diverse ecological scenarios.

Games such as arms races, Prisoner's Dilemma, chicken, battle of the sexes, and wars of attrition have become standard bases for considering the evolution of many social behaviors (any issue of *animal behavior* offers examples of these or variants of these games). These games, however, are not unique to evolutionary ecology. They are products of and recurrent themes in economics, engineering, sociology, and political science. It is from these disciplines that game theory first emerged as the mathematical tools for understanding and solving conflicts of interest.

We begin with a discussion of conventional or classical game theory that had its origins (in the 1930s) in the fields of economics, military sciences, engineering, and political science. In particular, evolution by natural selection is also a game. We highlight the similarities and differences between classical game theory and evolutionary game theory. Two novel features of evolutionary games distinguish them from classical games. In the evolutionary game, organisms (the players) inherit rather than choose their strategies. In the evolutionary game, payoffs (as fitness) determine directly the dynamics governing temporal changes in strategy frequencies. In the discussion of classical games, we introduce matrix games, continuous games, symmetry in games, and various solution concepts.

There is a special kind of symmetry in the evolutionary game. Collecting individuals that make up a population into **evolutionarily identical** groups permits the definition of a **fitness generating function** ($G$-function). The $G$-function eliminates the need to explicitly identify a different fitness function for every individual (of an evolutionarily identical group) that happens to be using a different strategy. In addition, the $G$-function plays a key role in the development of a theory for evolutionary games that by its nature lies outside of classical game theory.

Finally, we relate the $G$-function to major concepts and principles of ecology and evolution. From ecological principles, we know that all populations have the capacity to grow exponentially under ideal conditions. And no population can grow exponentially forever – there are limits to growth. The potential for exponential population growth and limits to growth generate the Malthusian **Struggle for Existence** (Gasue, 1934). From evolutionary principles, we know that like tends to beget like (an organism's traits are heritable), and that this process produces variation and somewhat inexact descendants as a consequence of mutations and Mendel's Laws of segregation and independent assortment. The near-faithful transmission of parental genes to their offspring produces **heritable variation**. As a principle of evolutionary ecology, we know that an organism's heritable phenotype (=strategy) may influence its mortality and fecundity. In this way, heritable variation influences the struggle for existence. The struggle for existence, heritable variation, and the influence of heritable variation on the struggle for existence are the three postulates Darwin used (see Subsection 1.1.2) to understand and explain evolution by natural selection. Because these postulates can be couched in terms of a game, evolution by natural selection can be modeled and understood as an evolutionary game.

Together the mathematical tools of Chapter 2 and the conceptual tools of this chapter allow for a mathematical modeling of evolution by natural selection. When done, the diversity and characteristics of life emerge as the combination of what is evolutionarily feasible and what is ecologically acceptable.

## 3.1 Classical games

von Neumann and Morgenstern (1944) conceived game theory as a mathematical tool for solving conflicts of interest. They envisioned two or more players

## 3.1 Classical games

each with their own set of choices (strategy set) and their own objective. A player tries to choose a strategy from its strategy set that maximizes its personal payoff. What separates a game problem from a standard optimization problem?[1] In a game with more than one player, any player's best strategy choice depends on the strategy choices made by all of the other players and the choices of other players do not necessarily favor a given player's goal. Such a situation is transparent when players are opponents in a **zero-sum game** where a player can increase his/her own payoff only at the expense of another's. Examples include most card games, parlor games, and individual or team sports. Even in games where individuals can maximize their collective payoff by coordinating strategies, an individual may be encouraged to cheat by choosing a strategy that maximizes his/her individual payoff at the expense of payoffs to others in the group. By cheating in a potentially cooperative, non-zero-sum game, an individual may increase his/her payoff but decrease the sum of payoffs among all players (e.g., Prisoner's Dilemma). In non-zero-sum games an individual's actions simultaneously influence the size of the pie and the individual's share of that pie.

In classical game theory as defined by von Neumann and Morgenstern, there are players, strategies, strategy constraint sets, payoffs, and rules for determining how the strategies employed by the players result in their respective payoffs. Classical game theory has produced a variety of solution concepts for predicting the game's outcomes and the players' optimal choice of strategies; e.g., min-max (von Neumann and Morgenstern, 1944), no-regret or Nash solution (Nash, 1951), Pareto optimality (Pareto, 1896), and Stackelberg solution (von Stackelberg, 1952). See Vincent and Grantham (1981) for further discussion on the relationships between these various solutions, concepts, and optimization theory.

### 3.1.1 The optimization problem

Economists often assume that corporations strive to maximize profits, and consumers strive to maximize utility. An engineer strives to maximize an aircraft's performance in terms of speed, load, range, and fuel efficiency. A physician tries to maximize a patient's likelihood of full recovery from an illness. People on vacation try to maximize a sense of fun, relaxation, or recreation. In all aspects of life, we adjust behavior to make choices that maximize some objective or set of objectives subject to constraints that limit our freedom of choice.

---

[1] If there is only one player in a game, then the game problem and the optimization problem are the same. In this way, all of standard optimization theory can be viewed as a one-player game.

The **optimization problem** deals with these situations. It is defined in terms of a payoff function (objective), a set of strategies, and a strategy constraint set that places limits upon the feasible choice of strategies. The objective may be scalar- or vector-valued and the set of strategies may be scalar- or vector-valued. Constraints can play an important role in determining the optimal solution. For example, suppose a strategy set allows only the discrete choices A, B, and C that yield payoffs of 7, 11, and 2, respectively. If an organism's objective is to maximize its payoff then it should select strategy B. But suppose strategy D offers a payoff of 13. Such a payoff is preferable to the organism but if it is constrained to choose among A, B, or C only then its optimal strategy must remain C. In general, strategies used in the determination of a payoff may be either continuous or discrete.[2] A good example of an optimization problem with continuous strategies is given by Smith and Fretwell (1974) who consider the trade-off between offspring size and offspring number.

**Example 3.1.1 (offspring size versus number)** *Suppose a mother has $y$ resources to commit to producing offspring, and she can choose how many offspring, $u_1$, to produce and the size of each offspring, $u_2$. The quality of each offspring increases with size, and the objective of the mother is to maximize the product of offspring number and offspring quality. The problem can be written as:*

*Maximize $E(u_1, u_2) = u_1 Q(u_2)$ subject to $u_1 u_2 = y$*

*where $Q(u_2)$ describes the relationship between offspring size and offspring quality. In this problem the payoff function is given by $E$, which is a function of a vector strategy composed of two components, $u_1$ and $u_2$. The constraint set is defined by the requirement that $u_1$ and $u_2$ lie on the curve defined by $u_1 u_2 = y$. In this case, the strategy set is continuous because the strategy set is a continuous function of $u_1$ and $u_2$. To find the values of $u_1$ and $u_2$ that maximize $E$ one can either use the tools of constrained optimization (Vincent and Grantham, 1981), or one can reduce the problem to a single strategy, $u_1$, by substituting the constraint into the objective function by noting that $u_2 = y/u_1$. For this example, let offspring quality increase with offspring size according to*

$$Q(u_2) = -a + \sqrt{u_2}$$

*where $u_2 > a^2$ and $a > 0$ represents the threshold size for offspring to have a positive influence on fitness. The original objective function is now*

---

[2] Recall that the models used to represent the population dynamics may also be either continuous (differential equations) or discrete (difference equations). Either type of strategy may be used with either type of population–dynamic model.

## 3.1 Classical games

*reformulated in terms of an unconstrained scalar-valued strategy*

$$E(u_1) = -au_1 + \sqrt{u_1 y}.$$

*Necessary conditions for the optimal value of $u_1^*$ that maximizes E are given by*

$$\frac{\partial E}{\partial u_1} = 0 \text{ and } \frac{\partial^2 E}{\partial u_1^2} < 0.$$

*Performing these operations we find that*

$$u_1^* = \frac{y}{2a^2} \text{ and hence } u_2^* = 2a^2.$$

In the above example, the mother's payoff does not depend on the choices made by other mothers. This need not be the case. Suppose the payoff to an individual selecting a strategy also depended upon what another individual selected and vice versa. Then we have a game. As noted in Section 2.3, an ecological model of population dynamics describes fitness and population growth rates. Fitness provides the payoff in the evolutionary game. In constructing an evolutionary game, the functional relationship between strategies and fitness must be defined. We will look first at matrix game theory to provide this relationship between strategies and payoff. In the traditional matrix game, the set of strategy choices is finite and discrete, and individuals pair-up to determine each other's rewards. A much broader class of matrix games is considered in Chapter 9.

### 3.1.2 Matrix games

Matrix games are those in which the payoff to an individual can be determined from a matrix of payoffs. The payoffs are assigned to each element of the matrix assuming that interactions among players are pairwise. One player chooses a row of the matrix and the other chooses a column of the matrix. The intersection of the row and the column determines a unique element of the matrix. For example, if player A's strategy is to choose the second row and player B's strategy is to choose the fourth column, the resultant payoff to player A is the value in the second row and fourth column of the matrix. As a consequence of this construction, the number of strategies available to each player is finite (often just two strategies) and discrete. The matrix game is asymmetric if each player has a different strategy set and/or if players experience different payoffs from playing the same strategy against opponents using a particular strategy. The matrix game is symmetric if players possess the same set of strategies and experience the same consequences of using a given strategy against an opponent with a particular strategy. See von Neumann and Morgenstern (1944) and Luce and Raiffa (1957) for formal definitions of symmetry in the context of games.

A two-player **asymmetric matrix game** requires a different payoff matrix **A** for the first player from the payoff matrix **B** for the second player. The first player chooses strategy $i$ and the second player chooses strategy $j$. The payoff to the first player is given by $a_{ij}$ located at the intersection of row $i$ and column $j$ in the matrix **A**. The payoff to the second player is given by $b_{ji}$ located at the intersection of row $j$ and column $i$ in the matrix **B**.

The payoffs for a two-player **symmetric matrix game** can be represented by a single square matrix and the payoffs to each player can be found by reversing focus. The entries in the matrix give the payoff to the focal player whose strategy is found along the rows and its opponent's strategy is found along the columns. Hence, if the first player uses strategy $i$ and the second chooses strategy $j$ then the payoffs to the players are $a_{ij}$ and $a_{ji}$ respectively.

**Example 3.1.2 (battle of the sexes)** *This is an example of an asymmetric game between a male and female couple. She prefers ice hockey, he prefers opera, and each prefers to do an activity together rather than alone. Each has his or her own payoff matrix and each has the same set of strategies, namely attend a hockey game or attend the opera.*

*Female payoff matrix*

|        | hockey | opera |
|--------|--------|-------|
| hockey | $a_f$  | $b_f$ |
| opera  | $c_f$  | $d_f$ |

*Male payoff matrix*

|        | hockey | opera |
|--------|--------|-------|
| hockey | $a_m$  | $b_m$ |
| opera  | $c_m$  | $d_m$ |

*In its most general form, the battle of the sexes has the following structure to the payoff elements. The woman prefers hockey, $a_f > d_f$, $b_f > c_f$, and the man prefers opera, $a_m < d_m$, $b_m < c_m$. They would prefer to do things together rather than alone: $a_f > b_f > c_f$ and $d_m > c_m > b_m$. The relationship between $b_f$ and $d_f$ determines whether the woman's preference for hockey is dominant ($b_f > d_f$) or subordinate ($b_f < d_f$) to her preference for doing things as a couple; and similarly $a_m$ and $c_m$ determine whether his preference for opera ($a_m < c_m$) or togetherness ($a_m > c_m$) dominates.*

**Example 3.1.3 (Prisoner's Dilemma; figure 3.1)** *This is an example of a symmetric, two-player, two-strategy matrix game. It is formulated in terms*

## 3.1 Classical games

**Figure 3.1** The Prisoner's Dilemma.

*of a single matrix:*

*Symmetric payoff matrix*

|   | A | B |
|---|---|---|
| A | $a_{11}$ | $a_{12}$ |
| B | $a_{21}$ | $a_{22}$ |

*Two individuals are being held in a prison in separate, isolated cells and they have each been told that, if they defect (from their partner in crime) by confessing, they will go free while the other will get five years in prison. However, if they cooperate (with their partner in crime) by not confessing, there is enough evidence to send them both to prison for two years. If they both defect, by confessing, they will each get four years. The potential exists to go free. This occurs if the focal individual defects (confesses) while its partner cooperates (remains mum). What would you do in this situation? If strategy A represents cooperate and B defect, then the elements of the payoff matrix take the form $a_{11} > a_{22}, a_{21} > a_{11}, a_{22} > a_{12}$, and $2a_{11} > a_{21} + a_{12}$. Strategy B always yields a higher payoff to a player than strategy A; yet both players using strategy A obtain a higher payoff than both players using B. The prisoner's dilemma forms the foundation for many inquiries into the likelihood and persistence of cooperative behaviors in economics, sociology, politics, and evolutionary ecology.*

**Example 3.1.4 (game of chicken; figure 3.2)** *The game of chicken (also known as the hawk-dove game introduced by Maynard Smith (1974)) has the*

**Figure 3.2** The game of chicken.

*same 2 × 2 matrix as the prisoner's dilemma but with a different relationship between the elements. Two children on bikes race toward each other and the first to swerve is the chicken. Let A be a strategy of swerving to avoid a collision (dove) and B the strategy of not swerving (hawk). The payoffs are scaled so that: $a_{11} > a_{22}$, $a_{21} > a_{11}$, $a_{12} > a_{22}$, and $a_{12} + a_{21} > 2a_{22}$. In this game, A is the best response to B, and B is the best response to A. The game of chicken brings back memories (unpleasant for us more gentle types) of the grade school playground.*

### 3.1.2.1 Continuous strategies

While the matrix games introduced above possess discrete strategies (choose a row or column) there are also versions of these games in which strategies are continuous. A **continuous strategy** is one in which choices are made over a possibly infinite continuum of values. These choices may be scalar- or vector-valued. An example of a scalar strategy is the amount of money a player in a gambling game decides to bet. An example of a vector strategy is the amount to bet on each of the 15–19 horses racing in the Kentucky derby.[3]

A matrix game, such as the prisoner's dilemma or the game of chicken, may be re-formulated as a continuous game with the introduction of the mixed-strategy concept. A **mixed strategy** is one in which the individual plays any one of the discrete strategies of a matrix game with a continuous probability between zero and 1. In a 2 × 2 symmetric game, $u_i$ may represent the probability that player $i$ uses strategy A and $(1 - u_i)$ is the probability of using strategy B. During any given play of the game, an individual must use either strategy A or strategy B, but with mixed strategies the actual strategy played

---

[3] Payoffs, known as odds, are determined by the collective amounts that all individuals bet on each horse.

has an element of uncertainty. An individual with a **pure strategy** always uses the same strategy A or B with certainty.

An individual using a mixed strategy will have a payoff function based on the probabilities of its mixed strategy. The **expected payoff** $E_i(u_i, u_j)$ to an individual $i$ using the mixed strategy $u_i$ against an individual using the mixed strategy $u_j$ can be determined from a matrix. For a mixed-strategy $2 \times 2$ matrix game

$$E_i(u_i, u_j) = \begin{bmatrix} u_i & 1 - u_i \end{bmatrix} \begin{bmatrix} a_{11} & a_{12} \\ a_{21} & a_{22} \end{bmatrix} \begin{bmatrix} u_j \\ 1 - u_j \end{bmatrix}.$$

where $i, j = 1, 2$. The matrix

$$\mathbf{A} = \begin{bmatrix} a_{11} & a_{12} \\ a_{21} & a_{22} \end{bmatrix}$$

could be from one of the games described above. For example, a mixed-strategy game of chicken would have the same elements in $\mathbf{A}$ as the game of chicken described above. Multiplying the matrix $\mathbf{A}$ by the two strategy vectors yields

$$E_i(u_i, u_j) = u_i u_j a_{11} + u_i(1 - u_j)a_{21} + (1 - u_i)u_j a_{12} + (1 - u_i)(1 - u_j)a_{22}.$$

We see that the game is **bi-linear**. The expected payoff to a player is linear in its own strategy $u_i$ and linear in the strategy of its opponent $u_j$. Furthermore, the effect of the opponent's strategy on a player's payoff is the product of the player's strategy and that of its opponent.

### 3.1.3 Solution concepts: max-min, Nash equilibrium, etc.

How should an individual go about choosing the "best" strategy? This question is central to both classical and evolutionary game theory. In thinking about how a given individual should choose a strategy, it becomes apparent that there is no one single approach to this problem. Consider the following two-player, four-strategy by four-strategy, payoff matrix for a symmetric game:

|   | A | B | C | D |
|---|---|---|---|---|
| A | 3 | 6 | 5 | 3 |
| B | 5 | 9 | 1 | 2 |
| C | 3 | 11 | 5 | 1 |
| D | 2 | 7 | 6 | 4 |

Each player has four strategy choices (choose row A, B, C, or D) and there are merits to each of them. Strategy A is a **max-min strategy**. It is the pessimist's strategy: "since I am not sure what my opponent is going to play, I am going to assume that it will be the strategy that minimizes my payoff!". Strategy A maximizes the lowest payoff that an individual can receive from playing an opponent that plays the least desirable strategy for that individual. The max-min strategy maximizes the row minima. However, if everyone plays strategy A an individual would do well to use another strategy such as B. Strategy B is a **group-optimal strategy**. It is attractive in that it provides the highest overall payoff given all individuals use the same pure strategy. As such, strategy B represents the maximum of the diagonal elements. However, if everyone plays B an individual would be tempted to use strategy C. Strategy C is attractive for several reasons. It represents the **max-max strategy** under the optimistic assumption that the opponent plays the most desirable strategy for that individual. Also, since row C has the highest average payoff, strategy C maximizes a player's expected payoff under the assumption that the other player selects his/her strategy at random. However, if everyone plays strategy C it behooves an individual to play strategy D. At first glance, strategy D has little to commend it. It is not max-min, max-max, nor does it maximize the value of the diagonal elements when played against itself. It is, however, a **no-regret strategy**. Such strategies are **Nash solutions** (Nash, 1951). If all individuals use strategy D, then an individual has no incentive to unilaterally change his/her strategy. If individuals are free to alter their strategies, a Nash solution is an equilibrium solution in the sense that if everyone uses strategy D no one should want to unilaterally change his/her strategy.

### 3.1.4 Continuous games

The introduction of the mixed-strategy concept for the matrix game changed the nature of the matrix game from one where the strategy set is discrete to one in which it is continuous. The mixed-strategy matrix game is a special case of the larger class of **continuous games** in which the payoff to each player is given by a continuous function of the strategies used by each of the players. In the general case, the payoffs $H_i$ may be an expected payoff (as in the case of mixed-strategy matrix games) or $H_i$ may be the actual payoff obtained from a single play of the game. Let there be $n_s$ players each using a scalar strategy. The payoffs to each of the separate players are given by the functions

$$H_1 = f_1(\mathbf{u})$$
$$\vdots$$
$$H_{n_s} = f_{n_s}(\mathbf{u})$$

where $\mathbf{u} = \begin{bmatrix} u_1 & \cdots & u_{n_s} \end{bmatrix}$.

**Example 3.1.5 (ant wars)** *Two species of ants live side by side in near equal conditions. However, species 2 likes to invade the space occupied by species 1. In this situation they have fitness functions*

$$H_1 = -u_2 K_2 + u_1 u_2 K_2 - u_1$$
$$H_2 = u_2 - \alpha u_1 u_2$$

*where $u_1$ is the fraction of time species 1 spends repelling species 2 and $u_2$ is the fraction of time species 2 spends in species 1's territory. For species 1, $-u_2 K_2$ is the loss of fitness due to invasion, $u_1 u_2 K_2$ is the return in fitness due to repelling and $-u_1$ is the loss of fitness due to the time spent repelling. For species 2, $u_2$ is the gain in fitness due to invasion and $-\alpha u_1 u_2$ is the loss of gain due to repelling. The parameter $K_2$ is the equilibrium population of species 2 and $\alpha$ is a conversion factor. Assume that $K_2 \geq 1$ and $\alpha \geq 1$. A Nash solution may be obtained for this game by setting*

$$\frac{\partial H_1}{\partial u_1} = \frac{\partial H_2}{\partial u_2} = 0$$

*yielding $u_1 = 1/\alpha$ and $u_2 = 1/K_2$ (see exercise 5.8, p.186 Vincent and Grantham, 1981).*

Some games with continuous strategies may not have continuous payoff functions. Such an example can be found in various forms of the **war of attrition**.

**Example 3.1.6 (war of attrition)** *In one form of this game, two individuals contribute either time or resources in hopes of outlasting their opponent. If one of the individuals has a strategy (measured as time or resources) that is greater than his/her opponent's then a prize is collected from the other player of value a. In playing the game, each individual pays a cost proportional to the strategy used. In a symmetric form of this game the expected payoff function has the following form for $i, j = 1, 2$, and $i \neq j$*

$$E\left(u_i, u_j\right) = -c u_i \text{ when } u_i < u_j$$
$$E\left(u_i, u_j\right) = a - c u_j \text{ when } u_i > u_j.$$

*A discontinuity in the payoff function occurs when $u_i = u_j$. At this point it jumps in value by the amount a. See Chapter 3 of Maynard Smith (1976) for a discussion of and solutions to games of this type.*

## 3.2 Evolutionary games

In this book, we are interested in solution concepts applicable to evolutionary games. While Darwin's theory of natural selection, dating from the 1850s, has all of the elements of a "game" it could not be formulated as such until the development of game theory nearly a century later. During the interim, fields such as population and quantitative genetics provided the lingua franca of evolutionary thinking up until the time that Maynard Smith began developing an evolutionary game theory in the early 1970s. We have seen that classical game theory is formulated in terms of payoff functions for each of the players. There is a connection between the payoff functions of game theory and fitness functions for individual players that links Darwin's ideas to game theory. However, the connection is not exact; that is, fitness is not a classical payoff function. The evolutionary game involves more than just fitness functions, it also contains an evolutionary dynamic that translates payoffs to individuals into strategy frequencies in the next generation.

**Example 3.2.1 (turning a matrix game into an evolutionary game)** *A mixed-strategy $2 \times 2$ matrix game has an expected payoff function for each player given by*

$$E_i(u_i, u_j) = u_i u_j a_{11} + u_i (1 - u_j) a_{21} + (1 - u_i) u_j a_{12} + (1 - u_i)(1 - u_j) a_{22}.$$

*Once the number of players has been specified, any of the solution concepts from classical game theory can be applied to yield a game theoretic solution. However, none of these need be the solution to the evolutionary game since payoff functions alone do not define the evolutionary game. We need to identify a dynamical relationship between expected payoffs and fitness. Let us use differential equation dynamics (Subsection 2.3.2) and let the change in the number of individuals using strategy $u_i$ be given as*

$$\dot{x}_i = x_i H_i (\mathbf{x}, \mathbf{u}) \tag{3.1}$$

*where $\mathbf{u} = \begin{bmatrix} u_1 & \cdots & u_{n_s} \end{bmatrix}$, $\mathbf{x} = \begin{bmatrix} x_1 & \cdots & x_{n_s} \end{bmatrix}$. Because there are $n_s$ different strategies, it is reasonable to assume that the fitness of strategy $i$ is the sum of the expected payoffs of playing $u_i$ against all strategies in proportion to their numbers in the population, that is*

$$H_i (\mathbf{x}, \mathbf{u}) = w_0 + \sum_{j=1}^{n_s} E\left(u_i, u_j\right) \frac{x_j}{N} \tag{3.2}$$

*where*

$$N = \sum_{k=1}^{n_s} x_k$$

## 3.2 Evolutionary games

*and $w_0$ describes the fitness of an individual in the absence of interactions with others. Population regulation can be built into this example by making $w_0$ a declining function of total population size, $N$. With fitness defined by $H_i$ (**x**, **u**), equations (3.1) and (3.2) together define the population dynamics for an evolutionary game.*

The above example illustrates the fact that an evolutionary game lies outside of classical game theory, but it lies firmly within the domain of Darwinian evolution. Because the evolutionary game is distinctly different from the classical game, evolutionary game theory requires new and appropriate solution concepts. Like the classical game, evolution by natural selection has players, strategies, strategy sets, and payoffs. The players are the individual organisms at any level of biological organization that manifests separate payoffs and separate strategies. Under Mendelian inheritance, individual genes cannot be the players, because their expected payoffs are the same as that of the whole organism. However, when studying the evolutionary consequences of phenomena such as meiotic drive, individual genes may be players and have "personal" payoffs divergent from other genes within the organism. Similarly, as symbioses (e.g., lichens) or social structures (eusocial insects) become increasingly tight, the individual organisms may cease being separate players. The resultant symbiotic relationship or supra-organism may become the player which manifests the strategy and receives the payoff. In fact, the presence or absence of an individual objective within a tight social network may provide the best means for defining and separating social systems that are highly **despotic** (individuals retain personal objectives and strategies within a social context) versus **eusocial** (the social unit manifests the objectives and strategies). For instance, the social foraging of ants supports the notion of a supra-organism (Anderson *et al.*, 2002; Portha *et al.*, 2002). Ants seem to completely subordinate any individual objectives for the good of the group. On the other hand, the social foraging of hyenas demonstrates individual agendas within a tight-knit social group (Hofer and East, 2003). As evolutionary games, one would ascribe strategies and payoffs to the ant colony, while ascribing strategies and payoffs to the individual hyenas of a pack. It's not a comforting thought to us humans, but cancer represents the moment that an individual cell of a multicellular organism breaks free and begins to manifest its own strategies and payoff function (Subsection 10.2.2). Depending upon the circumstances, the players that possess strategies and payoffs may be an ant colony, an individual hyena, or a cell.

**Strategies** are heritable phenotypes that have consequences for the players' payoffs. A strategy may be a fixed or variable trait. **Fixed strategies** represent invariant morphological or physiological traits. **Variable strategies** are contingent strategies that involve traits of assessment and responce. A player's strategy set is the set of all evolutionarily feasible strategies. In its narrowest sense, "evolutionarily feasible" is interpreted as the extant genetic variability within a population. Here, we intend a broader interpretation. A strategy is evolutionarily feasible if it either exists in the population or reoccurs regularly as a mutation. All breeds of domestic dog may provide a better indicator of the strategy set of *Canis* than the extant genetic variability found among wild members of the genus (i.e., wolf, coyote, jackal, dingo). And even domestic dogs may represent a rather dull subset of what would be truly evolutionarily feasible for domestic dogs, if we so desired. Payoffs in the evolutionary game come in the form of fitness; where fitness is defined as the expected per capita growth rate of a strategy given the ecological circumstances. Fitness is not defined as a property of an individual or of a group. Fitness is defined as the per capita growth rate in population density. Thus fitness directly influences changes in each strategy's frequency within the population. In *The Selfish Gene*, Dawkins (1976) recognizes that neither individuals nor groups have Darwinian fitness. Rather, individuals are carriers of the unit of selection. We prefer the term "Selfish Strategy" rather than "Selfish Gene." Substituting "strategy" for "gene" in most of Dawkins's famous book works well and likely serves the author's philosophy and perspective just as successfully. In other words, strategies have fitness associated with them.

Several features distinguish evolutionary games from classical games. In classical game theory, the focus is on the players. The players, more so than the game or strategies, persist through time. In the evolutionary game, the focus is on the strategies. Through births and deaths, the players come and go, but their strategies persist through time. In classical game theory, the players choose their strategies from their strategy sets. In the evolutionary game, players generally inherit their strategies, and occasionally acquire a novel strategy from the strategy set as a mutation. In classical game theory, every player can have a different strategy set and a different payoff function. In the evolutionary game, there will be populations of players who are evolutionarily identical in that they have the same strategy set and experience the same expected payoffs from using the same strategies (Vincent and Brown, 1984*b*; Brown and Vincent, 1987*c*). In classical game theory, rationality and self-interest provide the optimizing agent that encourages players to select sensible strategies. In the evolutionary game, natural selection serves as the agent of optimization. This is because evolutionary game theory has an inner and an outer game (Vincent and Brown, 1988). The inner game is that of classical game theory. This is the arena in which players receive payoffs in accord with their own and others' strategies. The outer game represents a dynamical link, via inheritance and fitness, in which

## 3.2 Evolutionary games

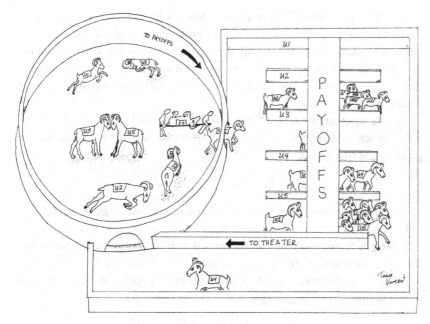

**Figure 3.3** The ecological theater and evolutionary play.

the payoffs to players with particular strategies become translated into changes in strategy frequencies. The **inner game** (Hutchinson's **ecological theater**; figure 3.3) and the **outer game** (his **evolutionary play**) combine to produce evolution by natural selection (Hutchinson, 1965).

### 3.2.1 Collapsing a population's fitness functions into a single $G$-function

Consider a population of organisms, in a given environment, that, by virtue of inheritance and common ancestry, have the same strategy set (=set of evolutionarily feasible phenotypes) and the same fitness consequences of possessing a particular strategy. Individuals within this population are playing a symmetric game with each other. However, if two or more populations of individuals come from different lineages and possess different strategy sets and different fitness consequences of possessing particular strategies then these populations are playing an asymmetric game. When individuals have the same set of strategies and the same set of fitness functions governing the consequences of strategies for per capita growth rates, we can refer to these individuals as evolutionarily identical and it becomes possible to collapse their fitness functions into a single fitness generating function (defined later), which we call the $G$-function. In what follows, we show for the simplest problem (see Subsection 1.4.1) that when individuals are evolutionarily identical, the fitness functions, $H_i$, introduced in Subsection 2.3.2 can be reformulated as a single fitness generating function ($G$-function). The same procedure is readily adapted for the more complex problems as well.

For the simplest problem, the population is composed of individuals with $n_s$ different scalar strategies. All of the strategies currently in the population are given by the row vector

$$\mathbf{u} = \begin{bmatrix} u_1 & \cdots & u_{n_s} \end{bmatrix}$$

where the strategies $u_i$ ($i = 1, \cdots, n_s$) represent a heritable characteristic such as body size.[4]

Different strategies in the population define differences among individuals. Thus, for each strategy in the population $u_i$, there is an associated number of individuals $x_i$ who possess $u_i$. All individuals in the population are represented by the vector

$$\mathbf{x} = \begin{bmatrix} x_1 & \cdots & x_{n_s} \end{bmatrix}$$

where $x_i$ make up the $n_s$ components.

Using our notation from Chapter 2, the individual fitness $H_i$ for the species $i$ (of density $x_i$ using strategy $u_i$) will be a function of $\mathbf{u}$ and $\mathbf{x}$ which we write as $H_i(\mathbf{u}, \mathbf{x})$. We also noted in Chapter 2 that there can be a functional dependence on a resource vector $\mathbf{y}$, which is not included as part of the model for the simplest problem. In fact, $\mathbf{y}$ will be included in the $G$-function only when resources are an explicit part of the ecological model. However, $\mathbf{u}$ and $\mathbf{x}$ are always essential elements in a Darwinian game.

Once the fitness functions have been defined, the population dynamics for each of the species is given as equations. If $H_i(\mathbf{u}, \mathbf{x})$ defines per capita growth rate, then the population dynamics is expressed as difference equations or as differential equations. The alternatives from Subsection 2.3.2 are of the form

$$\begin{array}{ll} \text{Difference:} & x_i(t+1) = x_i\left[1 + H_i(\mathbf{u}, \mathbf{x})\right] \\ \text{Exp. Difference:} & x_i(t+1) = x_i \exp H_i(\mathbf{u}, \mathbf{x}) \\ \text{Differential:} & \dot{x}_i = x_i H_i(\mathbf{u}, \mathbf{x}). \end{array}$$

---

[4] More generally, the full characterization of an organism requires a vector-valued strategy whose dimension equals the number of independent, heritable traits comprising the organism such as the shape of a fish in terms of length, depth, and breadth (Brown et al., 2005). Since vector strategies add their own notational burden, we consider them later.

## 3.2 Evolutionary games

For the simplest problem, we assume that, for a given strategy $\mathbf{u}$, there exists at least one non-zero equilibrium solution $\mathbf{x}^*$ (i.e., not every component of $\mathbf{x}^*$ is zero). There are two ways we can find a non-zero equilibrium solution for $\mathbf{x}^*$. One way is to solve for $\mathbf{x}^*$ from the system of equations

$$H_i\left(\mathbf{u}, \mathbf{x}^*\right) = 0, \quad i = 1, \cdots, n_s.$$

In general, a solution to this system of equations will require a numerical procedure. If more than one equilibrium solution exists, then the particular numerical solution obtained will depend on the initial guess made for $\mathbf{x}^*$. A second method for determining an equilibrium solution (when such a solution is asymptotically stable), for a given $\mathbf{u}$, is to choose an initial condition $\mathbf{x}(0)$ and simply iterate the difference equations or integrate the differential equations until a solution is reached.

At equilibrium, there will be one or more species with a non-zero population size. Does this solution predict the outcome of evolution by natural selection and is this an efficient way to model evolution within a population? The answer to both questions is NO. The equilibrium solution to the above system of equations considers only the outcome for those strategies already resident in the population. The solution does not consider, nor can it consider, the potentially infinite number of evolutionarily feasible strategies that will likely occur in the future via selection and/or mutation. We need to include the fact that each individual is not a distinct evolutionary lineage but rather shares an evolutionary history and context with others in the population via common ancestry and interbreeding.

The fact that there will be one or more groups of evolutionarily identical individuals in the population creates a special type of symmetry in the evolutionary game. In a sense, evolutionarily identical individuals are completely interchangeable. A population of evolutionarily identical individuals requires only a single fitness generating function to describe the fitness of all individuals (resident or otherwise) within the population (Vincent and Brown, 1984*b*, 1988).

**Definition 3.2.1 (fitness generating function)** *A function $G(v, \mathbf{u}, \mathbf{x})$ is a fitness generating function (G-function) for the population dynamics if and only if*

$$G(v, \mathbf{u}, \mathbf{x})|_{v=u_i} = H_i(\mathbf{u}, \mathbf{x}), \quad i = 1, \cdots, n_s \qquad (3.3)$$

*where $\mathbf{u}$ and $\mathbf{x}$ in $G$ are exactly the same vectors as in $H_i$. This is the G-function for the simplest problem with scalar strategies. Similar definitions for other situations are given in Chapter 4.*

The scalar $v$ is a place holder in the $G$-function. As a variable in the $G$-function, it is a **virtual strategy**. One can obtain the fitness function for any individual using a strategy $u_i$ by substituting it for the virtual strategy and evaluating $G$ at $v = u_i$. Use of a virtual strategy in the above definition may at first seem non-intuitive, but it plays a very important role in the development of an evolutionary game theory. The resident strategies **u**, their population sizes **x**, and the level of resources **y** (when applicable) describe the current biotic environment. The argument $v$ within the $G$-function determines the fitness that would accrue to a focal individual using any strategy of the strategy set were it to face this particular biotic environment. By changing $v$ to any strategy of the strategy set, one can determine the consequence for the focal individual using that strategy whether the strategy is actually present in the population or not.

The $G$-function is similar to certain fitness functions from population genetic, quantitative genetic (Charlesworth, 1990) and other formulations from evolutionary game theory (Auslander *et al.*, 1978; Abrams *et al.*, 1993a; Geritz, 1998). In such quantitative and game theory formulations, a function is constructed that describes the fitness of a small population with a mutant strategy that is pitted against a resident population that has a fixed genetic or strategy composition.

The evolutionary analyses of the 1970s (Lawlor and Maynard Smith, 1976; Auslander *et al.*, 1978; Mirmirani and Oster, 1978) used functions similar to but not identical to the $G$-function defined above. For example, Roughgarden (1976) used functions that do not include the strategy of the focal individual, $v$. Such a formulation results in frequency independence among individuals with the same strategy. When used in a frequency-dependence setting Roughgarden's formulation becomes a model of **group selection** (Abrams, 1987; Brown and Vincent, 1987a; Taper and Case, 1992). Under group selection one seeks a strategy that maximizes the growth rate of the entire group. In contrast, **individual selection** seeks a strategy that maximizes the growth rate of a focal individual. This distinction illustrates the need for a fitness generating function that allows for individual selection. This is precisely the role of the virtual variable in (3.3) where the strategy of the individual, $v$, is separated from the strategies used by that individual and all others, **u**. This particular formalization of the $G$-function was developed in the 1980s (Vincent and Brown, 1984b; Brown and Vincent, 1987b; Rosenzweig *et al.*, 1987; Vincent and Brown, 1988) with extensive application in the 1990s in works on adaptation, coevolution, and adaptive dynamics (Rees and Westoby, 1997; Schoombie and Getz, 1998; Cohen *et al.*, 1999; Mitchell, 2000). Furthermore, there have been numerous formulations

## 3.2 Evolutionary games

that converge on or become a $G$-function when the fitness function for a small population of mutants is redefined as a focal individual (Marrow et al., 1992; Abrams et al., 1993a; Rand et al., 1994; Dieckmann et al., 1995; Abrams and Harada, 1996; Metz et al., 1996; Kisdi and Geritz, 1999).

The fitness generating function $G(v, \mathbf{u}, \mathbf{x})$ determines the expected fitness of an individual using strategy $v$ as a function of its biotic environment that includes the extant strategies found among the different species within the population $\mathbf{u} = \begin{bmatrix} u_1 & \cdots & u_{n_s} \end{bmatrix}$, and their population sizes $\mathbf{x} = \begin{bmatrix} x_1 & \cdots & x_{n_s} \end{bmatrix}$. Sometimes, it is more useful to think of the biotic environment in terms of the sum of the population size of all species, $N = \Sigma x_i$, and the vector of strategy frequencies, $\mathbf{p} = \begin{bmatrix} p_1 & \cdots & p_{n_s} \end{bmatrix}$ where $p_i = x_i/N$. In this case,[5] the fitness generating function is written as $G(v, \mathbf{u}, \mathbf{p}, N)$. Either way, the fitness generating function not only represents the formulation of the inner game, but directly influences the changes in strategy frequencies (or population sizes) that represent the outer game.

Given a non-empty strategy vector $\mathbf{u}$, natural selection is **density dependent** if

$$\frac{\partial G(v, \mathbf{u}, \mathbf{p}, N)}{\partial N} \neq 0$$

and **frequency dependent** if

$$\frac{\partial G(v, \mathbf{u}, \mathbf{p}, N)}{\partial p_i} \neq 0, \quad i = 1, \cdots, n_s.$$

Whether the fitness generating function is represented in terms of the vector of population sizes, $G(v, \mathbf{u}, \mathbf{x})$, or in terms of the frequency of each strategy within a population, $G(v, \mathbf{u}, \mathbf{p}, N)$, is a matter of convenience and a matter of emphasis. When the vector of strategies represents potentially distinct species the former is preferable as it emphasizes the ecological perspective of considering the population sizes of each species separately. If the different components of $\mathbf{u}$ are interpreted as different strategies within a population, then the latter is preferable as it emphasizes the proportional representation of each strategy within a single-species population $N$. Either formulation will manifest the frequency and density dependence of the model. The frequency representation using $\mathbf{p}$ is more in line with traditional evolutionary perspectives of defining evolution as change in strategy frequency. A population vector representation, $\mathbf{x}$, is more in the spirit of ecological perspectives where the basis for fitness is per capita growth rates of populations or subpopulations.

---

[5] Further details for this case are presented in Section 4.7.

**Example 3.2.2 (density to frequency)** *Consider the differential equation model*

$$\dot{x}_i = x_i \, G(v, \mathbf{u}, \mathbf{x})|_{v=u_i}.$$

*The total number of individuals in a population is given by*

$$N = \sum_{i=1}^{n_s} x_i$$

*with the frequency of individuals*

$$p_i = \frac{x_i}{N} \quad \text{where} \quad \sum_{i=1}^{n_s} p_i = 1.$$

*Replacing $x_i = p_i N$ in the G-function, and noting that*

$$\frac{\dot{N}}{N} = \sum_{i=1}^{n_s} p_i \, G(v, \mathbf{u}, \mathbf{p}, N)|_{v=u_i}$$

*we define average fitness*

$$\overline{G}(v, \mathbf{u}, \mathbf{p}, N)\big|_{v=u_i} = \sum_{i=1}^{n_s} p_i \, G(v, \mathbf{u}, \mathbf{p}, N)|_{v=u_i}.$$

*It follows that*

$$\dot{p}_i = \frac{N \dot{x}_i - x_i \dot{N}}{N^2} = p_i \, G(v, \mathbf{u}, \mathbf{p}, N)|_{v=u_i} - p_i \, \overline{G}(v, \mathbf{u}, \mathbf{p}, N)\big|_{v=u_i}.$$

*We then have the equivalent population dynamics in terms of frequency*

$$\dot{p}_i = p_i \left[ G(v, \mathbf{u}, \mathbf{p}, N)|_{v=u_i} - \overline{G}(v, \mathbf{u}, \mathbf{p}, N)\big|_{v=u_i} \right]$$

$$\dot{N} = N \, \overline{G}(v, \mathbf{u}, \mathbf{p}, N)\big|_{v=u_i}.$$

*As a specific example, consider the simple G-function in the density format given by*

$$G(v, \mathbf{u}, \mathbf{x}) = f(v) - \sum_{i=1}^{n_s} u_i x_i.$$

*The population dynamics is written as*

$$\dot{x}_i = x_i \left[ f(u_i) - \sum_{i=1}^{n_s} u_i x_i \right].$$

*Converting to the frequency format we have the G-function*

$$G(v, \mathbf{u}, \mathbf{p}, N) = f(v) - N \sum_{i=1}^{n_s} u_i p_i$$

*with the frequency dynamics*

$$\dot{p}_i = p_i \left[ f(u_i) - N \sum_{i=1}^{n_s} u_i p_i - \sum_{i=1}^{n_s} p_i \left( f(u_i) - N \sum_{i=1}^{n_s} u_i p_i \right) \right] \quad (3.4)$$

$$= p_i \left[ G(v, \mathbf{u}, \mathbf{p}, N)|_{v=u_i} - \overline{G}(v, \mathbf{u}, \mathbf{p}, N)|_{v=u_i} \right]$$

$$\dot{N} = N \sum_{i=1}^{n_s} p_i \left( f(u_i) - N \sum_{i=1}^{n_s} u_i p_i \right) \quad (3.5)$$

$$= N \overline{G}(v, \mathbf{u}, \mathbf{p}, N)|_{v=u_i}.$$

*The formats are equivalent and will yield exactly the same results. The only difference is in terms of perspective. Sometimes (especially in matrix games) the total population size N is assumed constant (or ignored), leaving only frequency dynamics, (3.4). In such a case there would be no explicit density dependence. Alternatively the frequency could be assumed constant leaving only population dynamics (3.5) with no frequency dependence.*

### 3.2.2 Bauplans, G-functions, and taxonomic hierarchies

The German term **bauplan** is an old descriptor for classifying organisms according to common design features or design rules. It comes from the recognition that many groups of organisms differ only with respect to a small subset of their characteristics. The bauplan as a concept recognized that the fixity of many traits is more the rule than the exception among organisms. Today we might view this as developmental constraints on traits, or as phylogenetic constraints. The bauplan as a concept has obvious applications to taxonomy, in which organisms are ordered by characteristics in a hierarchical fashion. However, modern taxonomy went through a period of using small subsets of traits as the keys that define taxa in a somewhat dichotomous fashion – a leaning towards a dichotomous key approach to systematics. Modern taxonomy aims to group organisms based on strict evolutionary relationships. The means towards this end are the identification of shared derived characters as the tool for grouping organisms into mono-phyletic groupings. One seeks to define "branches" of the evolutionary tree by the points at which one group of organisms differs from another based upon one or several character shifts. Traits that are "conserved" become important in finding the branch points. Interestingly, this grouping of organisms in modern taxonomy by shared, heritable traits parallels the bauplan concept, even though the term had no strict evolutionary interpretation and would be viewed as archaic in modern phylogenetics.

We can think of the $G$-function as incorporating a bauplan along with its physical environment. The bauplan component of a $G$-function includes the strategy set of a group of evolutionarily identical individuals, and it includes the ways in which these characteristics of the bauplan interact with physical environments to influence fitness. A $G$-function and its associated strategy set provide a useful formalization of a bauplan as some set of design rules that define and constrain a particular group of species or higher taxa. The $G$-function is a mathematical construct that defines a group of individuals as evolutionarily identical. Individuals of a bauplan have the same set of evolutionarily feasible trait values, the same constraints on traits, and the same ecological consequences of actually possessing a particular trait value. While a bauplan is integral to a $G$-function (change the bauplan and the $G$-function must change), the actual $G$-function emerges from combining a bauplan with a particular physical environment (change the environment and the $G$-function likely changes even though the bauplan does not). We reserve the term bauplan to describe a $G$-function's strategy set and the associated propensities of these strategies to influence fitness when faced with particular ecological circumstances. Only when the actual environment is specified does the bauplan have an associated $G$-function.

We were tempted to let the $G$-function define a taxonomic grouping based on shared and derived characteristics rather than associate it with a bauplan. For instance, a $G$-function could describe the species within **families.** The taxonomic level of family may bring together all individuals that are evolutionarily identical in the sense that each could evolve quite readily the trait values of any species within the family (e.g., Canidae is represented by a dog $G$-function and Felidae by a cat $G$-function). For instance, artificially selecting a coyote to be a wolf would likely result in a "true" wolf rather than a coyote-like wolf. But, selecting a coyote to be a tiger would likely produce a coyote-like tiger rather than a tiger. Whether the taxonomic level of family roughly corresponds to a break point in bauplans and $G$-functions remains highly debatable and an untested empirical question. Further, our use of family or some other taxonomic grouping for a $G$-function would be co-opting terms and definitions that were not intended or developed for the $G$-function. These groupings have important and current usage in modern taxonomy. Still, the $G$-function needs a term to describe its categorizing of organisms based on evolutionarily identical individuals. We feel that bauplan is an excellent term for re-use in this modern context.

It successfully conveys the idea of grouping organisms by common evolutionary design rules. Within a given physical environment, one $G$-function means one bauplan, two $G$-functions mean two bauplans.

## 3.3 Evolution by natural selection

Darwin's reasoning was simple, elegant, and revolutionary. He started with the postulate of descent with change: like tends to beget like (heredity) and there is heritable variation associated with each type of organism. This by itself was neither novel nor revolutionary. Ancient programs of plant and animal breeding as well as cultural patterns of mate selection recognized (more or less accurately) "good" versus "bad" blood and the advantages of a good pedigree. However, the pool of variability was generally viewed as a narrow halo around a fixed, often "best" form. Novel variation constituted a degeneration of the blood line rather than novel opportunities for improved or different types of organisms. Darwin's postulate of heritable variation is an evolutionary principle that permits improvement through trial and error. But trial and error for what purpose? What constitutes an improvement?

Darwin's second postulate was relatively well known and unobjectionable at the time. It was based on the Malthusian principle that among organisms there is a struggle for existence. It follows directly from two principles of population ecology. First, all populations have the capacity to grow exponentially under ideal conditions. Second, no population can grow exponentially forever: there are limits to growth. Combining the principles of exponential population growth with limits to growth yields the struggle for existence. Darwin's first novel contribution was to see in the struggle for existence a means for directing and discriminating among heritable variation. It provides the opportunity for heritable traits to aid an organism in the struggle.

Darwin's last postulate, that heritable variation influences the struggle for existence, is a simple and elegant conjunction of the first two. And its consequence, evolution by natural selection, was intellectually revolutionary. Heritable variation provided the means for the natural process of trial and error to shape the traits of an organism. The struggle for existence provided the objective for which the organism's traits are judged superior or inferior.

### 3.3.1 Tautology and teleology in Darwinian evolution

Evolution by natural selection, and by extension evolutionary game theory, must be defensible against the twin accusation of tautology and teleology. Natural selection can seem like a truism or a tautology: natural selection is the survival of the fittest where the fittest are defined as those that survive – leading one to the conclusion that natural selection is all about the survival of the survivors. And, depending upon the formulation of Darwin's postulates, natural selection can indeed become vacuous by virtue of tautology (Lewontin, 1974; Endler, 1986). For any process that leads to the "erosion" and/or "accretion" of items can lead to a selective sieve that alters the character of a collection of things. This can include the composition of grain sizes on a sand dune, the sculpting of stream pebbles through abrasion, or the composition of household goods following a spring cleaning. If one does not know the "intention" or "objectives" of the sieving process then indeed all one knows are those that survive. Rosenberg (1985) makes the useful distinction between "what has been selected" and "what has been selected for" (see Mitchell and Valone (1990) for a fuller discussion). Simply knowing what has been selected falls into tautology as a predictive mechanism. However, knowing the objective of the selection process breaks the tautology and gives a selection process predictive powers. A wire mesh sieve *selects for* size, a winnowing process *selects for* specific gravity, etc.

When natural selection is viewed as winnowing among variation according to some criteria or objectives, natural selection risks becoming teleological. By requiring that a trait have a purpose are we presupposing a conscious intent or pre-meditated design? Fortunately, the answer is no (Mitchell and Valone, 1990; Reeve and Sherman, 1993). A trait does not need to have a purpose to be selected for, it need only have a function. The wings of many birds and insects are for flying, not in a purposeful sense, but in a functional sense. Hence, to avoid tautology and to avoid teleology, natural selection involves some set of evolutionarily feasible strategies (heritable variation) and an objective function that defines success in the struggle for existence. The strategy set and the fitness generating function serve our purposes in the context of evolutionary game theory. Traditionally this role has been played by the genetics and the relative fitnesses of phenotypes.

### 3.3.2 Darwin's postulates in evolutionary game theory

The fitness generating function encapsulates Darwin's three postulates as adapted from Lewontin (1961): (1) heritable variation, (2) struggle for existence, and (3) heritable variation influences the struggle for existence.

Consider the translation of Darwin's postulates into an evolutionary game setting:

1. Heritable variation: the individual organism possesses a heritable strategy, $u_i$, that may be any element in its strategy set: $u_i \in \mathcal{U}$.
2. Struggle for existence: the individual has a per capita growth rate determined from $G(v, \mathbf{u}, \mathbf{x})_{v=u_i}$ that is dependent on the densities and strategies of others.
3. Heritable variation influences the struggle: the per capita growth rate of an individual varies according to the choice of $v \in \mathcal{U}$.

### 3.3.3 Heritable variation and fitness

The Modern Synthesis (Dobzhansky, 1937; Huxley, 1942) was a major triumph of evolutionary thinking. The rediscovery of Mendel's Laws at the turn of the nineteenth century, the understanding of genes as the fundamental units of inheritance, and the development of population genetics by Wright, Fisher, and Haldane showed how an atomized and particulate recipe of inheritance (notions of discrete alleles at loci, and discrete loci packaged on chromosomes) was compatible with the production, persistence, and to some extent continuity of heritable variation required by Darwin's theory of natural selection. It is not surprising that population and quantitative genetics have become the lingua franca of evolutionary thinking and why the recipe of inheritance is considered central to any current theory of natural selection.

Genetics seemed to solve two additional issues in thinking about natural selection. First, Darwin was mostly wrong in his views on inheritance. With respect to heritable variation, Darwin was rather pluralistic and accepted blending inheritance (a kind of weighted averaging of the parental characteristics), **pangenesis** (all of the organs of the body contribute their own appropriate piece of the blueprint), and consequently some forms of inheritance of acquired traits. Mendelian and modern genetics corrected these errors and provided a rigorous means of modeling the trajectory of natural selection from some initial mix of genes.

Natural selection invites one to view evolution as the replacement of less fit characters by those that are more fit, perhaps reaching an evolutionary equilibrium, at which point one has the most-fit characteristics. In this way, the expected outcomes of natural selection have been associated with optimality and the production of the best or optimal traits. Done casually, this led Darwin and many others since to ask "Why does a plant (or animal) possess such and such a trait or characteristic?" This "explain a trait" approach to natural selection probably works remarkably well for traits with obvious survival functions or functions that are necessary for the organism's very existence. Remove a mammal's heart and it dies – a heart seems like a good idea for survival of a mammal-like organism. It also drove deep insights and thinking regarding traits such as altruism that seemed non-optimal or counter-productive to the individual. Sloppier forms of adaptation thinking led to sharp and often accurate criticisms of applying optimality to the outcomes of natural selection (Gould and Lewontin, 1979; Pierce and Ollason, 1987), but see Queller (1995). Here again, genetics was seen as the solution to these problems. Evolution could be defined as change in gene frequency. Fitness differences among individuals possessing different combinations of genes produced natural selection. And the resulting genetic composition of the population following this genic and genotypic fitness dynamics represented the predicted outcome of natural selection.

Population genetics and quantitative genetics seemed to resolve several issues raised by the adaptation approach. Under density-independent and frequency-independent selection, the population acquired a genetic composition that maximized the population's growth rate. Under density-dependent and frequency-independent selection, the population's genetic composition evolved so as to maximize equilibrium population size. Unfortunately, under frequency-dependent selection changes in gene frequency produce populations that do not appear to optimize any feature of the population's ecology. With concepts such as linkage, **epistasis** (a single gene influences several traits), and **pleiotropy** (several genes interact to produce a single trait), population genetics seemed to explain the maintenance of genetic variability, the persistence of what appear to be maladaptive phenotypes, and correlated changes in traits. With Hamilton's (1963) rule and the concept of inclusive fitness, population genetics could use the concept of genes being identical by descent to understand the evolution of **altruism** whereby an individual may sacrifice personal fitness to contribute to the survival or fecundity of a relative. So was Darwin in his ignorance of the recipe of inheritance merely lucky in coming up with a powerful idea for the wrong reason? Or does natural selection supervene the underlying genetics (Mitchell and Valone, 1990)?

A full appreciation of strategy dynamics (Meszena *et al.*, 1997) and the objective of natural selection requires a focus on heritable phenotypes (=strategies) as well as the fundamental units of inheritance. At its core, evolution by natural selection draws on both evolutionary and ecological principles. Yet evolutionary ecology, that harmonious blend of what is evolutionarily feasible and what is ecologically acceptable, has often been difficult to achieve. Combining the genes of population genetics with the individuals of population ecology is difficult at best. On the other hand, the heritable phenotypes of individuals

## 3.3 Evolution by natural selection

place the evolutionary and ecological contexts into the same currency. In the $G$-function, the evolutionary and ecological principles underlying evolution by natural selection can be described in the context of game theory. Evolutionary game theory becomes a modeling tool for predicting the trajectory of evolution (strategy dynamics) as well as the outcomes of natural selection (adaptation and optimality).

# 4
# $G$-functions for the Darwinian game

A bauplan for a group of evolutionarily identical individuals together with their environment represents the essential elements needed to construct a $G$-function. The bauplan has two aspects. First, it describes a set of evolutionarily feasible strategies, and, second, it specifies the intrinsic ecological properties, aptitudes, trade-offs, and limitations of this group. The environment provides a setting within which the bauplan produces species that evolve, diversify, and persist. For instance, hornbills, a frugivorous bird of African forests, differ in size, wing morphology, and bill characteristics. However, all members of the group are easily identifiable as hornbills quite distinct from other birds. It is reasonable to assume that all species of hornbills share the same bauplan and, hence, within the same environment their fitness is determined from a single $G$-function. Toucans of Central and South America occupy similar ecological niches to hornbills. These birds have radiated along similar morphological lines to hornbills. Yet they have a distinct bauplan from hornbills and from other bird groups. In an evolutionary game involving both hornbills and toucans, two $G$-functions would be required.

When modeling evolution, one usually has some taxa (such as hornbills or toucans) along with an environmental setting in mind. In this chapter, we undertake the practical task of bringing together the mathematical notation of Chapter 2 and the $G$-function, Definition 3.2.1, to formulate the required fitness generating functions. We will consider a number of different evolutionary games. We start by discussing the general procedure and then illustrate the method by developing $G$-functions for systems of increasing complexity, starting with the simplest biological models that can be described by a single $G$-function with scalar strategies. The same basic method is then used to determine $G$-functions for more complex systems.

## 4.1 How to create a $G$-function

It is usually quite easy to formulate a $G$-function for most biological situations of interest. Usually the $G$-function can be written as an analytical expression[1] by means of the following three steps.

1. Select an appropriate ecological model for the population dynamics. The model may be for a single population or species, it may be a life-history model with different age and stage classes, or it may be a model of population interactions that includes growth equations for competitors, resources, predators, etc.
2. Select strategies and strategy sets associated with the population, species, or community under consideration. The strategy set may be continuous and/or discrete. The strategy set is determined from hypotheses concerning genetic, developmental, physiological, and physical constraints on the set of evolutionarily feasible strategies. Determine feasible combinations of strategies based on equality constraints (e.g., heterozygosity, dominance, penetrance, etc. from Mendelian genetics) and/or upper and lower bounds by inequality constraints (akin to some quantitative genetic constraints). When defining strategies and strategy sets, one must decide whether the model calls for one, two, or more distinct sets of evolutionarily identical individuals. For instance, in a single model of density-dependent population growth, all individuals might be considered evolutionarily identical, in which case there is a single strategy set that will become associated with a single $G$-function. However, in a model of trophic interactions, it may be conjectured that the prey represent one set of evolutionarily identical individuals (a small mammal for instance) and the predators another (a raptorial bird). In this case, there will be a separate strategy set and $G$-function associated with each group of evolutionarily identical individuals.
3. Create the $G$-function(s) by hypothesizing how the individual's strategy, $\mathbf{v}$, as well as all strategies in the population, $\mathbf{u}$, influences the values of parameters in the ecological models of population dynamics. As soon as key parameters of a population model become functions of $\mathbf{v}, \mathbf{u}, \mathbf{x}, \mathbf{y}$, the ecological model becomes a $G$-function.

---

[1] There are situations involving stochasticity in which the $G$-function is defined within the context of a simulation where numerical iterations create the biotic environment against which a focal individual is compared (e.g., Schmidt et al., 2000).

**Example 4.1.1 (symmetric competition game)** *For step 1, let us reconsider the Lotka–Volterra model introduced in Subsection 2.4.5*

$$H_1(\mathbf{x}) = \frac{r_1}{K_1}\left(K_1 - \sum_{j=1}^{n_s} a_{1j} x_j\right)$$

$$\vdots = \vdots$$

$$H_{n_s}(\mathbf{x}) = \frac{r_{n_s}}{K_{n_s}}\left(K_{n_s} - \sum_{j=1}^{n_s} a_{n_s j} x_j\right).$$

*For step 2, we note that many of the parameters used in the above models could be either adaptive parameters or functions of adaptive parameters. For example, $r_i$, $K_i$, and $a_{ij}$ may all depend on metabolic rates and conversion efficiencies and if these more basic (adaptive) parameters were changed to increase say $r_i$ it is likely that $K_i$ would decrease, with $a_{ij}$ changing as well. Consider the situation where every $r_i$ is equal to the same constant*

$$r_i = r,$$

*every function $K_i$ is of the same functional form depending only on the scalar adaptive parameter $u_i$*

$$K_i = K(u_i),$$

*and every function $a_{ij}$ is of the same functional form depending on the adaptive parameters $u_i$ and $u_j$*

$$a_{ij} = a(u_i, u_j).$$

*The Lotka–Volterra model is then expressed as*

$$H_1(\mathbf{u}, \mathbf{x}) = \frac{r}{K(u_i)}\left(K(u_i) - \sum_{j=1}^{n_s} a(u_i, u_j) x_j\right)$$

$$\vdots = \vdots$$

$$H_{n_s}(\mathbf{u}, \mathbf{x}) = \frac{r}{K(u_{n_s})}\left(K(u_{n_s}) - \sum_{j=1}^{n_s} a(u_i, u_j) x_j\right).$$

*In order to complete the model, specific relationships for $K$ and $a$ must be given. In Subsection 2.4.5 we used the following distribution functions*

$$K(u_i) = K_m \exp\left(-\frac{u_i^2}{2\sigma_k^2}\right)$$

$$a(u_i, u_j) = \exp\left[-\frac{(u_i - u_j)^2}{2\sigma_a^2}\right].$$

For step 3 we note from the symmetry of the fitness functions that the Lotka–Volterra G-function for this system is given by

$$G(v, \mathbf{u}, \mathbf{x}) = \frac{r}{K(v)} \left[ K(v) - \sum_{j=1}^{n_s} a(v, u_j) x_j \right], \qquad (4.1)$$

where

$$K(v) = K_\mathrm{m} \exp\left(-\frac{v^2}{2\sigma_k^2}\right) \qquad (4.2)$$

and

$$a(v, u_j) = \exp\left[-\frac{(v - u_j)^2}{2\sigma_a^2}\right]$$

*as may be verified by direct substitution. The term $K_\mathrm{m}$ is the maximum value for the carrying capacity, $\sigma_k$ is related to the "range of resources," $\sigma_a$ is related to a species "niche width." Carrying capacity takes on a maximum value at $v = 0$. The variance of this distribution, $\sigma_k^2$, determines the severity with which an individual loses carrying capacity as its strategy deviates from $v = 0$. With a larger variance, the individual suffers less from a deviation. The competition term is a normal distribution with respect to $v$ and takes on a maximum when $v = u_j$. Its variance, $\sigma_a^2$, determines how quickly the competition coefficient changes as competitors deviate in their strategy values. A large variance means that the competition coefficient changes slowly with changes in $v$.*

Frequency-dependent selection enters the above model through the symmetric competition coefficients. As noted by Brown and Vincent (1987a), a result of this symmetry is that frequency dependence is lost as a factor determining the ESS. This limits the usefulness of this particular example.

## 4.2 Types of $G$-functions

The form and complexity of the $G$-function will depend on the complexity of the community under consideration. It will also depend upon the modeler's view of the system's population ecology, the suite of simplifying assumptions, and the presumed relationships between ecological parameters and evolutionary strategies. In the above example, we have illustrated how the $G$-function can be used to model a community of competitors under a number of simplifying assumptions. In particular, the model assumes that all individuals are evolutionarily identical, the strategies are scalars, and that there is no age or stage structure. However, by using appropriate notation, $G$-functions can be determined for systems with vector-valued strategies, groups of individuals that are not all evolutionarily identical, and life histories with explicit stages or age classes. Unfortunately, the notation becomes horrendous if we attempt to do everything at once. Rather, we will look at each of these cases with as much generality as possible without creating a notational overload.

The remainder of this chapter is devoted to **categorizing** increasingly complex $G$-functions that are discussed in more detail again in later chapters. We have attempted to make the categories correspond to classes of problems that we have studied and found to be useful. If a particular problem does not fall within a given category, it should be apparent how to modify the results given here. For example, if one is interested in multistage $G$-functions with vector strategies, then one can use the notation of both the multistage $G$-functions and the $G$-functions with vector strategies to handle this case.

## 4.3 $G$-functions with scalar strategies

This category includes any $G$-function that is used to model systems having **one unique bauplan** with a single **scalar strategy**. For this to be the case, each population's growth equation must be dependent on the same evolving trait. For example, flowering time has been identified as an important adaptive parameter in the modeling of annual plants (Cohen, 1971; Vincent and Brown, 1984a).

With $n_s$ species, their population densities and strategies are represented by the vectors

$$\mathbf{x} = \begin{bmatrix} x_1 & \cdots & x_{n_s} \end{bmatrix}$$
$$\mathbf{u} = \begin{bmatrix} u_1 & \cdots & u_{n_s} \end{bmatrix}.$$

Each strategy, $u_i$, is distinct and drawn from the same set of evolutionarily feasible strategies (as is required by a single bauplan – all individuals must be evolutionarily identical)

$$u_i \in \mathcal{U}, \quad i = 1, \cdots, n_s, \tag{4.3}$$

where $\mathcal{U}$ is a subset of a one-dimensional strategy space that represents the feasible strategy choices after all (if any) constraints have been imposed. For quantitative traits, the constraints are simply upper and lower bounds placed on the components of the strategy vector. As a shorthand, we use

$$\mathbf{u} \in \mathcal{U}$$

in place of (4.3).

The definition of the $G$-function for this case is given by (see Definition 3.2.1)

$$G(v, \mathbf{u}, \mathbf{x})|_{v=u_i} = H_i(\mathbf{u}, \mathbf{x}), \quad i = 1, \cdots, n_s.$$

In terms of the $G$-function, the **population dynamics** for the three dynamical systems introduced in Subsection 2.3.2 are of the following form

Population dynamics available for $G$-functions with scalar strategies

| Difference: | $x_i(t+1) = x_i \left[1 + G(v, \mathbf{u}, \mathbf{x})|_{v=u_i}\right]$ |
| Exp. Difference: | $x_i(t+1) = x_i \exp G(v, \mathbf{u}, \mathbf{x})|_{v=u_i}$ |
| Differential: | $\dot{x}_i = x_i G(v, \mathbf{u}, \mathbf{x})|_{v=u_i}$ |

where $i = 1, \ldots, n_s$.

**Example 4.3.1 (L–V competition game)** *Example 4.1.1 is one of many models based on the Lotka–Volterra system. We can obtain many variants of this game by simply changing the functional relationships for $r$, $K$, or $a$. For the L–V competition game, we again use (4.1) with the symmetric distribution for the carrying capacity as given by (4.2). However, the model differs from Example 4.1.1 by replacing the symmetrical distribution function $a(v, u_j)$ with a non-symmetric one*

$$a(v, u_j) = 1 + \exp\left\{-\frac{(v - u_j + \beta)^2}{2\sigma_a^2}\right\} - \exp\left\{-\frac{\beta^2}{2\sigma_a^2}\right\}.$$

*The term $\beta$ introduces an asymmetry into the competition term.*

Unlike Example 4.1.1, frequency dependence has a strong influence on the ESS for the L–V competition game. It has been analyzed in several papers (Brown and Vincent, 1987a; Vincent and Brown, 1987a; Vincent et al., 1993). We will show in Chapter 5 that, with this model, it is possible for ESS coalitions to form that involve the co-existence of more than one species.

## 4.4 $G$-functions with vector strategies

This category includes any $G$-function used to model systems with **one unique bauplan** and with **vector strategies**. It extends the scalar case by permitting heritable phenotypes with more than one evolving trait. For example, flowering time, root–shoot ratio, and height could make up a vector of adaptive parameters used in an annual plant model.

With $n_s$ species, the scalar population densities are represented by the vector

$$\mathbf{x} = \begin{bmatrix} x_1 & \cdots & x_{n_s} \end{bmatrix}.$$

The vector strategy of a given species $i$ is given by

$$\mathbf{u}_i = \begin{bmatrix} u_{i1} & \cdots & u_{in_u} \end{bmatrix}$$

where $n_u$ is the number of traits in the vector $\mathbf{u}_i$ (all strategy vectors are assumed to be of the same dimension). Each strategy vector represents a heritable phenotype with as many traits as the dimension of the strategy vector and it is distinct and drawn from the same set of evolutionarily feasible strategies

$$\mathbf{u}_i \in \mathcal{U}, \quad i = 1, \cdots, n_s. \tag{4.4}$$

where $\mathcal{U}$ is a subset of an $n_u$-dimensional strategy space that represents the feasible strategy choices after all constraints have been imposed. As a shorthand, we again use

$$\mathbf{u} \in \mathcal{U}$$

in place of (4.4).

The vector of all strategies present in the community is given by

$$\mathbf{u} = \begin{bmatrix} \mathbf{u}_1 & \cdots & \mathbf{u}_{n_s} \end{bmatrix}$$

where $\mathbf{u}$ is the **concatenation** of the vector strategies of all the species. That is, the first $n_u$ components of this vector belong to $x_1$, the second $n_u$ components belong to $x_2$ and so on. Collectively they form a vector of $n_s$ partitions with a total length determined by the product $n_s \cdot n_u$. For example, suppose that $n_s = 3$ and $n_u = 2$. The three strategy vectors are given by

$$\mathbf{u}_1 = \begin{bmatrix} u_{11} & u_{12} \end{bmatrix}$$
$$\mathbf{u}_2 = \begin{bmatrix} u_{21} & u_{22} \end{bmatrix}$$
$$\mathbf{u}_3 = \begin{bmatrix} u_{31} & u_{32} \end{bmatrix}$$

with

$$\mathbf{u} = \begin{bmatrix} u_{11} & u_{12} & | & u_{21} & u_{22} & | & u_{31} & u_{32} \end{bmatrix}$$

where the vertical bar is used for clarity to show the **partitioning** of each species' strategy vector.

The $G$-function definition for this case is an obvious extension of Definition 3.2.1

$$G(\mathbf{v}, \mathbf{u}, \mathbf{x})|_{\mathbf{v}=\mathbf{u}_i} = H_i(\mathbf{u}, \mathbf{x}), \quad i = 1, \cdots, n_s.$$

Clearly, the strategy of the focal individual as given by the virtual variable in $G(\mathbf{v}, \mathbf{u}, \mathbf{x})$ is also a vector of dimension $n_u$. The population dynamics for these

## 4.4 G-functions with vector strategies

systems are of the following form

Population dynamics available for $G$-functions with vector strategies

$$\begin{array}{ll} \text{Difference:} & x_i(t+1) = x_i\left[1 + G\left(\mathbf{v}, \mathbf{u}, \mathbf{x}\right)|_{\mathbf{v}=\mathbf{u}_i}\right] \\ \text{Exp. Difference:} & x_i(t+1) = x_i \exp G\left(\mathbf{v}, \mathbf{u}, \mathbf{x}\right)|_{\mathbf{v}=\mathbf{u}_i} \\ \text{Differential:} & \dot{x}_i = x_i\, G\left(\mathbf{v}, \mathbf{u}, \mathbf{x}\right)|_{\mathbf{v}=\mathbf{u}_i} \end{array} \quad (4.5)$$

where $i = 1, \ldots, n_s$.

**Example 4.4.1 (L–V big bully game)** *We again use the Lotka–Volterra competition model (4.1), but now introduce a vector-valued strategy that has two components. The first component influences carrying capacity*

$$K(\mathbf{v}) = \left(1 - v_2^2\right) K_{\max} \exp\left(-\frac{v_1^2}{2\sigma_k^2}\right)$$

*and the competition coefficients*

$$a\left(\mathbf{v}, \mathbf{u}_j\right) = 1 + B_j \exp\left[-\frac{\left(v_1 - u_{j1} + \beta\right)^2}{2\sigma_a^2}\right] - \exp\left[-\frac{\beta^2}{2\sigma_a^2}\right]$$

*in the same way as in Vincent et al. (1993) and Cohen et al. (1999). That is, an individual's carrying capacity fits a normal distribution with respect to the first component of its strategy, $v_1$. The competition experienced by an individual from another individual of its own or different species is influenced by the difference between the first component of the individual's strategy and the first component of another species's strategy. The factor $\beta$ describes the level of asymmetry in competition. When $\beta > 0$, an individual with a larger value for $v_1$ has a larger negative effect on an individual with a smaller $v_1$ than the smaller value has on the larger. The competition function follows a normal distribution with respect to $v_1$. Furthermore, the competition coefficient, by means of an additive adjustment term, takes on a value of 1 when individuals share the same value for the first component. The second component of an individual's strategy, $v_2$, influences carrying capacity, as indicated, and the competition coefficients via a "bully" function*

$$B_j = 1 + B_{\max}\left(u_{j2} - v_2\right).$$

The **bully function**, used in the above example, describes forms of competition where being slightly larger than your neighbor confers a competitive advantage by reducing the negative effects of others and increasing one's own negative effect on others. Height in trees provides an obvious example. Being taller than one's neighbor increases one's own access to light at the expense of shorter individuals that are now shaded. This favors a kind of arms race in which trees gain by evolving a height advantage against neighbors. However, this advantage is nullified as soon as others adopt a taller height. This arms race has a price. While the total amount of available sunlight remains unchanged whether the trees are short or tall, all of the trees must now produce and support the non-productive woody trunk that achieves height. This reduces the availability of resources for productive tissues such as roots, stems, and ultimately seeds. In trees, via tree trunks, competition for light produces a **tragedy of the commons** (Hardin, 1968). The advantage for being taller than one's neighbors provides a small individual benefit that is smaller than the collective loss. And this small advantage is eliminated as soon as others evolve a similar height; but the collective cost remains.

As a tragedy of the commons, we let the bully function, $B$, scale the competitive effect that others have on an individual. If others are larger than you then their negative effect is amplified; if smaller, then their negative effect is diminished. When individuals have the same value for $v_2$ then $B = 1$ and the effect of the bully function on the competition coefficient vanishes. But the individual pays a price in terms of its own carrying capacity by increasing $v_2$. An individual loses carrying capacity proportional to $(1 - v_2^2)$. This effect and functional form effectively restricts the reasonable values for this second component to $v_2 \in [0, 1)$. The function $B$ and its effect on the competition coefficients introduce an evolutionary arms race: bigger values for $v_2$ are better for competition and expensive in terms of $K$.

We will use the bully game later to illustrate convergent-stable saddle point solutions, adaptive speciation and ESS solutions composed of coalitions with more than two species.

## 4.5 $G$-functions with resources

This category includes any $G$-function used to model systems that explicitly include the dynamics of resources used by and influenced by the population of organisms. In particular, this includes all systems having **one unique bauplan** with **vector strategies** and **resource dynamics**. This class of system is useful when dealing with plants or animals feeding on an explicit, depletable resource that is not part of the evolving system.

In addition to the population density vector

$$\mathbf{x} = \begin{bmatrix} x_1 & \cdots & x_{n_s} \end{bmatrix}$$

there is a resource vector $\mathbf{y}$

$$\mathbf{y} = \begin{bmatrix} y_1 & \cdots & y_{n_y} \end{bmatrix}$$

where $n_y$ is the number of resources. The strategy vector satisfies the same

## 4.5 G-functions with resources

conditions as required for $G$-functions with vector strategies.

$$\mathbf{u}_i = \begin{bmatrix} u_{i1} & \cdots & u_{in_u} \end{bmatrix}.$$

The $G$-function in this case is defined by

$$G(\mathbf{v}, \mathbf{u}, \mathbf{x}, \mathbf{y})|_{\mathbf{v}=\mathbf{u}_i} = H_i(\mathbf{u}, \mathbf{x}, \mathbf{y}), \quad i = 1, \cdots, n_s.$$

The population dynamics for these systems are given by the following

Population dynamics available for $G$-functions with resources

| | |
|---|---|
| Difference: | $x_i(t+1) = x_i \left[1 + G(\mathbf{v}, \mathbf{u}, \mathbf{x}, \mathbf{y})|_{\mathbf{v}=\mathbf{u}_i}\right]$ |
| Exp. Difference: | $x_i(t+1) = x_i \exp G(\mathbf{v}, \mathbf{u}, \mathbf{x}, \mathbf{y})|_{\mathbf{v}=\mathbf{u}_i}$ |
| Differential: | $\dot{x}_i = x_i G(\mathbf{v}, \mathbf{u}, \mathbf{x}, \mathbf{y})|_{\mathbf{v}=\mathbf{u}_i}.$ |

The dynamical equations for the resource vector $\mathbf{y}$ are expressed in terms of a resource function $\mathbf{N}(\mathbf{u}, \mathbf{x}, \mathbf{y})$ also of dimension $n_y$

$$\mathbf{N} = \begin{bmatrix} N_1 & \cdots & N_{n_y} \end{bmatrix}.$$

In terms of $\mathbf{N}$, the resource dynamics are given by the following

Resource dynamics

| | |
|---|---|
| Difference: | $\mathbf{y}(t+1) = \mathbf{y} + \mathbf{N}(\mathbf{u}, \mathbf{x}, \mathbf{y})$ |
| Differential: | $\dot{\mathbf{y}} = \mathbf{N}(\mathbf{u}, \mathbf{x}, \mathbf{y}).$ |

The difference equation for resource dynamics is used with either of the difference equations for population dynamics.

**Example 4.5.1 (Bergmann's rule)** *Bergmann's rule notes how the body size of a mammal species or of a closely related group of mammal species (species of the same genus) increases with latitude. This increase in body size as one moves away from the equator has been interpreted as an adaptive consequence of colder temperatures and the utility of having a smaller surface-area-to-volume ratio. Size is viewed as mitigating the thermoregulatory costs of cold temperatures. In this example, we will develop and explore a model in which Bergmann's rule will emerge as a consequence of temperature-dependent consumer resource dynamics, and the consequence of body size for searching for and handling food items. We use a simple consumer-resource model in which the resource renews according to the Monod model. The consumers must search for and handle resources in a fashion modeled by the disk equation (Holling, 1965). Resource dynamics for y is described by subtracting consumer-induced mortality from resource renewal*

$$\dot{y} = r(K - y) - \sum_{i=1}^{n_s} \frac{a_i y x_i}{1 + a_i h_i y}$$

where r regulates the speed of resource population growth, K is the resource carrying capacity, a is the encounter probability of a consumer individual on resources, and h is the consumer's handling time on a resource item. Let the rate at which consumers increase in numbers be a function of net energy gain from consuming resources minus a foraging cost $c_i$

$$\dot{x}_i = x_i \left[ \frac{a_i y}{1 + a_i h_i y} - c_i \right].$$

**Allometry** considers the relationship between body size and important physiological, behavioral, morphological, and demographic parameters. For our purposes, we will let body size be the evolutionary strategy $u$. Hence we have the constraint set

$$\mathcal{U} = \{u \in \mathcal{R} \mid u > 0\}.$$

(One reads this equation as: the set $\mathcal{U}$ is composed of all $u$, an element of the one-dimensional real number space $\mathcal{R}$ such that $u$ is a positive number.) We will assume that there are allometric relationships between body size and encounter probability, handling time, and metabolic costs

$$a_i = A u_i^\alpha, \quad h_i = H u_i^{-\beta}, \quad c_i = C u_i^\gamma.$$

where $\alpha$ scales how encounter probability increases with body size, $\beta$ scales how handling time declines with body size, and $\gamma$ scales how foraging costs increase with body size. Relative to foraging costs, we will assume that body size represents a trade-off between being able to find and being able to handle resource items. Relative to foraging costs we will assume that smaller animals have a search advantage (e.g., it is advantageous for groups to split up when searching randomly for something) and larger animals have a handling advantage (there may be an economy of scale to the mouths and masticatory apparatus of animals). This tradeoff requires $\alpha < \gamma < \beta$. We can now formulate an evolutionary game for the evolution of body size by considering resource dynamics and the G-function for the consumer species

$$G(v, \mathbf{u}, \mathbf{x}) = \frac{A v^\alpha y}{1 + A H v^{(\alpha - \beta)} y} - C v^\gamma$$

$$\dot{y} = r(K - y) - \sum_{i=1}^{n_s} \frac{A u_i^\alpha y x_i}{1 + A H u_i^{(\alpha - \beta)} y}.$$

In this model, because competition among consumers is merely exploitative, the consumer species do not directly influence each other's fitness. The strategies and population sizes of other consumers enter the G-function only through their effect on resource abundance, y. In Chapter 7 we will show how Bergmann's rule is obtained using this consumer-resource model.

## 4.6 Multiple $G$-functions

This category includes any systems having **two or more bauplans** with **vector strategies**. Multiple $G$-functions are required when modeling organisms defined by a single bauplan in different environmental settings, when modeling organisms with different bauplans in the same environmental setting, or some combination of both.

An island world with only a single $G$-function would be a simple one. Among birds, suppose that Hawaii had only Hawaiian honeycreepers or that the Galápagos had only Darwin's finches. Through adaptation, species could come and go as the environment changes, but under the single $G$-function they are all evolutionarily identical. In our observations of this world, hopefully, at least one species always survives (a most dull world otherwise!).

In a more complicated world, not all individuals are evolutionarily identical and/or the environmental setting is not the same for everyone. These situations require more than one $G$-function to describe the evolutionary ecology. We let $n_g$ denote the number of distinct $G$-functions within the system. In dealing with this case, we take the view that $n_g$ is a fixed number; however, it is not required that the number of individuals modeled by a given $G$-function is non-zero, only that at least one $G$-function has at least one species at a non-zero population size. The reason for this is that the ESS need not involve individuals associated with every $G$-function. Consider a system in which there are two prey $G$-functions and one predator $G$-function. In the absence of predators, it is possible that prey from each $G$-function could arrive at an ESS. However, introduction of a predator could result in an ESS with the loss of all species associated with one of the prey $G$-functions. For example, in our island world of Hawaii, suppose that some white-eyes arrive and are able to co-evolve with the honeycreepers to form an ESS coalition of two or more. Then the rats arrive and eliminate all of the white-eyes, who have no defense against rats (purely hypothetical!).

Assume that each $G$-function has $n_{s_i} \overset{>}{\sim} 0, i = 1, \cdots, n_g$ different species. We use the notation, $\overset{>}{\sim}$ to denote that at least one $n_{s_i}$ must be greater than zero (with possibly all others zero). Let the **rising number of species**[2] $r_i$ be the number obtained when the number of species in the first $G$-function is added

---

[2] The notation for the rising number r (sans serif r) is an exception from our usual notation for number of indices (e.g., $n_r$). This is done to simplify notation when multiple subscripts are required).

to that in the second $G$-function, to that in the third $G$-function, etc., up to $G$-function $n_g$. We start with a rising number of zero and then add species going from one $G$-function to the next ($r_0$ is introduced for notational convenience)

$$r_0 = 0$$

$$r_1 = n_{s_1} = \sum_{j=1}^{1} n_{s_j}$$

$$r_2 = \sum_{j=1}^{2} n_{s_j}$$

$$\vdots$$

$$r_{n_g} = \sum_{j=1}^{n_g} n_{s_j} = n_s \geq 1$$

or more compactly

$$r_i = \sum_{j=1}^{i} n_{s_j} \text{ for } i = 1, \cdots, n_g$$

where $n_s$ is now the total number of species from all the bauplans. This use of $n_s$ is compatible with its previous use. We may now conveniently order the species according to

$$\mathbf{x}_1 = \begin{bmatrix} x_1 & \cdots & x_{r_1} \end{bmatrix}$$
$$\mathbf{x}_2 = \begin{bmatrix} x_{r_1+1} & \cdots & x_{r_2} \end{bmatrix}$$
$$\mathbf{x}_3 = \begin{bmatrix} x_{r_2+1} & \cdots & x_{r_3} \end{bmatrix}$$
$$\vdots$$
$$\mathbf{x}_{n_g} = \begin{bmatrix} x_{r_{(n_g-1)}+1} & \cdots & x_{r_{n_g}} \end{bmatrix}.$$

Taking note of the definition for $r_0$, we may write this more compactly as

$$\mathbf{x}_i = \begin{bmatrix} x_{r_{(i-1)}+1} & \cdots & x_{r_i} \end{bmatrix}$$

for $i = 1, \ldots, n_g$. If a $G$-function has zero species, it is still given a placeholder, by assigning it one species with a density of zero. In this way, all $G$-functions are accounted for. Using this notation we define the total density vector as

$$\mathbf{x} = \begin{bmatrix} \mathbf{x}_1 & | & \cdots & | & \mathbf{x}_{n_g} \end{bmatrix}.$$

The same $\mathbf{x}$ notation is used as in the single $G$-function case, with the understanding that $\mathbf{x}$ is now an extended vector that includes the catenation of densities of all the species from all the different $G$-functions.

## 4.6 Multiple G-functions

The members of each $G$-function have strategies drawn from a strategy set that is a property of the bauplan and generally unique to each $G$-function. As in the vector strategy case above, we require a double subscript notation in order to specify a particular strategy within a given $G$-function. Strategies used by the $i$th species are given by

$$\mathbf{u}_i = \begin{bmatrix} u_{i1} & \cdots & u_{in_{u_j}} \end{bmatrix}$$

where $n_{u_j}$ is the number of strategies in the vector $\mathbf{u}_i$ (strategy vectors from different $G$-functions may have different dimensions).

For example, suppose that $n_g = 3$, $n_{s_1} = 2$, $n_{s_2} = 1$, $n_{s_3} = 3$. In this case then, $r_1 = 2$, $r_2 = 3$, $r_3 = n_s = 6$ with

$$\mathbf{x} = \begin{bmatrix} x_1 & x_2 \mid x_3 \mid x_4 & x_5 & x_6 \end{bmatrix}$$

where we use the bar | to partition $\mathbf{x}$ corresponding to the various $G$-functions. Let $n_{u_1} = 3$ be the number of traits in the strategy vector of the first $G$-function; then the vector of strategies used by each of the species in the first $G$-function is given by

$$\mathbf{u}_1 = \begin{bmatrix} u_{11} & u_{12} & u_{13} \end{bmatrix}$$
$$\mathbf{u}_2 = \begin{bmatrix} u_{21} & u_{22} & u_{23} \end{bmatrix}$$

where $u_{ij}$ is the $j$th strategy vector of the $i$th population. Note that each strategy vector, under a given $G$-function, will always have the same number of components. If $n_{u_2} = 2$ then the vector of strategies in the second $G$-function is given by

$$\mathbf{u}_3 = \begin{bmatrix} u_{31} & u_{32} \end{bmatrix}$$

and if $n_{u_3} = 1$, then the third $G$-function has scalar strategies

$$u_4 = u_{41}$$
$$u_5 = u_{51}$$
$$u_6 = u_{61}.$$

The generalization is now fairly obvious: we define the vector of all strategies in the population by

$$\mathbf{u} = \begin{bmatrix} \mathbf{u}_1 & \cdots & \mathbf{u}_{n_g} \end{bmatrix}.$$

This gives us a notation consistent with previous use. In the above example, the vector of all strategies is given by

$$\mathbf{u} = \begin{bmatrix} u_{11} & u_{12} & u_{13} & u_{21} & u_{22} & u_{23} \mid u_{31} & u_{32} \mid u_{41} & u_{51} & u_{61} \end{bmatrix}.$$

The set of constraints for each $G$-function may be different; in general there will be $n_g$ different strategy sets. Thus

$$\mathbf{u}_i \in \mathcal{U}_j \tag{4.6}$$

where

$$i = r_0 + 1, \cdots, r_1 \quad \text{for} \quad j = 1$$
$$i = r_1 + 1, \cdots, r_2 \quad \text{for} \quad j = 2$$
$$\vdots \qquad \qquad \vdots \qquad \vdots \qquad \qquad (4.7)$$
$$i = r_{j-1} + 1, \cdots, n_s \quad \text{for} \quad j = n_g.$$

More compactly, instead of (4.7) we may write

$$i = r_{j-1} + 1, \cdots, r_j \quad \text{for } j = 1, \cdots, n_g. \qquad (4.8)$$

Once again we use

$$\mathbf{u} \in \mathcal{U}$$

as a shorthand for (4.6) and (4.8). For those situations in which the strategies used in each $G$-function are scalars, notation is simplified by dropping the double subscript. In this case we use $u_1$ for $u_{11}$, $u_2$ for $u_{21}$, etc. with

$$\mathbf{u} = \begin{bmatrix} u_1 & \cdots & u_{n_g} \end{bmatrix}.$$

Each $G$-function is defined by

$$G_j(\mathbf{v}, \mathbf{u}, \mathbf{x})\big|_{\mathbf{v}=\mathbf{u}_i} = H_i(\mathbf{u}, \mathbf{x})$$

where $i$ and $j$ are determined according to (4.8), $\mathbf{u}$, and $\mathbf{x}$ are defined as above and $\mathbf{v}$ is the virtual variable to be drawn from the $\mathcal{U}_j$ strategy set according to (4.8). For example, if there are two $G$-functions in which the first has two species with a two-component strategy vector and the second has one species with a three-component strategy vector, then $\mathbf{v}$ is set equal to $\begin{bmatrix} u_{11} & u_{12} \end{bmatrix}$ and $\begin{bmatrix} u_{21} & u_{22} \end{bmatrix}$ when evaluating $G_1$ and $\mathbf{v}$ is set equal to $\begin{bmatrix} u_{31} & u_{32} & u_{33} \end{bmatrix}$ when evaluating $G_2$.

The population dynamics for this class of systems are as follows

Population dynamics available for multiple $G$-functions

| | |
|---|---|
| Difference: | $x_i(t+1) = x_i \left[ 1 + G_j(\mathbf{v}, \mathbf{u}, \mathbf{x})\big|_{\mathbf{v}=\mathbf{u}_i} \right]$ |
| Exp. Difference: | $x_i(t+1) = x_i \exp G_j(\mathbf{v}, \mathbf{u}, \mathbf{x})\big|_{\mathbf{v}=\mathbf{u}_i}$ |
| Differential: | $\dot{x}_i = x_i\, G_j(\mathbf{v}, \mathbf{u}, \mathbf{x})\big|_{\mathbf{v}=\mathbf{u}_i}$ |

where $i$ and $j$ are determined according to (4.8).

**Example 4.6.1 (predator–prey coevolution)** *Predator–prey coevolution provides examples of evolutionary games with two $G$-functions (Brown and Vincent, 1992; Marrow et al., 1992; Abrams and Harada, 1996). Consider the predator–prey system introduced in Subsection 2.4.4. Assume that both the prey and the predators have scalar strategies, so we can use the simplified* strategy notation. Assume that the intrinsic growth rate for all prey is the same constant $r_1$ and the intrinsic growth rate of all predators is the same constant $r_2$. Assume that the carrying capacity of the prey $K(v)$ is a function of the individual prey's strategy only. The interaction term $a(v, u_j)$ is a function of the individual prey's strategy as well as the strategies of the other prey. The predation term $b(v, u_j)$ in $G_1$ is a function of the individual prey's strategy as well as the strategies of the predators. The predation term $b(v, u_j)$ in $G_2$ is a function of the individual predator's strategy as well as the strategies of the prey. Under these assumptions we have

$$\text{Prey}: G_1(v, \mathbf{u}, \mathbf{x}) = \frac{r_1}{K(v)} \left[ K(v) - \sum_{j=1}^{n_{s_1}} x_j a(v, u_j) \right] - \sum_{j=n_{s_1}+1}^{n_s} x_j b(v, u_j)$$

$$\text{Predator}: G_2(v, \mathbf{u}, \mathbf{x}) = r_2 \left[ 1 - \frac{\sum_{j=n_{s_1}+1}^{n_s} x_j}{c \sum_{j=1}^{n_{s_1}} x_j b(v, u_j)} \right]$$

where $n_{s_1}$ is the number of prey species and $n_s$ is the number of prey plus predator species. The following functional forms are assumed for the carrying capacity, competition coefficients, and capture probabilities

$$K(v) = K_{\max} \exp\left(-\frac{v^2}{\sigma_k^2}\right)$$

$$a(v, u_j) = \exp\left[-\frac{(v - u_j)^2}{\sigma_a^2}\right]$$

$$b(v, u_j) = b_{\max} \exp\left[-\frac{(v - u_j)^2}{\sigma_b^2}\right].$$

The fact that $G_1$ and $G_2$ are indeed $G$-functions for the given model may be easily tested using the $G$-function definition.

## 4.7 $G$-functions in terms of population frequency

This category includes any $G$-function that models systems having **one unique bauplan** with **vector strategies** when the $G$-function is of the form $G(\mathbf{v}, \mathbf{u}, \mathbf{p}, N)$ where $\mathbf{p}$ is a population frequency vector and $N$ is the total population size. This category is exactly the same as $G$-functions with vector strategies except $\mathbf{x}$ is replaced by $\mathbf{p}$ and $N$.

This formulation is more consistent with that used in the early days of evolutionary game theory when matrix games were the focus (Maynard Smith, 1974). Such games were generally formulated in terms of strategy frequency rather than population density. While matrix games generally do not have explicit population dynamics and considered only the frequency of resident strategies within the population, we do not take that approach here. Rather, we simply replace population dynamics with frequency dynamics. The results obtained are applicable to both matrix games (Chapter 9) and continuous games formulated in terms of frequency.

The total number of individuals in a population $N$ is given by

$$N = \sum_{i=1}^{n_s} x_i.$$

A measure of how well any given species is doing at time $t$ is given by their corresponding population size $x_i$. However, if we wish to measure how well one species is doing relative to others, then the **frequency** of those individuals

$$p_i = \frac{x_i}{N}$$

provides a metric. It is obvious from these definitions that

$$\sum_{i=1}^{n_s} p_i = 1.$$

Using the above definitions, we re-write the population dynamics for the vector case

Population dynamics in terms of fitness functions

| | |
|---|---|
| Difference: | $x_i(t+1) = x_i\left[1 + H_i(\mathbf{u}, \mathbf{x})\right]$ |
| Exp. Difference: | $x_i(t+1) = x_i \exp H_i(\mathbf{u}, \mathbf{x})$ |
| Differential: | $\dot{x}_i = x_i H_i(\mathbf{u}, \mathbf{x})$ |

(4.9)

in terms of frequency (Vincent and Fisher, 1988)

Frequency dynamics in terms of fitness functions

| | |
|---|---|
| Difference: | $p_i(t+1) = p_i \frac{1 + H_i(\mathbf{u},\mathbf{p},N)}{1 + \bar{H}}$ |
| Exp. Difference: | $p_i(t+1) = p_i \frac{\exp H_i(\mathbf{u},\mathbf{p},N)}{\sum_{i=1}^{n_s} p_i \exp H_i(\mathbf{u},\mathbf{p},N)}$ |
| Differential: | $\dot{p}_i = p_i\left[H_i(\mathbf{u}, \mathbf{p}, N) - \bar{H}\right]$ |

where $\bar{H}$ is the average fitness of the population as a whole

$$\bar{H} = \sum_{i=1}^{n_s} p_i H_i(\mathbf{u}, \mathbf{p}, N). \quad (4.10)$$

The equations for the total population size are

## 4.7 G-functions in terms of population frequency

**Total population size dynamics**

Difference: $N(t+1) = N\left(1+\bar{H}\right)$
Exp. Difference: $N(t+1) = N \sum_{i=1}^{n_s} p_i \exp H_i(\mathbf{u}, \mathbf{p}, N)$
Differential: $\dot{N} = N\bar{H}$

The above equations are equivalent to (4.9) and will yield exactly the same results. The only difference is viewpoint. In this form, frequency dependence (in terms of $\mathbf{p}$) and density dependence (in terms of $N$) are made explicit. Often (especially in matrix games) the total population size is ignored and the $N$ dependence is dropped from $H$.

Thus we have a vector of population frequencies

$$\mathbf{p} = \begin{bmatrix} p_1 & \cdots & p_{n_s} \end{bmatrix}$$

with the strategy vector satisfying exactly the same conditions as the vector case above. However, since the exponential difference equations cannot be expressed in terms of $\bar{H}$ they must always be treated as a special case. In order to avoid undue complexity, we drop this system from this category of $G$-functions.

The $G$-function in terms of population frequency is defined by

$$G(\mathbf{v}, \mathbf{u}, \mathbf{p}, N)|_{\mathbf{v}=\mathbf{u}_i} = H_i(\mathbf{u}, \mathbf{p}, N) \quad i = 1, \cdots, n_s.$$

The frequency dynamics, in terms of the $G$-function, is given by

**Frequency dynamics in terms of a $G$-function**

Difference: $p_i(t+1) = p_i \dfrac{1+G(\mathbf{v},\mathbf{u},\mathbf{p},N)|_{\mathbf{v}=\mathbf{u}_i}}{1+\bar{G}}$
Differential: $\dot{p}_i = p_i\left[G(\mathbf{v}, \mathbf{u}, \mathbf{p}, N)|_{\mathbf{v}=\mathbf{u}_i} - \bar{G}\right]$

(4.11)

where $i = 1, \ldots, n_s$ and

$$\bar{G} = \sum_{i=1}^{n_s} p_i\, G(\mathbf{v}, \mathbf{u}, \mathbf{p}, N)|_{\mathbf{v}=\mathbf{u}_i}.$$

The total population dynamics is given by the following

**Total population size dynamics**

Difference: $N(t+1) = N\left(1+\bar{G}\right)$
Differential: $\dot{N} = N\bar{G}$

(4.12)

**Example 4.7.1 (L–V competition game in terms of frequency)** *This game (Example 4.3.1) reformulated in terms of frequency is given by*

$$G(v, \mathbf{u}, \mathbf{p}, N) = \frac{r}{K(v)}\left(K(v) - N \sum_{j=1}^{n_s} a(v, u_j) p_j\right),$$

with

$$K(v) = K_m \exp\left\{-\frac{v^2}{2\sigma_k^2}\right\}$$

$$\alpha(v, u_i) = 1 + \exp\left\{-\frac{(v - u_i + \beta)^2}{2\sigma_\alpha^2}\right\} - \exp\left\{-\frac{\beta^2}{2\sigma_\alpha^2}\right\}.$$

While the reformulation of this game is trivial, the sets of equations required to solve them are quite different. Compare (4.5) with (4.10), (4.11), and (4.12). Unless one wants a frequency–density viewpoint there is no advantage to using this formulation. However, this is the viewpoint most often used in matrix game theory (Chapter 9).

## 4.8 Multistage $G$-functions

This category includes any $G$-function that models systems having **one unique multistage bauplan** with **scalar strategies**. The multistage case requires a matrix **G**-function of the form $\mathbf{G}(v, \mathbf{u}, \mathbf{x})$. The ecological community is composed of $n_s$ species and all species have a similar life history that has $n_h$ **life history stages**. These stages may represent age classes (as in a standard life history table; Deevey, 1947), developmental stages (as in models with ontogenetic niche shifts; Werner and Gilliam, 1984), states within a structured population (as in individuals occupying different habitats or places within a metapopulation; Hanski, 1991), or different habitats or places that are coupled by migration. Species are identified by their respective population sizes, $\mathbf{x}_i$, and strategies, $u_i$. A species's population size includes the number of individuals of that species within each life history stage. The species's scalar strategy is drawn from some relevant set of heritable traits. In the following, we show how these elements combine to model each species's population dynamics as a **population projection matrix**. The matrix entries represent the transition processes among life-history stages. Some entries may be probabilities and represent the likelihood of surviving from one age class to the next. Other entries may be values that reflect the number of offspring produced by adult classes. Frequency and density dependence can potentially enter through every element of the matrix.

We need to restate the definition for **x**. It follows that, with only one multistage bauplan, all species will have the same number of life-history stages. Thus the density of any species is a vector made up of $n_h$ life-history stages.

$$\mathbf{x}_i = \begin{bmatrix} x_{i1} & \cdots & x_{in_h} \end{bmatrix}.$$

For example, $x_{23}$ is the population density of the third stage of species 2. Let

## 4.8 Multistage G-functions

the vector

$$\mathbf{x} = \begin{bmatrix} \mathbf{x}_1 \mid \cdots \mid \mathbf{x}_{n_s} \end{bmatrix}$$

be densities associated with $n_s$ species in a community. It follows that $\mathbf{x}$ is a vector whose length is determined by the product $n_s \times n_h$.

No special notation is needed for the strategy vector as it satisfies exactly the same conditions as specified for $G$-functions with scalar strategies. However, defining the **G-matrix** is more complicated. We are aided in this regard by again using the rising number concept. In this case, the rising number count the number of life stages as we add species starting from zero. For example, if there are three species, $n_s = 3$, each with two life-history stages $n_h = 2$, then

$$r_1 = 0$$
$$r_2 = n_h = 2$$
$$r_3 = 2n_h = 4$$

more generally

$$r_i = (i-1)n_h.$$

We use this notation to model the dynamics of the $i$th species population densities, $\mathbf{x}_i$, in terms of fitness functions of the form $H_{ij}(\mathbf{u}, \mathbf{x})$ as given in the following table (exponential difference equations are not included since they are not derivable from differential form equations, see Subsection 2.3.3). For notational clarity the function arguments $(\mathbf{u}, \mathbf{x})$ are not always included

Multistage dynamics in terms of fitness functions

| | |
|---|---|
| Difference: | $x_{i1}(t+1) = x_{i1}\left[1 + H_{(r_i+1)1}\right] + x_{i2}H_{(r_i+1)2} + \cdots$ <br> $\quad + x_{in_s}H_{(r_i+1)n_h}$ <br> $x_{i2}(t+1) = x_{i1}H_{(r_i+2)1} + x_{i2}\left[1 + H_{(r_i+2)2}\right] + \cdots$ <br> $\quad + x_{in_s}H_{(r_i+2)n_h}$ <br> $\vdots = \vdots$ <br> $x_{in_h}(t+1) = x_{i1}H_{(r_i+n_h)1} + x_{i2}H_{(r_i+n_h)2} + \cdots$ <br> $\quad + x_{in_s}\left[1 + H_{(r_i+n_h)n_h}\right]$ |
| Differential: | $\dot{x}_{i1} = x_{i1}H_{(r_i+1)1} + x_{i2}H_{(r_i+1)2} + \cdots + x_{in_s}H_{(r_i+1)n_h}$ <br> $\dot{x}_{i2} = x_{i1}H_{(r_i+2)1} + x_{i2}H_{(r_i+2)2} + \cdots + x_{in_s}H_{(r_i+2)n_h}$ <br> $\vdots = \vdots$ <br> $\dot{x}_{in_h} = x_{i1}H_{(r_i+n_h)1} + x_{i2}H_{(r_i+n_h)2} + \cdots + x_{in_s}H_{(r_i+n_h)n_h}$ |

(4.13)

where $i = 1, \cdots, n_s$. For example, suppose that there are two species, $n_s = 2$, each of which has three life history stages, $n_h = 3$. The equations for the differential case would be written as

$$\dot{x}_{11} = x_{11}H_{11} + x_{12}H_{12} + x_{13}H_{13}$$
$$\dot{x}_{12} = x_{11}H_{21} + x_{12}H_{22} + x_{13}H_{23}$$
$$\dot{x}_{13} = x_{11}H_{31} + x_{12}H_{32} + x_{13}H_{33}$$
$$\dot{x}_{21} = x_{21}H_{41} + x_{22}H_{42} + x_{23}H_{43}$$
$$\dot{x}_{22} = x_{21}H_{51} + x_{22}H_{52} + x_{23}H_{53}$$
$$\dot{x}_{23} = x_{21}H_{61} + x_{22}H_{62} + x_{23}H_{63} \; .$$

In matrix form, (4.13) is written as

$$\boxed{\begin{array}{ll} \text{Difference:} & \mathbf{x}_i(t+1) = \mathbf{x}_i \left[ \mathcal{I} + \mathbf{H}_i^{\mathrm{T}}(\mathbf{u}, \mathbf{x}) \right] \\ \text{Differential:} & \dot{\mathbf{x}}_i = \mathbf{x}_i \mathbf{H}_i^{\mathrm{T}}(\mathbf{u}, \mathbf{x}) \end{array}}$$

where $\mathcal{I}$ is the $n_h \times n_h$ identity matrix and

$$\mathbf{H}_i(\mathbf{u}, \mathbf{x}) = \begin{bmatrix} H_{(r_i+1)1}(\mathbf{u}, \mathbf{x}) & H_{(r_i+1)2}(\mathbf{u}, \mathbf{x}) & \cdots & H_{(r_i+1)n_h}(\mathbf{u}, \mathbf{x}) \\ H_{(r_i+2)1}(\mathbf{u}, \mathbf{x}) & H_{(r_i+2)2}(\mathbf{u}, \mathbf{x}) & \cdots & H_{(r_i+2)n_h}(\mathbf{u}, \mathbf{x}) \\ \vdots & \vdots & \ddots & \vdots \\ H_{(r_i+n_h)1}(\mathbf{u}, \mathbf{x}) & H_{(r_i+n_h)2}(\mathbf{u}, \mathbf{x}) & \cdots & H_{(r_i+n_h)n_h}(\mathbf{u}, \mathbf{x}) \end{bmatrix}$$

is the **fitness matrix** for individuals of type $i$ and $\mathbf{H}_i^{\mathrm{T}}$ is the **transpose** of $\mathbf{H}_i$ obtained by interchanging the rows and columns of $\mathbf{H}_i$ (see Example 2.4.1). We assume that $\mathbf{H}_i[\mathbf{u}, \mathbf{x}]$ has continuous partial derivatives with respect to $\mathbf{x}$, and $\mathbf{u}$.

For a given $\mathbf{u}$ and $\mathbf{x}$ it is possible to define a (scalar) fitness function for the multistage case by using a special function that is evaluated using one of the eigenvalues of $\mathbf{H}_i$. This function is related to the concept of a dominant eigenvalue.

**Definition 4.8.1 (dominant eigenvalue)** *The eigenvalue $\lambda_i$ of a matrix $\mathbf{A}$ is called the dominant eigenvalue of $\mathbf{A}$ if $|\lambda_i| > \lambda_j$ for all eigenvalues $\lambda_j$ where $i$ is not equal to $j$.*

Unfortunately the dominant eigenvalue does not provide a usable general definition of fitness. One problem is that a dominant eigenvalue need not even exist (e.g., consider the matrix $\mathbf{H}_i(\mathbf{u}, \mathbf{x})$ that has only the eigenvalues $-2$ and $+2$). Even when the eigenvalues are distinct, a different definition is needed when the population dynamics is given by differential equations. A more useful general definition is provided by the concept of a **critical value**.

## 4.8 Multistage G-functions

**Definition 4.8.2 (critical value)** *Given $\mathbf{u}$ and $\mathbf{x}$, let $\boldsymbol{\lambda}_i = \begin{bmatrix} \lambda_{i1} \cdots \lambda_{in_s} \end{bmatrix}$ be the vector of eigenvalues corresponding to $\mathbf{H}_i(\mathbf{u}, \mathbf{x})$. Let $\mathrm{abs}(\boldsymbol{\lambda}_i)$ be the vector of absolute values and $\mathrm{Re}\,(\boldsymbol{\lambda}_i)$ be the vector of real parts. Let $\max\,(\mathrm{abs}(\boldsymbol{\lambda}_i))$ and $\max\,(\mathrm{Re}\,(\boldsymbol{\lambda}_i))$ be the values of the maximum components of the vector (the maximum value is unique; however, there may be more than one solution with this value). Those components of the vector of eigenvalues that have this maximum value are called critical values. The following notation is used*

> Difference:   $\mathrm{crit}\,\mathbf{H}_i(\mathbf{u}, \mathbf{x}) = \max\,(\mathrm{abs}(\boldsymbol{\lambda}_i))$
> Differential: $\mathrm{crit}\,\mathbf{H}_i(\mathbf{u}, \mathbf{x}) = \max\,(\mathrm{Re}\,(\boldsymbol{\lambda}_i))$.

In general the critical value need not be an eigenvalue. Only when all eigenvalues are real, distinct, and non-negative (or non-positive) will the critical value be the dominant eigenvalue for the difference equation case. The critical value definition leads us to the following definition of fitness.

**Definition 4.8.3 (fitness for multistage G-functions)** *Given $\mathbf{u}$ and $\mathbf{x}$, and a fitness matrix $\mathbf{H}_i(\mathbf{u}, \mathbf{x})$, the fitness of individuals, associated with a multistage G-function, using the strategy $u_i$ is defined by*

$$H_i(\mathbf{u}, \mathbf{x}) = \mathrm{crit}\,\mathbf{H}_i(\mathbf{u}, \mathbf{x}).$$

We thus have both a **G-matrix**

$$\mathbf{G}(v, \mathbf{u}, \mathbf{x})|_{v=u_i} = \mathbf{H}_i(\mathbf{u}, \mathbf{x})$$

and a *G*-function

$$G(v, \mathbf{u}, \mathbf{x})|_{v=u_i} = H_i(\mathbf{u}, \mathbf{x})$$

for the multistage systems. Working in terms of the scalar *G*-function has advantages in several situations over using the **G**-matrix directly. In some cases $G(v, \mathbf{u}, \mathbf{x})$ can be used to solve analytically for a potential ESS and it can be used to plot the **adaptive landscape** introduced in Chapter 5. However, when solving for the population dynamics, the **G**-matrix must be used. The system dynamics, in terms of the **G**-matrix, is given by the following

> Population dynamics in terms of the **G**-matrix
>
> Difference:   $\mathbf{x}_i(t+1) = \mathbf{x}_i \left[ \mathcal{I} + \mathbf{G}^\mathrm{T}(v, \mathbf{u}, \mathbf{p}) \big|_{v=u_i} \right]$
> Differential: $\dot{\mathbf{x}}_i = \mathbf{x}_i\,\mathbf{G}^\mathrm{T}(v, \mathbf{u}, \mathbf{p})\big|_{v=u_i}$

**Example 4.8.1 (life cycle game)** *The life cycle game in the following example may not conform directly to an actual organism, but it does illuminate* *some features of multistage G-functions. In this life cycle, the creature has two stages: a competitive reproductive stage, and a non-competitive non-reproductive stage. An organism in the reproductive stage produces offspring (immediately recruited into this reproductive stage), suffers mortality that increases with the density of reproductives, and has some probability of becoming non-reproductive. Individuals in the non-reproductive stage suffer mortality that increases with the density of reproductives, and have some probability of becoming reproductive (births, deaths, and transition probabilities are independent of the density of non-reproductives). Let a scalar-valued strategy of an individual affect two aspects of the life cycle. First, an individual's likelihood of shifting from reproductive to non-reproductive increases linearly with its strategy. Second, this same strategy of the individual influences its reproduction rate when it is in the reproductive stage. We let the strategies of others weight the mortality effect that reproductives have on those in the reproductive stage. We incorporate these assumptions into the following model of population growth. The model describes the rate of change in population density of stage 1 (reproductives) and stage 2 (non-reproductives) of the ith species (as influenced by* **u** *and* **x**)

$$\dot{x}_{i1} = \left( f(u_i) - \sum_{j=1}^{n_x} u_j x_{j1} \right) x_{i1} + u_i x_{i2}$$

$$\dot{x}_{i2} = u_i x_{i1} - \left( \sum_{j=1}^{n_x} x_{j1} \right) x_{i2}.$$

*From the definition of the* **G***-matrix it follows that*

$$\mathbf{G}(v, \mathbf{u}, \mathbf{x}) = \begin{bmatrix} f(v) - \sum_{j=1}^{n_x} u_j x_{j1} & v \\ v & -\left( \sum_{j=1}^{n_x} x_{j1} \right) \end{bmatrix}.$$

*Note how the* **G***-matrix highlights the effects on fitness of the focal individual's strategy, $v$, the strategies of others,* **u**, *and the effects of population densities,* **x**.

Since the **G**-matrix in the above example is symmetric, the eigenvalues are real. This will often be the case, but not always.

## 4.9 Non-equilibrium dynamics

Generally, for fixed strategies, there exist asymptotically stable equilibrium solutions for the population dynamics. The population sizes of individuals using

## 4.9 Non-equilibrium dynamics

particular strategies tend either towards zero or towards stable and positive values. However, this need not be the case and, for any of the categories introduced above, the possibility exists for non-equilibrium population dynamics including limit cycles (continuous), $n$-cycles (discrete), and chaos. We will examine non-equilibrium dynamics for $G$-functions with scalar strategies in the following chapters.

# 5
# Darwinian dynamics

**Darwinian dynamics** couples **population dynamics**[1] with **strategy dynamics**[2] to model the evolutionary process. So far, we have focused on ecological models by using the $G$-function to express the population dynamics. In this chapter we obtain strategy dynamics using the same $G$-function by assuming heritable variation as a distribution of strategies around the mean strategy used by each species in the population.

Any theory of evolution that includes natural selection is incomplete, unless it includes both population dynamics and strategy dynamics. The resulting Darwinian dynamics captures the full rich behavior of the evolutionary processes. Its use clarifies two important features of evolutionary stability: resistance to invasion and dynamic attainability (Eshel and Motro, 1981; Eshel, 1983; Taylor, 1989; Christiansen, 1991; Takada and Kigami, 1991; Taylor, 1997).

The concept of a strategy dynamic requires us to deal with issues of mutation, heritable variation, and whether species represent asexual lineages or populations of sexually interbreeding individuals. The strategy $\mathbf{u}_i$ of a species is no longer considered fixed. Rather, it describes a population's mean strategy value that contains some variability in value among the individuals of the population. The introduction of strategy dynamics with the population dynamics leads to a new time scale. In addition to the ecological time scale there is now an evolutionary time scale. Because population dynamics and strategy dynamics may occur on different time scales, it is useful to make this distinction. We will show that population dynamics generally, but not always, occur on a faster time scale than strategy dynamics. We will also show that strategy dynamics

---

[1] Population dynamics are those relationships between population density and the factors that affect density changes with time. Any of the population models of Chapter 2 represent population dynamics; see in particular Subsection 2.3.2.

[2] Strategy dynamics are relationships between strategy values and the factors affecting changes in these values with time.

## 5.1 Strategy dynamics and the adaptive landscape

can be visualized as occurring on an **adaptive landscape**. For a given **u** (existing strategies among species) and **x** (population sizes of each species), an adaptive landscape plots the per capita growth rate, $G(\mathbf{v}, \mathbf{u}, \mathbf{x})$, as a function of a focal individual's strategy, **v**. It is similar to Wright's **fitness landscape** (Wright, 1931, 1969). However, unlike Wright's rigid fitness landscape, the adaptive landscape is pliable. It readily changes shape with changes in population densities and strategies within and among the various species. This chapter concludes with additional material on the various forms of dynamical stability associated with Darwinian dynamics for $G$-functions with scalar strategies.

### 5.1 Strategy dynamics and the adaptive landscape

The $G$-functions developed in the previous chapter can determine the fates of any number of specific strategies played together. By alternatively setting **v** equal to each of the strategies, the fate of each strategy can be followed through time by means of the strategies' population dynamics. Some strategies will persist at positive population sizes while others will not due to a population size decline towards extinction ($x_i = 0$). Often, when several strategies are played against each other, only a small number (possibly just one) survives. This is true whether the strategies are very close to each other in value (relative to the entire strategy set) or whether the strategies are far apart in strategy space.

As a tutorial, consider a simple $G$-function derived from logistic population growth. In this model we let carrying capacity be a function of a scalar strategy, and we assume that fitness is influenced by the individual's strategy and the combined population sizes of all existing strategies

$$G(v, \mathbf{x}) = \frac{r}{K(v)} \left( K(v) - \sum_{j=1}^{n_s} x_j \right)$$

where

$$K(v) = K_m \exp\left(-\frac{v^2}{2\sigma_k^2}\right)$$

and

$$r = 0.1, K_m = 100, \sigma_k^2 = 2.$$

Consider four species with the following strategy values $\mathbf{u} = \begin{bmatrix} 0.2 & 0.5 & 1 & 2 \end{bmatrix}$ which result in four models of population dynamics with the following associated carrying capacities $K = \begin{bmatrix} 99 & 93.9 & 77.9 & 36.8 \end{bmatrix}$. Now let each population start at a density of $\mathbf{x}(0) = \begin{bmatrix} 1 & 1 & 1 & 1 \end{bmatrix}$, where $\mathbf{x}(0)$

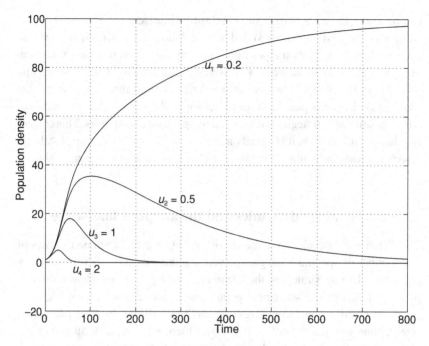

**Figure 5.1** The species with highest carrying capacity survives.

values are the strategies' population sizes at time 0. As illustrated in Figure 5.1, the resulting population dynamics, obtained using the differential equation dynamics,[3] starts with a rapid rise in the population sizes of all four strategies. But soon we see a fairly rapid decline in the third and fourth species, a slow decline in the second species, and the eventual sole survivorship of the first species. We have the beginnings of a crude but effective **strategy dynamic** for producing evolutionary change. While the strategy of each population remained fixed, the overall frequency of each strategy within the population changed with time. Eventually the species with the highest carrying capacity replaced all others. In this case, selection did not occur within a species but rather among species.

We refine the selection process by starting with species $x_1$ as the sole survivor ($u_1 = 0.2$) and then add two new species with strategies close to $x_1$'s strategy: $\mathbf{u} = \begin{bmatrix} 0.2 & 0.15 & 0.3 \end{bmatrix}$. We then solve the population dynamics using an initial population size of 1 for each species. The population dynamics is more rapid,

---

[3] For this and many examples we will use the continuous time description of population dynamics (differential equation form) rather than one of the discrete, difference equation forms. All of these three forms provide similar results as long as we avoid parameter values that produce population dynamics with limit cycles or chaos.

## 5.1 Strategy dynamics and the adaptive landscape

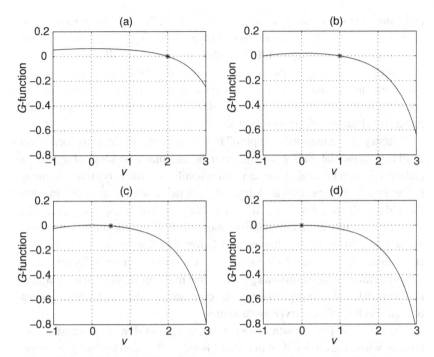

**Figure 5.2** The "star" locates the strategy at the equilibrium value for $x^*$.

but the selection process of replacing one strategy by another occurs much more slowly as the second species (with the highest carrying capacity of the three) replaces the other two. We can continue this process of selection by considering species $u_2 = 0.15$ as the sole survivor and adding two new strategies very close to and on either side of 0.15. With each iteration of this selection process, there will be a sole surviving species, the surviving species will have the strategy that has the highest carrying capacity, and this strategy will be the one closest to $v = 0$. This strategy maximizes fitness by virtue of maximizing $K$.

We can visualize how the selection process of replacing old strategies by new ones takes place by examining the adaptive landscape which plots $G(v, \mathbf{x}^*)$ versus $v$ where $\mathbf{x}^*$ is the equilibrium population vector. For this example, the adaptive landscape is hump shaped and always reaches a maximum at $v = 0$ as illustrated in Figure 5.2. This particular landscape changes shape only in response to changes in the combined population sizes of all species present at equilibrium: $\sum_{j=1}^{n_s} x_j^*$. In response to increases and decreases in population size, the landscape sinks and rises, respectively, as illustrated in the frames of Figure 5.2. For this example, there is only one species present at equilibrium and frame (a) corresponds to $u_1 = 2$, $x_1^* = 36.8$, frame (b) corresponds to $u_1 = 1$, $x_1^* = 77.9$, frame (c) corresponds to $u_1 = 0.5$, $x_1^* = 93.9$, and frame (d) corresponds to $u_1 = 0$, $x_1^* = 100$. Note that, at equilibrium, all extant strategies have zero per capita growth rates (fitness of zero). However, only at $u_1 = 0$ does the adaptive landscape take on a maximum at the strategy value producing zero fitness. Inspection of the landscape shows whether nearby strategies could invade by virtue of having higher fitness.

If strategies are introduced "uphill" from a surviving strategy not located on a hill top, an escalator effect occurs from the interplay of ecological dynamics and the selection process. A strategy that is uphill can invade by virtue of having higher fitness. But, as soon as it does so, population sizes change in a manner that reduces the new surviving strategy's fitness to zero. In the above example, a strategy that moves up the adaptive landscape will become the new survivor, but it will also result in a higher population size that ultimately causes the entire landscape to sink. As we shall see, this effect is general. Evolutionary dynamics favors changes in strategy values that move up the slope of the adaptive landscape while the ecological dynamics constantly pushes the landscape down beneath the feet of the surviving strategies.

The above logistic growth model is an example of **density-dependent selection**, where the fitness of an individual is only influenced by total population size, and the individual's own strategy, and not directly by the strategies of others. The strategies of others were only relevant to fitness insofar as their presence or absence influences total population sizes. In such a situation, evolution by natural selection always favors the strategy that maximizes equilibrium population size (Wright, 1931).

The continual introduction of novel strategies illustrates how selection results in an evolutionary process. This evolutionary process can proceed quickly when strategies are very different from each other and have very different fitness values. When different strategies are close to each other in value the evolutionary process can, in general, take a long time to sort the losing strategies from the surviving strategies. It becomes important in an evolutionary game to consider the source of heritable variation and the source of new strategies either within or among species.

## 5.2 The source of new strategies: heritable variation and mutation

The example of the previous section illustrates an evolutionary process based on the introduction of novel strategies. Missing from this procedure are

## 5.2 The source of new strategies: heritable variation and mutation

specific mechanisms for producing novel strategies. We have not included the mechanism of inheritance, the underlying genetics, the sources of novel strategies for invasion, or the occurrence of mutations within the population. These topics have been well covered in literature on population and quantitative genetics (Falconer, 1960; Wright, 1960; Crow and Kimura, 1970; Speiss, 1977; Wright, 1977). Also, there exists an excellent literature relating underlying genetics to the strategies of evolutionary games (Dieckmann et al., 1995; Hammerstein, 1996; Hofbauer and Sigmund, 1998; Kisdi and Geritz, 1999).

Here we examine two important issues that strongly influence strategy dynamics. The first issue concerns how mutations or novel strategies enter the population. Novel strategies can be introduced as heritable variation around the strategies of existing species, or as separate species starting out at small population sizes. The second issue concerns the consequences of having either asexual or sexual reproduction. With **asexual** reproduction it does not matter whether novel strategies are thought of as separate species or as heritable variation within a species. For this reason, strict game theoretic models have sometimes been seen as applicable only to species with asexual or haploid inheritance. With **sexual** reproduction it matters whether heritable variation exists within a species where mixing, genetic exchange, and recombination can occur, or whether heritable variation takes the form of non-interbreeding species. These issues will influence the resulting strategy dynamics, and the speed and extent to which evolution by natural selection can explore the strategy set and move along the adaptive landscape.

We use the term **invasion-driven heritable variation** to describe the situation where novel strategies are introduced as separate species that do not interbreed with the extant species (Roughgarden, 1987; Brown and Vincent, 1987a). Such a situation applies generally to any haploid or asexual species, regardless of whether the new strategy arises as a mutation within the extant population or whether it immigrates from outside the population. It applies only to sexual species within the context of a different species invading a community from elsewhere, or in the unusual event that a mutation within the population confers both a change of strategy and reproductive isolation (e.g., polyploidy in some plant species). When viewing the evolutionary process as primarily invasion-driven there is no need to specify heritable variation within a species. All of the relevant variation is presumed to exist among the species found in the vector **u**. Each time a new strategy invades, the dimension of **u** increases by one species, and each time a species becomes extinct it declines by one. This is roughly the procedure that we followed in the tutorial example of evolution under logistic population growth.

Under invasion-driven heritable variation, the modeler has considerable flexibility regarding the introduction of new strategies. Invading strategies will generally be introduced at small population numbers. However, this need not be the case. With regard to the timing of invasions, one or more novel strategies can be introduced frequently or infrequently at regular or irregular intervals. Or each new strategy can be introduced only after the previous invader has either been established or become extinct. With regard to the strategy value of the invader, novel strategies may, at one extreme, be drawn randomly from the entire strategy set, or at the other extreme drawn according to a narrower distribution from around the strategies of the extant species. Drawing invading strategies from across the strategy set allows for "hopeful monsters" (Goldschmidt, 1940) that greatly enhance the opportunity for natural selection to explore the strategy set. This may or may not be biologically reasonable given the recipe of inheritance and the source of invading strategies. Alternatively, a micro-mutation approach of selecting invaders' strategies within some small neighborhood of existing strategy values may conform more faithfully to constraints imposed by the genetics, or by the existing variability among closely related allopatric species. A micro-mutation approach restricts the ways natural selection explores the strategy set. Regardless of the rules placed upon the invaders' strategies, invasion-structured heritable variation occurs among species and not within species.

We use the term **strategy-driven heritable variation** to describe the situation where novel strategies are introduced as variability within extant species that may or may not interbreed. In this case, the population can be described as having a mean strategy value with variability in strategy value around this mean among different individuals within the population. Mutation, recombination, and other genetic changes are the presumed processes that continually introduce deviations from the mean in some individual's strategy value. It is often convenient to approximate the population's distribution of strategy values as a normal distribution. In this case, a population's strategy can be fully described by its mean and variance. Natural selection will reduce the variance in strategy values around the mean, and mutation will serve to increase this variance. Consequently, both the mean and the variance of a population's strategy value will change with time. If the population is sexual, the population's mean strategy will change with natural selection, with interbreeding, and with the underlying genetics. The interbreeding becomes part of the "environment" to which the strategy adapts, and the genetics must be accounted for as constraints on the strategy set. Mutation, interbreeding, and genetics can influence the variability among individuals in strategy values. Sexually reproducing species, by virtue of interbreeding, can produce more complex changes in both the mean and the variance of a population's strategy.

## 5.3 Ecological time and evolutionary time

Natural selection is both an ecological and an evolutionary process. Both strategy dynamics and population dynamics contribute to the sorting out of surviving strategies that constitutes evolution by natural selection. Consequently, game theory models of natural selection have two time scales. There is an **ecological time scale**, $T_{ec}$, associated with the time it takes for the density of individuals to return to an ecological equilibrium when perturbed from this equilibrium and an **evolutionary time scale**, $T_{ev}$, associated with the time it takes for the strategies to return to a strategy equilibrium when perturbed from this equilibrium. It may often be that the evolutionary time scale is much slower than the ecological time scale, but this need not be the case. Grant and Grant (2002) observed rapid natural evolution of beak size in Darwin's finches. Strong selective pressure is exerted by year to year variability in population sizes that can routinely reach 20–30%. During a particularly interesting ecological crunch period for the finches, beak morphology and body size changed on the order of 2–3%.

If we move the system away from an equilibrium state a measure of the ecological time scale $T_{ec}$ is determined by the amount that fitness has changed,[4] whereas the evolutionary time scale $T_{ev}$ is determined by the slope of the adaptive landscape as well as the distribution of available strategies.[5] The relative fitness among individuals with different strategies determines the evolutionary time scale. What would our (model) world be like if $T_{ec}$ = finite number and $T_{ev} = \infty$? This implies that there is no evolution of strategies (by any means!). All species present in the population would be determined from the initial conditions used to set up the ecological model, with no opportunity for changes in strategy values. However, even in this case, the population mimics an evolutionary process in the sense that the frequency at which strategies occur among species in the community changes with changing population sizes, demonstrating that ecological and evolutionary dynamics can never be entirely distinct. After an equilibrium state has been obtained, all of the species that will ever be can be identified and counted. If the environment were stable, that would be the end of it. If the environment were not stable (e.g., allowed mixing of one ecosystem with another), we would discover species becoming extinct. There are two possible effects due to environmental changes: loss of species and changes in equilibrium numbers of the remaining species. Clearly this is part of, but not all of, the real world, and only a small portion of the evolutionary process.

---

[4] $T_{ec}$ is related to the "slowest" eigenvalue associated with the population dyanamics equations linearized about the equilibrium solution.

[5] $T_{ev}$ is related to the "slowest" eigenvalue associated with the strategy dynamics equations linearized about the equilibrium solution.

Species come and go as a part of the evolutionary process. No doubt the most common situation is when $T_{ec} < T_{ev}$. In this case, over short periods of time we expect to see equilibrium population levels fluctuate due to environmental effects. Note that our assumption of the existence of an equilibrium population does not imply that the equilibrium level could not change on an ecological time scale due to changes in the environment. In the short term, we would likely see the extinction of species as in the $T_{ev} = \infty$ case. However, over the long term we would expect to see evolution taking place with species changing their strategy values and new species appearing. In this case, it is possible to study evolutionary effects by assuming that the ecological dynamics is always at or near equilibrium by setting $G(\mathbf{v}, \mathbf{u}, \mathbf{x}^*) = 0$ with the only time-dependent equation given by the strategy dynamics. Note that this assumption *does not eliminate the ecological dynamics*, it simply replaces a set of differential equations with algebraic equations describing equilibrium population sizes as a function of the species' strategy values.

It is also possible that $T_{ec} \approx T_{ev}$ (approximately equal). In this case, the population dynamic equations and the strategy dynamic equations are inseparably coupled and we might see non-equilibrium or even chaotic behavior for the Darwinian dynamics even when each system is by itself[6] asymptotically stable for an equilibrium solution. The case of $T_{ec} > T_{ev}$ seems unlikely due to the way population dynamics generates strategy dynamics; the sorting out of winning strategies is done on the basis of the ecological equations. However, this case should not be discounted for unnatural situations, involving genetic engineering for example.

The remainder of this chapter focuses on how heritable variation within a population drives strategy dynamics.

## 5.4 *G*-functions with scalar strategies

The study of natural selection using Darwinian dynamics needs an explicit mathematical formulation for the strategy dynamics. Strategy dynamics is derivable from the population dynamics equations provided that a distribution of strategies about some mean exists for each species. This distribution requires a distinction between phenotypes and species. A community of evolutionarily identical individuals is made up of both phenotypes and species.[7] In developing

---

[6] That is, **u** constant for the population dynamics equations and **x** constant in the strategy dynamics equations.

[7] When a distinction between the two is not important, we use the same notation for both. That is $u_i$ can refer to the strategy of a phenotype or species and $x_i$ can refer to the density of a phenotype or species.

## 5.4 G-functions with scalar strategies

strategy dynamics, the distinction between the two is important. We refer to a **species** as a set of evolutionarily identical individuals whose strategies, referred to as **phenotypes**, aggregate around a distinct mean strategy value. In Section 8.2 this species definition is developed further using a **strategy-species concept**.[8] We show below that, when the phenotypes of a particular species aggregate around some mean strategy value (interbreeding may facilitate this aggregation), the distribution of strategies results in an evolutionary dynamic. In this case, $\mathbf{u}_i$ is the mean strategy for the phenotypes of the species and $x_i$ is the species's population size (sum of the densities of all of the phenotypes of that species). That is, $\mathbf{u}_i$ and $x_i$ still refer to a species's strategy and density.

In order to simplify what follows we use the following short-cut notation

$$G|_w = G(v, \mathbf{u}, \mathbf{x})|_{v=w}$$

$$\left.\frac{\partial G}{\partial v}\right|_w = \left.\frac{\partial G(v, \mathbf{u}, \mathbf{x})}{\partial v}\right|_{v=w}$$

where $w$ is any scalar strategy. Using this notation the population dynamics for this category of $G$-functions becomes the following

Population dynamics

$$\boxed{\begin{array}{ll} \text{Difference:} & x_i(t+1) = x_i \left[1 + G|_{u_i}\right] \\ \text{Exp. Difference:} & x_i(t+1) = x_i \exp G|_{u_i} \\ \text{Differential:} & \dot{x}_i = x_i \, G|_{u_i}. \end{array}} \quad (5.1)$$

The individuals in the population $x_i$ will express the genetic variability inherent in that population as a whole. We previously introduced bounds on genetic variability in Chapter 4 with the definition of the strategy set $\mathcal{U}$. We characterize genetic variability by assuming that there are $n_p$ distinct phenotypes within each species. In the following development of strategy dynamics, interbreeding and/or mutation may produce and maintain the distribution of phenotypes within a species but only the differences in fitness among phenotypes as produced by the $G$-function influence the mean strategy of the species. The variable $u_i$ no longer refers to the strategy of any given individual in the population $x_i$, but rather we now define $u_i$ to be the **mean strategy** of all individuals in the population $x_i$. When the process of interbreeding or the actual genetic system strongly influences the mean of the strategy distribution, then these particulars of the breeding system and the genetic mechanisms producing the strategies must be explicitly incorporated into the $G$-function as an important component of a strategy's fitness.

---

[8] The strategy species concept has similarities to the **morphological species concept** and does not invoke the **biological species concept**. These concepts are discussed in more detail in Section 8.1.

### 5.4.1 Mean strategy dynamics

We need a double subscript notation to keep track of the different phenotypes in the population $x_i$. We use the notation $x_{ij}$ to designate a phenotype $j$ within the species $i$. It follows from the definition of the $G$-function that the population dynamics for the phenotypes are given by the following

Phenotype dynamics

$$
\begin{aligned}
\text{Difference:} \quad & x_{ij}(t+1) = x_{ij}\left[1 + G|_{u_{ij}}\right] \\
\text{Exp. Difference:} \quad & x_{ij}(t+1) = x_{ij} \exp G|_{u_{ij}} \\
\text{Differential:} \quad & \dot{x}_{ij} = x_{ij} \, G|_{u_{ij}}.
\end{aligned}
\qquad (5.2)
$$

The density of species $x_i$ is simply the sum of the densities of all the phenotypes of that species

$$x_i = \sum_{j=0}^{n_p} x_{ij}.$$

Likewise we use the notation $u_{ij}$ to designate the strategy used by phenotype $x_{ij}$. If we let $\delta u_{ij}$ be the difference between $u_{ij}$ and the mean value $u_i$, it follows that

$$u_{ij} = u_i + \delta u_{ij}. \qquad (5.3)$$

Note that $\delta$ is not a variable, but $\delta u_{ij}$ is. Equation (5.3) is thought of as representing the genetic variability within the population $i$. The **frequency of phenotypes** $x_{ij}$ in the population $x_i$ using the strategies $u_{ij}$ is simply the ratio of the population of $x_{ij}$ divided by the total population for that species and it is designated by

$$q_{ij} = \frac{x_{ij}}{x_i}.$$

By definition, the mean strategy $u_i$ is determined by

$$u_i = \sum_{j=0}^{n_p} q_{ij} u_{ij}. \qquad (5.4)$$

However, since

$$\sum_{j=0}^{n_p} q_{ij} u_{ij} = \sum_{j=0}^{n_p} q_{ij}\left(u_i + \delta u_{ij}\right) = u_i + \sum_{j=0}^{n_p} q_{ij} \delta u_{ij}$$

it follows that

$$\sum_{j=0}^{n_p} q_{ij} \delta u_{ij} = 0. \qquad (5.5)$$

## 5.4 G-functions with scalar strategies

It is possible to determine how the mean strategy $u_i$ evolves (i.e., a strategy dynamic for $u_i$) by simply considering the changes that must take place in the mean strategy as a result of the density dynamics. It is important to note that, while the mean strategy can and does change in time due to changes in frequency of the phenotypes using the strategies $u_{ij}$, the strategies $u_{ij}$ are fixed and do not change with time (however, the frequency of individuals using these strategies does change with time).

From (5.4) it follows that

$$u_i(t+1) = \sum_{j=0}^{n_p} q_{ij}(t+1) u_{ij}$$

for the difference equation models and

$$\dot{u}_i = \sum_{j=0}^{n_p} \dot{q}_{ij} u_{ij}$$

for the differential equation model. Using (5.2) for the determination of $q_{ij}(t+1)$ and $\dot{q}_{ij}$ we have for the first difference equation model

$$q_{ij}(t+1) = \frac{x_{ij}(t+1)}{x_i(t+1)} = \frac{x_{ij}\left(1 + G|_{u_{ij}}\right)}{x_i\left(1 + G|_{u_i}\right)}$$

$$= q_{ij} \frac{\left(1 + G|_{u_i}\right)}{\left(1 + G|_{u_i}\right)} + q_{ij} \frac{\left(G|_{u_{ij}} - G|_{u_i}\right)}{\left(1 + G|_{u_i}\right)}$$

$$= q_{ij}\left[1 + \frac{\left(G|_{u_{ij}} - G|_{u_i}\right)}{\left(1 + G|_{u_i}\right)}\right]$$

and for the second difference equation model

$$q_{ij}(t+1) = \frac{x_{ij}(t+1)}{x_i(t+1)} = \frac{x_{ij} \exp\left(G|_{u_{ij}}\right)}{x_i \exp\left(G|_{u_i}\right)}$$

$$= q_{ij} \exp\left(G|_{u_{ij}} - G|_{u_i}\right)$$

and for the differential equation model

$$\dot{q}_{ij} = \frac{d}{dt}\left(\frac{x_{ij}}{x_i}\right) = \frac{x_i \dot{x}_{ij} - x_{ij} \dot{x}_i}{x_i^2}$$

$$= \frac{x_i \, G|_{u_{ij}} x_{ij} - x_{ij} \, G|_{u_i} x_i}{x_i^2}$$

$$= q_{ij}\left(G|_{u_{ij}} - G|_{u_i}\right).$$

Thus the mean strategy dynamics for the three models are given by the following

Strategy dynamics

$$
\begin{aligned}
\text{Difference:} \quad & u_i(t+1) = u_i + \frac{1}{\left(1+G|_{u_i}\right)} \sum_{j=0}^{n_p} \left(G|_{u_{ij}} - G|_{u_i}\right) q_{ij} u_{ij} \\
\text{Exp. Difference:} \quad & u_i(t+1) = \sum_{j=0}^{n_p} \exp\left(G|_{u_{ij}} - G|_{u_i}\right) q_{ij} u_{ij} \\
\text{Differential:} \quad & \dot{u}_i = \sum_{j=0}^{n_p} \left(G|_{u_{ij}} - G|_{u_i}\right) q_{ij} u_{ij}
\end{aligned}
$$

(5.6)

Note that the only assumption used to obtain the results given in (5.6) is that a finite number of fixed strategies is available to the phenotypes. While these results may be used to determine mean strategy dynamics, they require keeping track of a large number of phenotypes. Some thought also needs to be given to how the original population is distributed among the phenotypes. A reasonable assumption would be to assign the majority of the population to phenotypes in the neighborhood of the mean strategy. However, all phenotypes must be assigned a fraction of the population, even though this fraction might be quite small or zero.

The equations given in (5.6) can be simplified by making some further assumptions. Consider using the first two terms of a Taylor series expansion that provide a **first-order approximation** for $\left(G|_{u_{ij}} - G|_{u_i}\right)$. Taking note of (5.3), small $\delta u_{ij}$ results in

$$G|_{u_{ij}} - G|_{u_i} \approx \left.\frac{\partial G}{\partial v}\right|_{u_i} \delta u_{ij}.$$

If we now treat this equation as an equality and substitute it into (5.6) we obtain

$$
\begin{aligned}
\text{Difference:} \quad & u_i(t+1) = u_i + \frac{1}{1+G|_{u_i}} \left.\frac{\partial G}{\partial v}\right|_{u_i} \sum_{j=0}^{n_p} \delta u_{ij} q_{ij} u_{ij} \\
\text{Exp. Difference:} \quad & u_i(t+1) = u_i + \left.\frac{\partial G}{\partial v}\right|_{u_i} \sum_{j=0}^{n_p} \delta u_{ij} q_{ij} u_{ij} \\
\text{Differential:} \quad & \dot{u}_i = \left.\frac{\partial G}{\partial v}\right|_{u_i} \sum_{j=0}^{n_p} \delta u_{ij} q_{ij} u_{ij}
\end{aligned}
$$

(5.7)

where the exponential function in the second difference equation model has been replaced by the first two terms of the expansion $e^x = 1 + x + \cdots$. The latter assumption, while not inconsistent with the small $\delta u_{ij}$ assumption, is more restrictive for the exponential model as it also requires $\left.\frac{\partial G}{\partial v}\right|_{u_i} \delta u_{ij}$ to be small.

## 5.4 G-functions with scalar strategies

These equations have the common term

$$\sum_{j=0}^{n_p} \delta u_{ij} q_{ij} u_{ij} = \sum_{j=0}^{n_p} \delta u_{ij} q_{ij} \left( u_i + \delta u_{ij} \right)$$

$$= u_i \sum_{j=0}^{n_p} \delta u_{ij} q_{ij} + \sum_{j=0}^{n_p} \delta u_{ij} q_{ij} \delta u_{ij}.$$

By virtue of (5.5), the first summation term to the right of the lower equals sign is zero and it follows that

$$\sum_{j=0}^{n_p} \delta u_{ij} q_{ij} u_{ij} = \sum_{j=0}^{n_p} \delta u_{ij} q_{ij} \delta u_{ij} = \sigma_i^2 \qquad (5.8)$$

where by definition $\sigma_i^2$ is the **variance** in $\delta u_{ij}$ from the mean $u_i$.

Using this result in (5.7) gives us approximate strategy dynamics for the three different models.

First-order strategy dynamics

| | |
|---|---|
| Difference: | $\Delta u_i = \frac{\sigma_i^2}{1+G\vert_{u_i}} \frac{\partial G}{\partial v}\vert_{u_i}$ |
| Exp. Difference: | $\Delta u_i = \sigma_i^2 \frac{\partial G}{\partial v}\vert_{u_i}$ |
| Differential: | $\dot{u}_i = \sigma_i^2 \frac{\partial G}{\partial v}\vert_{u_i}$ |

(5.9)

where

$$\Delta u_i = u_i \left( t + 1 \right) - u_i.$$

These equations are much easier to use than (5.6) because there is no need for summations. They also relate more directly to known biological processes. For example, the variance $\sigma_i^2$ scales the rate of evolutionary change. This coefficient has much in common with the way heritability, phenotypic variances, and additive genetic variances scale evolutionary rates in quantitative genetic models (Taper and Case, 1985).

**Fisher's Fundamental Theorem** of Natural Selection asserts that the rate of increase in fitness of any organism at any time is equal to its additive genetic variance in fitness at that time (Fisher, 1930). It may be interpreted in terms of the adaptive landscape as follows. A population's mean strategy will change in the direction of the upward slope of the adaptive landscape at a rate that is directly proportional to the slope of the adaptive landscape and the amount of heritable variation within the population (additive genetic variance). The slope of the adaptive landscape at any point represents the change in fitness for a given change in strategy. The greater the genetic variance, the greater the change in fitness and, by (5.9), the more rapidly an organism will evolve toward equilibrium. For this reason $\sigma_i^2$ is sometimes referred to as the **speed** term in these equations.

It should be noted that constraints on the strategy set

$$\mathbf{u} \in \mathcal{U}$$

must be accounted for when using strategy dynamics. This is done by including extra coding when solving the equations to ensure that constraints are satisfied (Vincent and Grantham, 1997). It is apparent that Fisher's Fundamental Theorem must also be modified accordingly. Without explicitly stating it for each case, it is understood that the strategy constraints must be satisfied when using strategy dynamics for all situations discussed below.

### 5.4.1.1 Large difference in time scales

After initial conditions have been specified, equations (5.1), (5.6) or (5.9) may be solved (iterated or integrated) to determine the outcome of the evolutionary game. However, in so doing, recall the two **time scales**: an ecological time scale, $T_{ec}$, associated with the population dynamics (5.1) and an evolutionary time scale, $T_{ev}$, associated with the strategy dynamics (5.6) or (5.9). The time scales measure the time it takes for a system to return to a fixed point or equilibrium solution after being displaced from such a solution. We see from the development of the strategy dynamics equations that the return time for the strategy dynamics must be slower than the return time for the population dynamics (since a change in mean strategy can only come about by the change in frequency and hence numbers of the phenotypes). In general, the evolutionary time scale could range from somewhat slower than the ecological time scale (bacteria) to an evolutionary time scale that is very much slower than the ecological time scale (turtles). We expect the latter to be the more common situation. However, because time scales can change rapidly under strong selective pressures, they should not be thought of as fixed.

When there is a large difference in time scales, the biological system will spend most of its time near a slowly changing ecological equilibrium solution and we can simplify the solution process by replacing the population dynamics by the fixed point or equilibrium conditions as given by the algebraic equations

$$G(v, \mathbf{u}, \mathbf{x})|_{v=u_i} = 0. \tag{5.10}$$

We may think of (5.10) as a set of equations that, for a given $\mathbf{u}$, may be solved for the fixed point or equilibrium value $\mathbf{x}^*$. In this case the Darwinian dynamics reduces to just the strategy dynamics. Rewriting (5.9) for this case (using the argument notation to emphasize that $\mathbf{x}$ is evaluated at $\mathbf{x}^*$) we obtain the following

**First-order strategy dynamics with equilibrium population x***

$$\text{Difference:} \quad \Delta u_i = \frac{\sigma_i^2}{1 + G(v,\mathbf{u},\mathbf{x}^*)|_{v=u_i}} \left.\frac{\partial G(v,\mathbf{u},\mathbf{x}^*)}{\partial v}\right|_{v=u_i}$$

$$\text{Exp. Difference:} \quad \Delta u_i = \sigma_i^2 \left.\frac{\partial G(v,\mathbf{u},\mathbf{x}^*)}{\partial v}\right|_{v=u_i} \quad (5.11)$$

$$\text{Differential:} \quad \dot{u}_i = \sigma_i^2 \left.\frac{\partial G(v,\mathbf{u},\mathbf{x}^*)}{\partial v}\right|_{v=u_i}$$

In the next example, we use these equations to examine evolution when there is a large difference between $T_{ec}$ and $T_{ev}$. It is illuminating to observe how the ESS strategy evolves, by plotting $G(v, \mathbf{u}, \mathbf{x}^*)$ as a function of $v$ at different times as evolution takes place.

**Example 5.4.1 (L–V competition game)** *Recall from Example 4.3.1 that this game has a G-function given by*

$$G(v, \mathbf{u}, \mathbf{x}) = \frac{r}{K(v)} \left( K(v) - \sum_{j=1}^{n_s} a(v, u_j) x_j \right),$$

with

$$K(v) = K_m \exp\left\{ -\frac{v^2}{2\sigma_k^2} \right\}$$

$$\alpha(v, u_i) = 1 + \exp\left\{ -\frac{(v - u_i + \beta)^2}{2\sigma_\alpha^2} \right\} - \exp\left\{ -\frac{\beta^2}{2\sigma_\alpha^2} \right\}.$$

*This model was developed to demonstrate how a **coalition**[9] of strategies can form that involves more than one species. We will explore this feature using the differential equation model by examining snapshots of the adaptive landscape while solving (5.11) from an arbitrary starting condition to an equilibrium solution under two different sets of parameters. Following Vincent et al., (1993), we use the parameters $r = 0.25$, $K_m = 100$, and $\sigma_k^2 = \sigma_\alpha^2 = \beta^2 = 4$ as a first case. Figure 5.3 illustrates the evolution of a single strategy ($n_s = 1$, $\sigma_1^2 = 0.5$) from an initial (arbitrary) value of $u_1 = -3$ to a final equilibrium value of $u_1 = 1.213$ with a corresponding equilibrium population density[10] of $x_1^* = 83.20$.*

---

[9] When $n_{s^*} > 1$ species have a non-zero population at equilibrium, their corresponding strategies are refered to as a coalition (in the sense of combination) of $n_{s^*}$ strategies. A formal definition is given in Subsection 6.2.3.

[10] This is the only equilibrium solution obtained under these conditions no matter what starting value is used for $u_1$ or how many species are used in the simulation. Exactly the same results are obtained using all three dynamical system models. Given $n_s$ different strategies, some (at least 1) will converge to $u_1$ with a finite equilibrium population. Others will have their population numbers converge to zero.

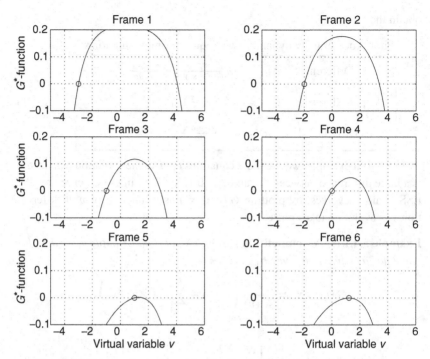

**Figure 5.3** Strategy dynamics on the adaptive landscape for the Lotka–Volterra model with $\sigma_k^2 = 4$.

At equilibrium, $G(v, \mathbf{u}, \mathbf{x}^*)$ takes on a maximum value at $u_1$. The adaptive landscape changes shape somewhat during this process. This feature becomes much more pronounced as $\sigma_k$ is increased. For example, increasing to $\sigma_k^2 = 12.5$ leads to a dramatically different result, as shown in Figure 5.4. In this case, a single strategy is again used starting at an initial value of $u_1 = -2$; however, as the strategy climbs the adaptive landscape, we see a valley appear in frame 4. This valley tracks the strategy and moves under it at the equilibrium point $u_1 = 3.79$, $x_1^* = 56.28$. We have a fascinating result. While the strategy always climbs upward on the adaptive landscape, when it reaches equilibrium it is at a local minimum! Could this solution possibly be the endpoint of evolution? We will now show that this solution is not stable with respect to changes in the number of species. The existence of the high peak in the last frame of Figure 5.4 suggests that an evolutionary equilibrium will require a coalition of strategies with more than one species. This is indeed the case. Furthermore, the existence of a coalition of one stable strategy at a local minimum allows for **speciation** to take place ($n_s = 1 \Rightarrow n_s = 2$), resulting in a stable coalition of two species. Figure 5.5 illustrates this process. The last frame of Figure 5.4 and the first

## 5.4 G-functions with scalar strategies

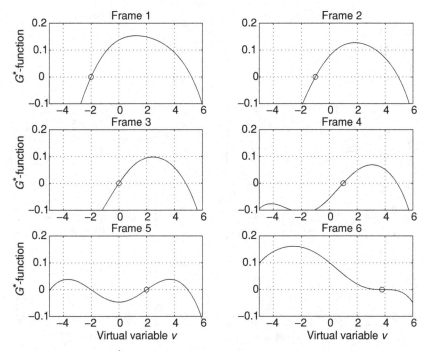

**Figure 5.4** With $\sigma_k^2 = 12.5$ and $n = 1$, strategy dynamics produces an equilibrium point that is a local minimum.

frame of Figure 5.5 differ only in that $n_s = 1$ has been replaced by $n_s = 2$. A second strategy has been added at a value very close to $u_1 = 3.79$. This could occur in a natural system through a process of **assortative mating** (like tends to mate with like); that is, in the distribution of phenotypes available to a given species through genetic variability, as modeled here by (5.3), there is a tendency for individuals in the tails of the distributions to mate with each other. The stable local minimum in the adaptive landscape created by the coalition of one solution creates an opportunity for two distinct types, close in strategy value, to co-exist, forming a coalition of two species. Each type is located on either side of the valley previously occupied by the coalition of one and hence will be facing a hill to climb via strategy dynamics. Under strategy dynamics, the two types climb their respective hills as seen in frame 2 of Figure 5.5. Frames 3–5 illustrate that as this process continues, the adaptive landscape changes shape, and at equilibrium (frame 6), the two peaks are distinct and separated enough to identify them as separate species. As demonstrated in Vincent et al. (1993) this same model can have coalitions greater than two species (using a larger value of $\sigma_k^2$).

130                    *Darwinian dynamics*

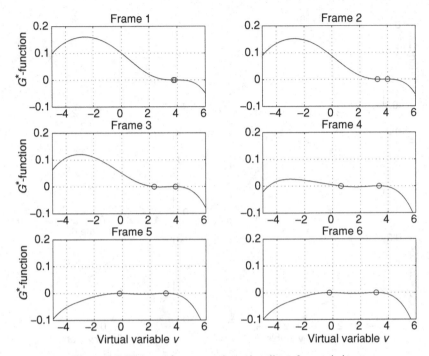

**Figure 5.5** With $n = 2$, strategy dynamics allows for speciation.

This example has important implications for how speciation can take place. It was Darwin's claim that evolution takes place through gradual changes as driven by natural selection. Each slight variation toward the final state is advantageous. This claim has often been used by his critics as a major problem with his theory. If we looked only at the final landscape as given by frame 6 of Figure 5.5, we might also have the same objection. How could speciation possibly take place in order to create the two isolated peaks? The two species in final form appear to have such a large difference in character that it is hard to imagine one slowly evolving from the other. However, we have seen exactly how it can be done in the way Darwin envisioned. The existence of a stable local minimum for one strategy provides an opportunity for two strategies nearly identical in mean value to co-exist. A minimum develops between the two strategies and natural selection drives the strategies, using small changes, to the final state as represented by the peaks. Speciation will be discussed in more detail in Chapter 8.

#### 5.4.1.2  Small difference in time scales
When the difference between the evolutionary time scale $T_{ev}$ and the ecological time scale $T_{ec}$ is not large, the complete set of Darwinian dynamics must be

## 5.5 G-functions with vector strategies

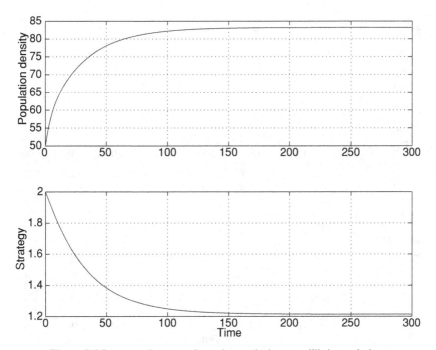

**Figure 5.6** Low-speed strategy dynamics results in an equilibrium solution.

used to predict the path of evolution. In this case, it increases the likelihood that solutions will result in non-equilibrium dynamics.

**Example 5.4.2 (L–V competition game)** *Let us reconsider the previous example using the difference equation model, all other parameters the same as the first case $(\sigma_k^2 = 4)$ and with two different speed terms. The population and strategy dynamics using $\sigma_1^2 = 0.5$ as illustrated in Figure 5.6 shows that we get the same equilibrium solution $(x_1^* = 83.2, u_1 = 1.213)$ as in the previous example. However, increasing the speed term to $\sigma_1^2 = 35$ (thus greatly increasing $T_{ev}$) we get the results of Figure 5.7. The fast evolutionary dynamics results in an equilibrium solution that is no longer asymptotically stable.*

## 5.5 G-functions with vector strategies

Strategy dynamics with vector strategies may be developed in a fashion similar to the scalar case. As before, $x_{ij}$ denotes the $j$th phenotype of species $i$ and $\mathbf{u}_i$ denotes the mean strategy vector for the $i$th species

$$\mathbf{u}_i = \begin{bmatrix} u_{i1} & \cdots & u_{i n_u} \end{bmatrix}.$$

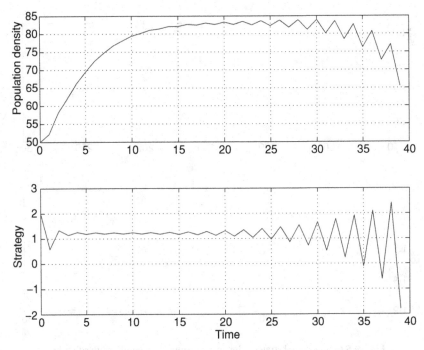

**Figure 5.7** High-speed strategy dynamics results in unstable Darwinian dynamics.

In general, the variability in strategy values among the phenotypes of species $i$ will be small relative to the strategy differences between the species. Assume that for each species $i$ there are $n_p$ phenotypes and that the $n_p$ phenotypes within a species mate assortatively (like mates with like). Each component of $\mathbf{u}_i$ is assumed to have a distribution of strategies. For example, the first component of the strategy vector is of the form

$$\begin{bmatrix} u_{i1} + \delta u_{i11} & u_{i1} + \delta u_{i12} & \cdots & u_{i1} + \delta u_{i1n_p} \end{bmatrix}.$$

Since each phenotype can have a vector strategy with $n_u$ components, notation can become complex. However, we can handle all strategy components and all phenotypes for each species by defining the strategy matrix

$$\mathbf{U}_i = \begin{bmatrix} u_{i1} + \delta u_{i11} & u_{i1} + \delta u_{i12} & \cdots & u_{i1} + \delta u_{i1n_p} \\ u_{i2} + \delta u_{i21} & u_{i2} + \delta u_{i22} & \cdots & u_{i2} + \delta u_{i2n_p} \\ \vdots & \vdots & u_{ij} + \delta u_{ijk} & \vdots \\ u_{in_u} + \delta u_{in_u 1} & u_{in_u} + \delta u_{in_u 2} & \cdots & u_{in_u} + \delta u_{in_u n_p} \end{bmatrix}$$

where each *row* represents the variations in a given component of the strategy vector and each *column* represents a phenotype. The first subscript represents the different species $i$. The second represents the trait within a strategy; there are $n_u$ such traits. The third subscript represents the phenotype. There are $n_p$ phenotypes, and each column represents the set of traits that a phenotype carries. A unique set of traits represent a phenotype's strategy.

The difference between the mean value and the actual strategy value of all extant phenotypes within species $i$ is represented by the matrix

$$\delta \mathbf{U}_i = \begin{bmatrix} \delta u_{i11} & \delta u_{i12} & \cdots & \delta u_{i1n_p} \\ \delta u_{i21} & \delta u_{i22} & \cdots & \delta u_{i2n_p} \\ \vdots & \vdots & \delta u_{ijk} & \vdots \\ \delta u_{in_u 1} & \delta u_{in_u 2} & \cdots & \delta u_{in_u n_p} \end{bmatrix}.$$

Thus

$$\mathbf{U}_i = \mathbf{u}_i^T \Omega + \delta \mathbf{U}_i$$

where

$$\Omega = \begin{bmatrix} 1 & 1 & \cdots & 1 \end{bmatrix}$$

is a row vector containing $n_p$ ones. That is, $\mathbf{u}_i^T \Omega$ is an $n_u \times n_p$ matrix. When it is added term by term to $\delta \mathbf{U}_i$ it results in the phenotype trait values for species $i$ with mean $\mathbf{u}_i$. Changes in the components of $u_{ij}$ are independent since the components of $\delta u_{ij}$ are independent. While genetic phenomena such as linkage, epistasis, and pleiotropy may force dependencies among the components of the strategy vector, we assume that such phenomena do not preclude the components from varying independently of each other. By making this assumption, any subsequent covariances in the components' strategy dynamics and among the values of components are due to natural selection, not forced genetic or physical constraints on the strategy set.

The density of the species $x_i$ may be written as the sum of the densities of all the phenotypes within species $i$

$$x_i = \sum_{j=1}^{n_p} x_{ij}.$$

The mean strategy for the population $x_i$ is defined as

$$\mathbf{u}_i = \sum_{j=1}^{n_p} \frac{x_{ij}}{x_i} \mathbf{u}_{ij} \tag{5.12}$$

where $\mathbf{u}_{ij}$ is the $j$th column vector of $\mathbf{U}_i$.

We can track how the mean strategy $\mathbf{u}_i$ evolves (i.e., a strategy dynamics for $\mathbf{u}_i$) by considering the changes that must take place in the mean strategy as a result of changes in phenotype densities within species $i$. If we define

$$q_{ij} = \frac{x_{ij}}{x_i}$$

and

$$\mathbf{q}_i = \begin{bmatrix} q_{i1} & \cdots & q_{in_p} \end{bmatrix}^T \tag{5.13}$$

then (5.12) may be written in the following compact matrix form

$$\mathbf{u}_i = \mathbf{U}_i \mathbf{q}_i. \tag{5.14}$$

We may now calculate a strategy dynamic for the mean strategy $\mathbf{u}_i$ using (5.14). Here we use the idea that the dynamics of $\mathbf{u}_i$ as it changes from its original nominal value is due to changes in the relative frequency of phenotypes (the $q_{ij}$). Even though $\mathbf{u}_i$ changes with time we may take $\mathbf{U}_i$ to be constant since $n_p$ can be a very large number, with only a fraction of the possible phenotypes having a non-zero $q_{ij}$ at any point in time. A cluster of non-zero and ever-changing $q_{ij}$ phenotype strategies are able to follow $\mathbf{u}_i$ as a distribution of strategies about the mean. In other words, the change in $\mathbf{u}_i$ is due only to a change in $\mathbf{q}_i$ as given by the following

Strategy dynamics in terms of $\mathbf{q}$

| | |
|---|---|
| Difference: | $\Delta \mathbf{u}_i = \mathbf{U}_i \Delta \mathbf{q}_i$ |
| Exp. Difference: | $\Delta \mathbf{u}_i = \mathbf{U}_i \Delta \mathbf{q}_i$ |
| Differential: | $\dot{\mathbf{u}}_i = \mathbf{U}_i \dot{\mathbf{q}}_i$ |

In terms of the $G$-function, the population dynamics for the species and the phenotypes is written as

Species dynamics

| | |
|---|---|
| Difference: | $x_i(t+1) = x_i \left[ 1 + G\vert_{\mathbf{u}_i} \right]$ |
| Exp. Difference: | $x_i(t+1) = x_i \exp G\vert_{\mathbf{u}_i}$ |
| Differential: | $\dot{x}_i = x_i G\vert_{\mathbf{u}_i}$ |

Phenotype dynamics

| | |
|---|---|
| Difference: | $x_{ij}(t+1) = x_{ij} \left[ 1 + G\vert_{\mathbf{u}_{ij}} \right]$ |
| Exp. Difference: | $x_{ij}(t+1) = x_{ij} \exp G\vert_{\mathbf{u}_{ij}}$ |
| Differential: | $\dot{x}_{ij} = x_{ij} G\vert_{\mathbf{u}_{ij}}$ |

## 5.5 G-functions with vector strategies

It follows that for the first difference equation model

$$q_{ij}(t+1) = \frac{x_{ij}(t+1)}{x_i(t+1)} = \frac{x_{ij}\left(1+G|_{\mathbf{u}_{ij}}\right)}{x_i\left(1+G|_{\mathbf{u}_i}\right)}$$

$$= q_{ij}\frac{\left(1+G|_{\mathbf{u}_i}\right)}{\left(1+G|_{\mathbf{u}_i}\right)} + q_{ij}\frac{\left(G|_{\mathbf{u}_{ij}}-G|_{\mathbf{u}_i}\right)}{\left(1+G|_{\mathbf{u}_i}\right)}$$

$$\Delta q_{ij} = q_{ij}\frac{\left(G|_{\mathbf{u}_{ij}}-G|_{\mathbf{u}_i}\right)}{\left(1+G|_{\mathbf{u}_i}\right)}$$

and for the second difference equation model

$$q_{ij}(t+1) = \frac{x_{ij}(t+1)}{x_i(t+1)} = \frac{x_{ij}}{x_i}\frac{\exp\left(G|_{\mathbf{u}_{ij}}\right)}{\exp\left(G|_{\mathbf{u}_i}\right)}$$

$$= q_{ij}\exp\left(G|_{\mathbf{u}_{ij}}-G|_{\mathbf{u}_i}\right) + q_{ij} - q_{ij}$$

$$\Delta q_{ij} = q_{ij}\left[\exp\left(G|_{\mathbf{u}_{ij}}-G|_{\mathbf{u}_i}\right)-1\right]$$

and for the differential equation model

$$\dot{q}_{ij} = \frac{d}{dt}\left(\frac{x_{ij}}{x_i}\right) = \frac{x_i\dot{x}_{ij}-x_{ij}\dot{x}_i}{x_i^2}$$

$$= \frac{x_i\,G|_{\mathbf{u}_{ij}}\,x_{ij}-x_{ij}\,G|_{\mathbf{u}_i}\,x_i}{x_i^2}$$

$$= q_{ij}\left(G|_{\mathbf{u}_{ij}}-G|_{\mathbf{u}_i}\right).$$

Writing these results in matrix form we have the following

Strategy dynamics

| | | |
|---|---|---|
| Difference: | $\Delta \mathbf{q}_i$ | $= \frac{1}{1+G|_{\mathbf{u}_i}}\mathbf{Q}_i\Delta\mathbf{G}_i$ |
| Exp. Difference: | $\Delta \mathbf{q}_i$ | $= \mathbf{Q}_i\exp\Delta\mathbf{G}_i$ |
| Differential: | $\dot{\mathbf{q}}_i$ | $= \mathbf{Q}_i\Delta\mathbf{G}_i$ |

(5.15)

where $\Delta\mathbf{q}_i$ and $\dot{\mathbf{q}}_i$ are column vectors as in (5.13) and $\mathbf{Q}_i$, is an $n_p \times n_p$ diagonal matrix

$$\mathbf{Q}_i = \begin{bmatrix} q_{i1} & 0 & \cdots & 0 \\ 0 & q_{i2} & \cdots & 0 \\ \vdots & \vdots & \ddots & 0 \\ 0 & 0 & 0 & q_{in_p} \end{bmatrix}.$$

By definition

$$\exp \Delta \mathbf{G}_i = \begin{bmatrix} \exp\left(G|_{\mathbf{u}_{i1}} - G|_{\mathbf{u}_i}\right) - 1 \\ \vdots \\ \exp\left(G|_{\mathbf{u}_{in_p}} - G|_{\mathbf{u}_i}\right) - 1 \end{bmatrix}$$

and

$$\Delta \mathbf{G}_i = \begin{bmatrix} G|_{\mathbf{u}_{i1}} - G|_{\mathbf{u}_i} \\ \vdots \\ G|_{\mathbf{u}_{in_p}} - G|_{\mathbf{u}_i} \end{bmatrix}.$$

This result may be used to determine the strategy dynamics. However, doing so requires keeping track of a large number of phenotypes, and one may need to know the form of the distribution of strategies around the mean strategy. A reasonable assumption assigns the majority of the population to phenotypes within the immediate neighborhood of the mean strategy. However, all phenotypes must occur as some fraction of the population, even though this fraction might be quite small or even zero.

Using a Taylor series expansion about $\mathbf{u}_i$ to approximate $\Delta \mathbf{G}_i$ in terms of first- and higher-order terms (HOT) we obtain

$$\begin{bmatrix} G|_{\mathbf{u}_{i1}} - G|_{\mathbf{u}_i} \\ \vdots \\ G|_{\mathbf{u}_{in_p}} - G|_{\mathbf{u}_i} \end{bmatrix}$$

$$= \begin{bmatrix} \frac{\partial G}{\partial v_1}\bigg|_{\mathbf{u}_i} \delta u_{i11} + \frac{\partial G}{\partial v_2}\bigg|_{\mathbf{u}_i} \delta u_{i21} + \cdots + \frac{\partial G}{\partial v_{n_u}}\bigg|_{\mathbf{u}_i} \delta u_{in_u 1} \\ \vdots \\ \frac{\partial G}{\partial v_1}\bigg|_{\mathbf{u}_i} \delta u_{i1n_p} + \frac{\partial G}{\partial v_2}\bigg|_{\mathbf{u}_i} \delta u_{i2n_p} + \cdots + \frac{\partial G}{\partial v_{n_u}}\bigg|_{\mathbf{u}_i} \delta u_{in_u n_p} \end{bmatrix} + \text{HOT}$$

$$= \begin{bmatrix} \delta u_{i11} & \delta u_{i21} & \cdots & \delta u_{in_u 1} \\ \delta u_{i12} & \delta u_{i22} & \cdots & \delta u_{in_u 2} \\ \vdots & \vdots & \ddots & \vdots \\ \delta u_{i1n_p} & \delta u_{i2n_p} & \cdots & \delta u_{in_u n_p} \end{bmatrix} \begin{bmatrix} \frac{\partial G}{\partial v_1} \\ \frac{\partial G}{\partial v_2} \\ \vdots \\ \frac{\partial G}{\partial v_{n_u}} \end{bmatrix}_{\mathbf{u}_i} + \text{HOT}.$$

For small $\delta \mathbf{u}_{ij}$ this gives

$$\Delta \mathbf{G}_i \approx \delta \mathbf{U}_i^T \frac{\partial G}{\partial \mathbf{v}}\bigg|_{\mathbf{u}_i}.$$

## 5.5 G-functions with vector strategies

If we treat this approximation as an exact equality and substitute it into (5.15) we obtain

$$
\begin{aligned}
\text{Difference:} \quad & \Delta \mathbf{u}_i = \frac{1}{1+G|_{\mathbf{u}_i}} \mathbf{U}_i \mathbf{Q}_i \delta \mathbf{U}_i^T \left. \frac{\partial G}{\partial \mathbf{v}} \right|_{\mathbf{u}_i} \\
\text{Exp. Difference:} \quad & \Delta \mathbf{u}_i = \mathbf{U}_i \mathbf{Q}_i \delta \mathbf{U}_i^T \left. \frac{\partial G}{\partial \mathbf{v}} \right|_{\mathbf{u}_i} \\
\text{Differential:} \quad & \dot{\mathbf{u}}_i = \mathbf{U}_i \mathbf{Q}_i \delta \mathbf{U}_i^T \left. \frac{\partial G}{\partial \mathbf{v}} \right|_{\mathbf{u}_i}
\end{aligned}
$$

where the exponential term has been replaced by the first two terms in its expansion. All equations have in common

$$
\begin{aligned}
\mathbf{U}_i \mathbf{Q}_i \delta \mathbf{U}_i^T \left. \frac{\partial G}{\partial \mathbf{v}} \right|_{\mathbf{u}_i} &= [\mathbf{u}_i \Omega + \delta \mathbf{U}_i] \mathbf{Q}_i \delta \mathbf{U}_i^T \left. \frac{\partial G}{\partial \mathbf{v}} \right|_{\mathbf{u}_i} \\
&= \left( \mathbf{u}_i \Omega \mathbf{Q}_i \delta \mathbf{U}_i^T + \delta \mathbf{U}_i \mathbf{Q}_i \delta \mathbf{U}_i^T \right) \left. \frac{\partial G}{\partial \mathbf{v}} \right|_{\mathbf{u}_i}.
\end{aligned}
$$

If we assume that $q_{ij}$ and $\delta \mathbf{u}_{ij}$ are symmetric ($\delta \mathbf{u}_{ij}$ has a mean of zero), we can now show that the first term is zero. That is

$$\mathbf{u}_i \Omega \mathbf{Q}_i \delta \mathbf{U}_i^T = \mathbf{0}$$

since

$$
\begin{bmatrix} u_{i1} & u_{i1} & \cdots & u_{i1} \\ u_{i2} & u_{i2} & \cdots & u_{i2} \\ \vdots & \vdots & \vdots & \vdots \\ u_{in_u} & u_{in_u} & \cdots & u_{in_u} \end{bmatrix} \begin{bmatrix} q_{i1} & 0 & \cdots & 0 \\ 0 & q_{i2} & \cdots & 0 \\ \vdots & \vdots & \vdots & \vdots \\ 0 & 0 & \cdots & q_{in_u} \end{bmatrix} \begin{bmatrix} \delta u_{i11} & \delta u_{i21} & \cdots & \delta u_{in_u 1} \\ \delta u_{i12} & \delta u_{i22} & \cdots & \delta u_{in_u 2} \\ \vdots & \vdots & \ddots & \vdots \\ \delta u_{i1n_p} & \delta u_{i2n_p} & \cdots & \delta u_{in_u n_p} \end{bmatrix}
$$

$$
= \begin{bmatrix} u_{i1} q_{i1} & u_{i1} q_{i2} & \cdots & u_{i1} q_{in_u} \\ u_{i2} q_{i1} & u_{i2} q_{i2} & \cdots & u_{i2} q_{in_u} \\ \vdots & \vdots & & \vdots \\ u_{in_u} q_{i1} & u_{in_u} q_{i2} & \cdots & u_{in_u} q_{in_u} \end{bmatrix} \begin{bmatrix} \delta u_{i11} & \delta u_{i21} & \cdots & \delta u_{in_u 1} \\ \delta u_{i12} & \delta u_{i22} & \cdots & \delta u_{in_u 2} \\ \vdots & \vdots & \ddots & \vdots \\ \delta u_{i1n_p} & \delta u_{i2n_p} & \cdots & \delta u_{in_u n_p} \end{bmatrix}.
$$

Multiplying this result yields a matrix

$$
\begin{bmatrix} u_{i1} \left( q_{i1} \delta u_{i11} + q_{i2} \delta u_{i12} + \cdots \right. & \cdots & u_{i1} \left( q_{i1} \delta u_{in_u 1} + q_{i2} \delta u_{in_u 2} + \cdots \right. \\ \left. + q_{in_u} \delta u_{i1n_p} \right) & & \left. + q_{in_u} \delta u_{in_u n_p} \right) \\ \vdots & \vdots & \vdots \\ u_{in_u} \left( q_{i1} \delta u_{i11} + q_{i2} \delta u_{i12} + \cdots \right. & \cdots & u_{in_u} \left( q_{i1} \delta u_{in_u 1} + q_{i2} \delta u_{in_u 2} + \cdots \right. \\ \left. + q_{in_u} \delta u_{i1n_p} \right) & & \left. + q_{in_u} \delta u_{in_u n_p} \right) \end{bmatrix}
$$

in which each term sums to zero due to the assumed symmetry. This leaves us with the result

$$\begin{aligned}
\text{Difference:} &\quad \Delta\mathbf{u}_i = \frac{1}{1+G|_{\mathbf{u}_i}} \delta\mathbf{U}_i \mathbf{Q}_i \delta\mathbf{U}_i^T \left.\frac{\partial G}{\partial \mathbf{v}}\right|_{\mathbf{u}_i} \\
\text{Exp. Difference:} &\quad \Delta\mathbf{u}_i = \delta\mathbf{U}_i \mathbf{Q}_i \delta\mathbf{U}_i^T \left.\frac{\partial G}{\partial \mathbf{v}}\right|_{\mathbf{u}_i} \\
\text{Differential:} &\quad \dot{\mathbf{u}}_i = \delta\mathbf{U}_i \mathbf{Q}_i \delta\mathbf{U}_i^T \left.\frac{\partial G}{\partial \mathbf{v}}\right|_{\mathbf{u}_i}
\end{aligned}$$

Finally if we define the covariance matrix

$$\mathcal{D}_i = \delta\mathbf{U}_i \mathbf{Q}_i \delta\mathbf{U}_i^T$$

we may write the first-order strategy dynamics as follows

First-order strategy dynamics

$$\begin{aligned}
\text{Difference:} &\quad \Delta\mathbf{u}_i = \frac{\mathcal{D}_i}{1+G|_{\mathbf{u}_i}} \left.\frac{\partial G}{\partial \mathbf{v}}\right|_{\mathbf{u}_i} \\
\text{Exp. Difference:} &\quad \Delta\mathbf{u}_i = \mathcal{D}_i \left.\frac{\partial G}{\partial \mathbf{v}}\right|_{\mathbf{u}_i} \\
\text{Differential:} &\quad \dot{\mathbf{u}}_i = \mathcal{D}_i \left.\frac{\partial G}{\partial \mathbf{v}}\right|_{\mathbf{u}_i}
\end{aligned} \quad (5.16)$$

For example, if $n_u = 2$, and $n_s = 2$, then these equations for the differential equation case are of the form

$$\dot{u}_{11} = \sigma_{11}^2 \left.\frac{\partial G}{\partial v_1}\right|_{v_1=u_{11}} + \delta_{11}^2 \left.\frac{\partial G}{\partial v_2}\right|_{v_2=u_{12}}$$

$$\dot{u}_{12} = \delta_{12}^2 \left.\frac{\partial G}{\partial v_1}\right|_{v_1=u_{11}} + \sigma_{12}^2 \left.\frac{\partial G}{\partial v_2}\right|_{v_2=u_{12}}$$

$$\dot{u}_{21} = \sigma_{21}^2 \left.\frac{\partial G}{\partial v_1}\right|_{v_1=u_{21}} + \delta_{21}^2 \left.\frac{\partial G}{\partial v_2}\right|_{v_2=u_{22}}$$

$$\dot{u}_{22} = \delta_{22}^2 \left.\frac{\partial G}{\partial v_1}\right|_{v_1=u_{21}} + \sigma_{22}^2 \left.\frac{\partial G}{\partial v_2}\right|_{v_2=u_{22}}$$

where $\sigma^2$ represents variance and $\delta^2$ represents covariance. Notation is simplified if $\mathcal{D}_i$ is the same for each species.

Because of interbreeding and heritable independence among the strategy components, the covariance terms will be small and made even smaller by the summation of both positive and negative terms. If the covariance terms are close to zero relative to the variance terms, we may write the strategy dynamics for

## 5.5 G-functions with vector strategies

each trait of species $i$ as follows

First-order strategy dynamics with small covariance

$$\text{Difference:} \quad \Delta u_{ik} = \frac{\sigma_{ik}^2}{1+G|_{u_i}} \left.\frac{\partial G}{\partial v}\right|_{u_i}$$

$$\text{Exp. Difference:} \quad \Delta u_{ik} = \sigma_{ik}^2 \left.\frac{\partial G}{\partial v}\right|_{u_i}$$

$$\text{Differential:} \quad \dot{u}_{ik} = \sigma_{ik}^2 \left.\frac{\partial G}{\partial v_k}\right|_{u_i}$$

where $\sigma_{ik}^2$ is the variance element corresponding to strategy component $k$. First-order strategy dynamics in this situation has an analog in the evolutionary dynamics of quantitative traits in quantitative genetic models.

For example, if $n_u = 2$ and $n_s = 2$ these equations for the differential equation model are of the form

$$\dot{u}_{11} = \sigma_{11}^2 \left.\frac{\partial G}{\partial v_1}\right|_{v_1 = u_{11}}$$

$$\dot{u}_{12} = \sigma_{12}^2 \left.\frac{\partial G}{\partial v_2}\right|_{v_2 = u_{12}}$$

$$\dot{u}_{21} = \sigma_{21}^2 \left.\frac{\partial G}{\partial v_1}\right|_{v_1 = u_{21}}$$

$$\dot{u}_{22} = \sigma_{22}^2 \left.\frac{\partial G}{\partial v_2}\right|_{v_2 = u_{22}}.$$

Note that, when using a scalar strategy, there is only one variance term and no covariance terms, it follows that

$$\mathcal{D}_i = \sigma_{i1}^2 = \sigma_i^2.$$

In examples dealing with scalar strategies we use $\sigma_i^2$ to denote strategy variance.

**Example 5.5.1 (L–V big bully game)** *Recall from Example 4.4.1, this game is defined by*

$$G(v, \mathbf{u}, \mathbf{x}) = \frac{r}{K(v)} \left( K(v) - \sum_{j=1}^{n_s} a(v, \mathbf{u}_j) x_j \right) \quad (5.17)$$

$$K(\mathbf{v}) = \left(1 - v_2^2\right) K_{\max} \exp\left(-\frac{v_1^2}{2\sigma_k^2}\right)$$

$$a(\mathbf{v}, \mathbf{u}_j) = 1 + B \exp\left[-\frac{(v_1 - u_{j1} + \beta)^2}{2\sigma_a^2}\right] - \exp\left[-\frac{\beta^2}{2\sigma_a^2}\right]$$

$$B = 1 + B_{\max} (u_{j2} - v_2).$$

In the above functions for $G$, $K$, $\alpha$, and $B$, both components of an individual's strategy influence its fitness via effects on competition coefficients and carrying capacity. The two components do this in very different ways. The first component has a value for $v_1$ that maximizes carrying capacity (introduces stabilizing selection) and competition is minimized by having $v_1$ much larger or much smaller than one's competitors (introduces elements of disruptive selection). The second component, $v_2$, is under directional selection to be larger with respect to competition and under directional selection to be smaller with respect to carrying capacity. Consider first a single species with $\mathbf{u}_1(0) = (0, 0)$ and an initial population of $x_1 = 100$. We set the model's parameters equal to the following values

$$K_{\max} = 100$$
$$r = 0.25$$
$$\sigma_\alpha^2 = 4$$
$$\sigma_k^2 = 2$$
$$\beta = 2$$
$$B_{\max} = 1.$$

Under Darwinian dynamics (population dynamics plus first-order strategy dynamics) with

$$\mathcal{D}_1 = \begin{bmatrix} 0.5 & 0.1 \\ 0.1 & 0.5 \end{bmatrix}$$

the single species's strategy and population size, using the differential equation model, converge on the following values

$$x^* = 84.08$$
$$\mathbf{u}^* = [0.6065 \quad 0.2796].$$

The same solution is obtained independently of the number of starting species and initial conditions. The same solution is also obtained if the covariance terms in $\mathcal{D}_1$ are set equal to zero.

In the next two chapters we will examine the properties and significance of the solutions obtained for this example.

## 5.6 $G$-functions with resources

Using the same approach as with vector strategies we obtain the following

First-order strategy dynamics

$$\begin{aligned}
\text{Difference:} \quad & \Delta u_i = \frac{\mathcal{D}_i}{1+G(\mathbf{v},\mathbf{u},\mathbf{x})|_{\mathbf{v}=u_i}} \left.\frac{\partial G(\mathbf{v},\mathbf{u},\mathbf{x},\mathbf{y})}{\partial v}\right|_{\mathbf{v}=u_i} \\
\text{Exp. Difference:} \quad & \Delta u_i = \mathcal{D}_i \left.\frac{\partial G(\mathbf{v},\mathbf{u},\mathbf{x},\mathbf{y})}{\partial v}\right|_{\mathbf{v}=u_i} \\
\text{Differential:} \quad & \dot{u}_i = \mathcal{D}_i \left.\frac{\partial G(\mathbf{v},\mathbf{u},\mathbf{x},\mathbf{y})}{\partial v}\right|_{\mathbf{v}=u_i}
\end{aligned} \quad (5.18)$$

**Example 5.6.1 (Bergmann's rule)** *From Example 4.5.1 we have the model*

$$G(v,\mathbf{u},\mathbf{x}) = \frac{Av^\alpha y}{1 + AHv^{(\alpha-\beta)}y} - Cv^\gamma$$

$$\dot{y} = r(K - y) - \sum_{i=1}^{n_s} \frac{Au_i^\alpha y x_i}{1 + AHu_i^{(\alpha-\beta)}y}.$$

*Consider a specific form of this model by assigning the following parameter values*

$$\begin{aligned}
r &= 0.25 \\
K &= 100 \\
A &= 0.05 \\
C &= 1.5 \\
H &= 0.3 \\
\alpha &= 0.5 \\
\beta &= 1 \\
\gamma &= 0.75.
\end{aligned}$$

*Under the Darwinian dynamics (5.18) and starting from the initial conditions*

$$x(0) = 50, \; y(0) = 20, \; u(0) = 1$$

*with a variance $\sigma^2 = 0.05$, we obtain the equilibrium solution for the differential and exponential difference models*

$$x^* = 10.517, \; y^* = 54.000, \; u^* = 0.6561.$$

*The difference equation model is unstable using these parameters.*

Additional solutions to this problem are given in Chapter 6. In Chapter 7 we obtain an analytical result that satisfies Bergmann's rule.

## 5.7 Multiple $G$-functions

Using the same methods as in the vector strategy case we obtain the following

## First-order strategy dynamics

$$\text{Difference:} \quad \Delta \mathbf{u}_i = \frac{\mathcal{D}_i}{1+G(\mathbf{v},\mathbf{u},\mathbf{x})|_{\mathbf{v}=\mathbf{u}_i}} \left. \frac{\partial G_j(\mathbf{v},\mathbf{u},\mathbf{x})}{\partial v} \right|_{\mathbf{v}=\mathbf{u}_i}$$

$$\text{Exp. Difference:} \quad \Delta \mathbf{u}_i = \mathcal{D}_i \left. \frac{\partial G_j(\mathbf{v},\mathbf{u},\mathbf{x})}{\partial v} \right|_{\mathbf{v}=\mathbf{u}_i} \quad (5.19)$$

$$\text{Differential:} \quad \dot{\mathbf{u}}_i = \mathcal{D}_i \left. \frac{\partial G_j(\mathbf{v},\mathbf{u},\mathbf{x})}{\partial v} \right|_{\mathbf{v}=\mathbf{u}_i}$$

where $i$ and $j$ are determined according to the rising number defined in Section 4.6

$$i = r_{j-1} + 1, \ldots, r_j \quad \text{for } j = 1, \ldots, n_b.$$

**Example 5.7.1 (predator–prey coevolution)** *This model as presented in Subsection 4.6.1 has the two bauplans*

$$\text{Prey}: G_1(v, \mathbf{u}, \mathbf{x}) = \frac{r_1}{K(v)} \left[ K(v) - \sum_{j=1}^{n_{s_1}} x_j a(v, u_j) \right] - \sum_{j=n_{s_1}+1}^{n_s} x_j b(v, u_j)$$

$$\text{Predator}: G_2(v, \mathbf{u}, \mathbf{x}) = r_2 \left[ 1 - \frac{\sum_{j=n_{s_1}+1}^{n_s} x_j}{c \sum_{j=1}^{n_{s_1}} x_j b(v, u_j)} \right]$$

*where*

$$K(v) = K_{\max} \exp\left(-\frac{v^2}{\sigma_k^2}\right)$$

$$a(v, u_j) = \exp\left[-\frac{(v-u_j)^2}{\sigma_a^2}\right]$$

$$b(v, u_j) = b_{\max} \exp\left[-\frac{(v-u_j)^2}{\sigma_b^2}\right].$$

*Consider a specific form of this model with the following constants:*

$$r_1 = r_2 = c = 0.25$$
$$K_{\max} = 100$$
$$b_{\max} = 0.15$$
$$\sigma_k^2 = 2$$
$$\sigma_a^2 = 4$$
$$\sigma_b^2 = 10.$$

Under Darwinian dynamics (using (5.19) with $\sigma_1^2 = \sigma_2^2 = 0.5$ and the covariance terms set equal to zero), assuming one species of each type, and with the initial conditions

$$x_1(0) = x_2(0) = 50, \quad u_1(0) = 2, \quad u_2(0) = -2$$

we obtain the equilibrium solution

$$x_1^* = 30.77, \quad x_2^* = 1.154, \quad u_1^* = u_2^* = 0.$$

This solution is obtained using either the differential equation or the exponential difference dynamical models. However, the initial conditions used are not within the domain of attraction (see Subsection 2.5.2) to the equilibrium solution when using the difference equation system. The initial conditions

$$x_1(0) = 35, \quad x_2(0) = 10, \quad u_1(0) = 0.2, \quad u_2(0) = -0.2$$

are within the domain of attraction for the difference equation model and the same solution is obtained. By varying the parameter $\sigma_b^2$ many other solutions are possible. We will investigate some of these in the next two chapters.

The above example illustrates how the stability of an equilibrium point for a non-linear system is often only local and how the domain of attraction for the same $G$-function may change from one dynamical system to the next.

## 5.8 $G$-functions in terms of population frequency

Recall from Chapter 4 that exponential difference equations are not included in this grouping of $G$-functions. Using the same approach as with vector strategies we obtain the following

First-order strategy dynamics

$$\boxed{\begin{aligned} \text{Difference:} \quad & \Delta \mathbf{u}_i = \frac{\mathcal{D}_i}{1 + G(\mathbf{v},\mathbf{u},\mathbf{p},N)|_{\mathbf{v}=\mathbf{u}_i}} \left.\frac{\partial G(\mathbf{v},\mathbf{u},\mathbf{p},N)}{\partial \mathbf{v}}\right|_{\mathbf{v}=\mathbf{u}_i} \\ \text{Differential:} \quad & \dot{\mathbf{u}}_i = \mathcal{D}_i \left.\frac{\partial G(\mathbf{v},\mathbf{u},\mathbf{p},N)}{\partial \mathbf{v}}\right|_{\mathbf{v}=\mathbf{u}_i} \end{aligned}} \quad (5.20)$$

**Example 5.8.1 (L–V competition game in terms of frequency)** *This game reformulated in terms of frequency is defined by*

$$G(v, \mathbf{u}, \mathbf{p}, N) = \frac{r}{K(v)} \left( K(v) - N \sum_{j=1}^{n_s} a(v, u_j) p_j \right),$$

with

$$K(v) = K_m \exp\left\{-\frac{v^2}{2\sigma_k^2}\right\}$$

$$\alpha(v, u_i) = 1 + \exp\left\{-\frac{(v - u_i + \beta)^2}{2\sigma_\alpha^2}\right\} - \exp\left\{-\frac{\beta^2}{2\sigma_\alpha^2}\right\}.$$

All of the results obtained from the density formulation of this game are again obtained. For example, using $K_m = 100$, $r = 0.25$, and $\sigma_k^2 = \sigma_\alpha^2 = \beta^2 = 4$, we obtain from the $p$ and $N$ equations in Section 4.7 along with (5.20) the equilibrium solution

$$u_1^* = 1.213, \quad p_1^* = 1, \quad N^* = 83.20.$$

This is the same result as obtained in the scalar $G$-function example and is applicable to both differential and difference equation dynamics.

## 5.9 Multistage $G$-functions

Recall from Section 4.8 that a scalar $G$-function for the multistage case is defined using the critical value concept

$$G(v, \mathbf{u}, \mathbf{x})|_{v=u_i} = \operatorname{crit} \mathbf{H}_i(\mathbf{u}, \mathbf{x}).$$

Because critical value is a function of $v$, the same procedure is used as for the $G$-functions with scalar strategies case. Only here we obtain first-order strategy dynamics in terms of the **gradient of critical value**

First-order strategy dynamics

| | |
|---|---|
| Difference: | $\Delta u_i = \frac{\sigma_i^2}{1+G\vert_{u_i}} \left.\frac{\partial G(v,\mathbf{u},\mathbf{x})}{\partial v}\right\vert_{v=u_i}$ |
| Differential: | $\dot{u}_i = \sigma_i^2 \left.\frac{\partial G(v,\mathbf{u},\mathbf{x})}{\partial v}\right\vert_{v=u_i}$ |

**Example 5.9.1 (life cycle)** Example 4.8.1 has the following **G**-matrix

$$\mathbf{G}(v, \mathbf{u}, \mathbf{x}) = \begin{bmatrix} f(v) - \sum_{j=1}^{n_x} u_j x_{j1} & v \\ v & -\left(\sum_{j=1}^{n_x} x_{j1}\right) \end{bmatrix}.$$

As in Vincent and Brown (2001) let the relationship between reproduction rate and the individual's strategy conform to a downward parabola that reaches a *peak value at $v = 2$*

$$f(v) = -1 + 4v - v^2.$$

*The strategy of $v = 2$ would maximize fitness if the model were independent of $x$ and $u$. However, in this example, the **G**-matrix is both density and frequency dependent. By using Darwinian dynamics for the differential equation case, we obtain the coalition of one equilibrium solution*

$$x_{11}^* = 1.781, \quad x_{12}^* = 4.562, \quad u_1^* = 4.562.$$

## 5.10 Non-equilibrium Darwinian dynamics

For any of the above categories of $G$-functions there need not be stable equilibrium solutions to the population dynamics equations and/or strategy dynamics equations. In the following example, an equilibrium solution is obtained for the strategy in spite of the fact that the population can be periodic or even chaotic.

**Example 5.10.1 (non-equilibrium L–V game)** *The $G$-function for the L–V competition game is defined by*

$$G(v, \mathbf{u}, \mathbf{x}) = \frac{r}{K(v)} \left( K(v) - \sum_{j=1}^{n_s} a(v, u_j) x_j \right)$$

where

$$K(v) = K_m \exp\left(-\frac{v^2}{2\sigma_k^2}\right)$$

$$a(v, u_i) = 1 + \exp\left\{-\frac{(v - u_i + \beta)^2}{2\sigma_a^2}\right\} - \exp\left\{-\frac{\beta^2}{2\sigma_a^2}\right\}$$

*and $r > 2$. Consider the same set of parameters used with the frequency $G$-function example except for increasing $r$. We use the parameters $K_m = 100$, $r = 2.5$, and $\sigma_k^2 = \sigma_\alpha^2 = \beta^2 = 2$. Under difference equation strategy dynamics, we once again obtain $u_1^* = 1.213$. However, as illustrated in Figure 5.8, the population density follows a 4 cycle. Increasing $r$ to $r = 2.8$ results in a chaotic solution for the population density as illustrated in Figure 5.9. Note, however, that difference equation strategy dynamics yields the same equilibrium solution for $u_1^*$.*

This example illustrates how non-equilibrium population dynamics does not necessarily imply non-equilibrium strategy dynamics. See Section 6.8 for further discussion.

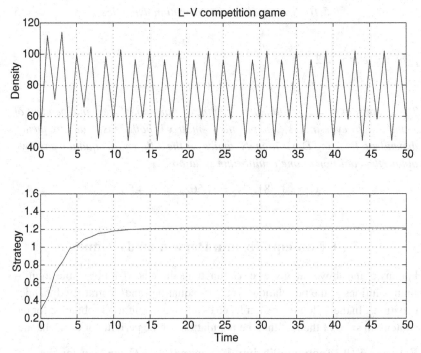

**Figure 5.8** With $r = 2.5$, strategy dynamics results in an equilibrium solution for $u_1$ and a 4-cycle solution for density.

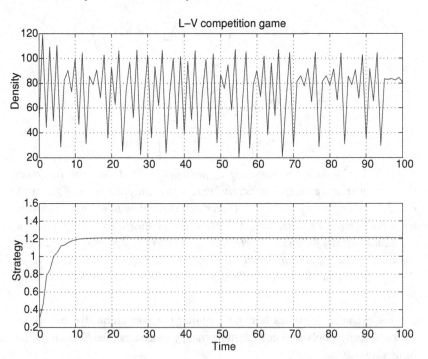

**Figure 5.9** Increasing to $r = 2.8$ results in a chaotic solution for the population density.

## 5.11 Stability conditions for Darwinian dynamics

Stability conditions for Darwinian dynamics are determined from the population dynamics and the first-order strategy dynamics. We develop these conditions using $G$-functions under differential equation dynamics with a single species, $x_1$, and a scalar strategy, $u_1$. As in Cohen et al. (1999) assume that there are no constraints on the strategy and that a unique equilibrium solution exists for (5.1) and (5.9). A non-trivial equilibrium ($x_i \neq 0$) must satisfy

$$G(v, u_1, x_1)|_{v=u_1} = 0 \tag{5.21}$$

$$\left.\frac{\partial G(v, u_1, x_1)}{\partial v}\right|_{v=u_1} = 0 \tag{5.22}$$

which yield the equilibrium solutions $x_1 = x_1^*$, $v = u_1 = u_1^*$. In order to examine the stability of these solutions in the neighborhood of $x_1^*$ and $u_1^*$ we obtain the **perturbation equations** associated with the total system dynamics by examining a first-order Taylor series expansion of the differential equations in terms of the perturbation variables defined by

$$\delta x_1 = x_1 - x_1^*$$
$$\delta u_1 = u_1 - u_1^*$$
$$\delta v = v - u_1^*$$

noting that, when $v$ is replaced by $u_1$, $\delta v = \delta u_1$. We obtain

$$\delta \dot{x}_1 = \left[G(v, u_1, x_1) + x_1^* \frac{\partial G(v, u_1, x_1)}{\partial x_1}\right]_* \delta x_1$$
$$+ x_1^* \left[\frac{\partial G(v, u_1, x_1)}{\partial v} + \frac{\partial G(v, u_1, x_1)}{\partial u_1}\right]_* \delta u_1$$
$$\delta \dot{u}_1 = \sigma_1^2 \left[\frac{\partial^2 G(v, u_1, x_1)}{\partial x_1 \partial v} \delta x_1 + \left(\frac{\partial^2 G(v, u_1, x_1)}{\partial v^2} + \frac{\partial^2 G(v, u_1, x_1)}{\partial u_1 \partial v}\right) \delta u_1\right]_*$$

where $[\ ]_*$ denotes that all of the arguments in the expression are replaced by $x_1 = x_1^*$, $v = u_1 = u_1^*$. Using the equilibrium conditions (5.21) and (5.22) the first equation becomes

$$\delta \dot{x}_1 = x_1^* \left[\frac{\partial G(v, u_1, x_1)}{\partial x_1} \delta x_1 + \frac{\partial G(v, u_1, x_1)}{\partial u_1} \delta u_1\right]_*.$$

Since $x_1^* > 0$ and $\sigma_1^2 > 0$, the stability of the system is determined by the eigenvalues of the matrix

$$\begin{bmatrix} \dfrac{\partial G(v, u_1, x_1)}{\partial x_1} & \dfrac{\partial G(v, u_1, x_1)}{\partial u_1} \\ \dfrac{\partial^2 G(v, u_1, x_1)}{\partial x_1 \partial v} & \dfrac{\partial^2 G(v, u_1, x_1)}{\partial v^2} + \dfrac{\partial^2 G(v, u_1, x_1)}{\partial u_1 \partial v} \end{bmatrix}_*.$$

The stability analysis can be simplified if we assume that there is a large difference in time scale between the ecological dynamics and the strategy dynamics. This implies that the local ecological dynamics for population size (5.1) occurs very fast relative to the evolutionary dynamics that determine strategy values (5.9). Thus from (5.1) we solve the algebraic equation

$$G(u_1, u_1, x_1) = 0$$

for $x_1$ to give a relationship of the form

$$x_1 = f(u_1)$$

and substitute into (5.9) to determine the strategy dynamics as a function of $v$ and $u_1$ only. In this case the perturbation equation for $u_1$ becomes

$$\delta \dot{u}_1 = \sigma_1^2 \left[ \frac{\partial^2 G(v, u_1, x_1)}{\partial v^2} + \frac{\partial^2 G(v, u_1, x_1)}{\partial u_1 \partial v} + \frac{\partial^2 G(v, u_1, x_1)}{\partial x_1 \partial v} \frac{\partial f(u_1)}{\partial u_1} \right]_* \delta u_1.$$

Thus, for stable strategy dynamics it is necessary that

$$\left[ \frac{\partial^2 G(v, u_1, x_1)}{\partial v^2} + \frac{\partial^2 G(v, u_1, x_1)}{\partial u_1 \partial v} + \frac{\partial^2 G(v, u_1, x_1)}{\partial x_1 \partial v} \frac{\partial f(u_1)}{\partial u_1} \right]_* < 0. \quad (5.23)$$

Equation (5.23) is a necessary condition for the strategy dynamics to be asymptotically stable under the assumption that the population dynamics is stable and that $T_{ec} \gg T_{ev}$. This result is closely related to the concept of **convergent stability** (Eshel and Motro, 1981) that we discuss in more general terms in Chapter 6. Briefly, convergent stability refers to the ability of an evolving population to return to an equilibrium point $x_1^*, u_1^*$ if it is displaced from it. Equation (5.23) is widely used (e.g., Abrams et al., 1993b; Metz et al., 1996; and others). Denote the first term in (5.23) by $A$, the second by $B$, and the third by $C$. Abrams et al. (1993b) ignored the term $C$ and consider $A + B < 0$ as the condition for convergent stability. Metz et al. (1996) subsume the term $C$ into term $B$. However, keeping term $C$ separate from $B$ highlights the interaction between strategy dynamics and population dynamics in influencing convergent stability.

The term $C = 0$ occurs in two special circumstances. In the first of these, population size may be independent of strategies and strategy dynamics: $\partial f / \partial u_1 = 0$. This is unlikely because strategies increase in frequency precisely because they have higher fitness in the current population. Thus, changes in fitness and strategy values will generally also change equilibrium population sizes. A second way for $C = 0$ occurs when population size is made implicit in the $G$-function by substituting $f(u_1)$ for $x_1^*$ directly into $G$ prior to evaluating the necessary conditions. In this way, the term $B$ will implicitly subsume $C$ and $A + B < 0$ becomes the correct necessary condition for convergent stability (Metz et al., 1996).

While the necessary condition for convergent stability contains $A$, which also figures in the necessary conditions for invasion resistance, the conditions for convergent stability and invasion resistance are somewhat independent of each other: one may be satisfied while the other is not. This leads to four possibilities when evaluating strategies that are stability candidates in the sense that $\left.\frac{\partial G(v,u_1,x_1)}{\partial v}\right|_{v=u_1} = 0$:

(i) An evolutionarily stable maximum results when $u_1$ is both resistant to invasion ($A < 0$) and convergent stable ($A + B + C < 0$).
(ii) An evolutionarily stable minimum (Abrams et al., 1993b) occurs when $u_1$ is not resistant to invasion ($A > 0$) but is convergent stable ($B + C < 0$ and $|B + C| > A$).
(iii) An evolutionarily unstable maximum (Eshel, 1983; Abrams et al., 1993b; Taylor, 1997) results when $u_1$ is resistant to invasion ($A < 0$) but is not convergent stable ($B + C > 0$ and $B + C > |A|$).
(iv) An evolutionarily unstable minimum results when $u_1$ is not resistant to invasion ($A > 0$) and not convergent stable ($A + B + C > 0$).

We have examined the stability of Darwinian dynamics for the simple cases that are common in the literature. The convergent stable conditions for a coalition of two or more are horrendous (Cohen et al., 1999) and it is not an approach that is suitable for the general case. In fact, it is not necessary, as we will see in Chapter 6.

## 5.12 Variance dynamics

It is of interest to determine how the variance in strategy value changes with time as Darwinian dynamics drives the system toward equilibrium. What we show here is that, under the assumptions used to obtain the first-order scalar strategy dynamics, the variance does not change with time. However, one should not attempt to generalize this result; as we will show in Section 8.3, through simulation, if we relax the assumption of a symmetric small variance, then the variance changes with time. We also observe that the variance becomes small as the system approaches equilibrium. In the absence of some process that will generate new variability, natural selection will eventually eliminate the variation about a mean strategy. In this section we only deal with the differential equation model, but the same result is obtained for the difference equation models as well (Vincent *et al.*, 1993).

From (5.8) the variance is given by

$$\sigma_i^2 = \sum_{j=0}^{n_p} \delta u_{ij} q_{ij} \delta u_{ij} \tag{5.24}$$

thus

$$\frac{d\sigma_i^2}{dt} = \sum_{j=0}^{n_p} \delta u_{ij} \dot{q}_{ij} \delta u_{ij},$$

where

$$\dot{q}_{ij} = \frac{\dot{x}_{ij}}{\dot{x}_i} = q_{ij} \frac{G(v_{ij}, \mathbf{u}, \mathbf{x})|_{u_{ij}}}{G(v, \mathbf{u}, \mathbf{x})|_{u_i}}. \tag{5.25}$$

Therefore

$$\frac{d\sigma_i^2}{dt} = \frac{\sum_{j=0}^{n_p} \delta u_{ij} q_{ij} \left[ G(v_{ij}, \mathbf{u}, \mathbf{x})|_{u_{ij}} \right] \delta u_{ij}}{G(v, \mathbf{u}, \mathbf{x})|_{u_i}}. \tag{5.26}$$

Equation (5.26) can be used to compute the dynamics of the phenotypic variance for the general case. However, because of the summation, it is not convenient to use. If we make the same small $\delta u_{ij}$ assumption as we did with first-order strategy dynamics, we can use the Taylor series expansion about $v_{ij} = u_i$ to approximate (5.25); namely

$$\dot{q}_{ij} = q_{ij} \frac{G(v, \mathbf{u}, \mathbf{x})|_{u_i} + \frac{\partial G(v, \mathbf{u}, \mathbf{x})}{\partial v}\Big|_{u_i} \delta u_{ij}}{G(v, \mathbf{u}, \mathbf{x})|_{u_i}},$$

using this to substitute into (5.24) yields

$$\frac{d\sigma_i^2}{dt} = \sum_{j=0}^{n_p} \delta u_{ij} q_{ij} \frac{\frac{\partial G(v, \mathbf{u}, \mathbf{x})}{\partial v}\Big|_{u_i}}{G(v, \mathbf{u}, \mathbf{x})|_{u_i}} \left(\delta u_{ij}\right)^2.$$

Assuming that the distribution of strategies about $u_i$ is symmetric (i.e., for every $j$ there is a $k$ such that $\delta u_{ij} q_{ij} = -\delta u_{ik} q_{ik}$) it follows that

$$\frac{d\sigma_i^2}{dt} = 0.$$

# 6
# Evolutionarily stable strategies

Natural selection produces strategies that are continually better than those discarded along the way to some evolutionary equilibrium. Intuitively this implies that eventually, natural selection should produce the "best" strategy for a given situation. The flowering time of a plant, the leg length of a coyote, or the filter feeding system of a clam should produce higher fitness than alternative strategies that are evolutionarily feasible (within the genetic, developmental, and physical constraints in the bauplan). In graphical form these products of natural selection should reside on peaks of the adaptive landscape. Yet we have seen in the previous chapter how, under Darwinian dynamics, natural selection may produce strategies that evolve to minimum points, maximum points, and saddlepoints on the adaptive landscape.

An evolutionary ecologist studying the traits of a species whose strategy has evolved to a convergent stable minimum on the adaptive landscape may, on reflection, be surprised. At this minimum, any individual with a strategy that deviates slightly from that produced by Darwinian dynamics has a higher, not lower, fitness than the resident strategy. While an evolutionarily stable minimum can result from Darwinian dynamics, this strategy is not the "correct" solution to the evolutionary game. In this chapter, we expand upon the original word definition of an **evolutionarily stable strategy** (ESS) as given by Maynard Smith: "An ESS is a strategy such that, if all members of a population adopt it, then no mutant strategy could invade the population under the influence of natural selection" (Maynard Smith, 1974). We update this definition by putting it into a mathematical context applicable to the modeling approach developed in the previous chapters. In Chapter 7, we show that ESS solutions occur only at maximum points on the adaptive landscape.

Because Darwinian dynamics is a combination of two dynamical processes (population dynamics and strategy dynamics) natural selection involves two stability processes. As discussed in Chapters 2 and 5, stability of a dynamical system is always in reference to an operating condition. Generally, the operating condition refers to an equilibrium solution to the dynamical equations.[1] These are points in **x**, **u** (and possibly **y**) space at which no change in state takes place. For now, assume that **y** is not part of the model and suppose that we have found an equilibrium point denoted by $\mathbf{x}^*$ and $\mathbf{u}^*$. By definition, at this equilibrium point the Darwinian dynamics produces zero change in **x** and **u**, requiring

$$G(v, \mathbf{u}^*, \mathbf{x}^*)\big|_{v=u^*} = 0$$

$$\frac{\partial G(v, \mathbf{u}^*, \mathbf{x}^*)}{\partial v}\bigg|_{v=u^*} = 0.$$

Perturbations in **x** and **u** from their equilibrium values are made by setting

$$\mathbf{x} = \mathbf{x}^* + \delta\mathbf{x}$$
$$\mathbf{u} = \mathbf{u}^* + \delta\mathbf{u}$$

where $\delta\mathbf{x}$ represents the amount **x** has been moved away from $\mathbf{x}^*$ ($\delta\mathbf{x}$ must be chosen so that the population density remains positive) and $\delta\mathbf{u}$ represents the amount **u** has been moved away from $\mathbf{u}^*$ ($\delta\mathbf{u}$ must be chosen so that $\mathbf{u} \in \mathcal{U}$).

Consider first the situation in which no strategy changes are allowed from $u = u^*$. If the population dynamics equation eventually returns the system to $\mathbf{x}^*$ as $t \to \infty$, we say that $\mathbf{x}^*$ is **ecologically stable**. This definition is useful for ecological studies when the evolutionary time scale is very much greater than the ecological time scale. Likewise we could define a type of stability based on keeping **x** fixed at $\mathbf{x}^*$ and examining the stability of $\mathbf{u}^*$ under strategy dynamics. However, such a definition is not generally useful for the study of evolution under natural selection because the time scales are rarely reversed.

A more appropriate definition is to allow both **x** and **u** to vary. The equilibrium points $\mathbf{x}^*$, $\mathbf{u}^*$ are **convergent stable** (Eshel, 1996) if, for any non-zero $\delta\mathbf{x}$ with $\mathbf{x} > \mathbf{0}$ and any $\delta\mathbf{u} \neq \mathbf{0}$ such that $\mathbf{u} \in \mathcal{U}$, the Darwinian dynamics (i.e., population dynamics plus strategy dynamics) eventually returns the system to $\mathbf{x}^*$ and $\mathbf{u}^*$ as $t \to \infty$. This definition might seem sufficient for defining an evolutionarily stable strategy; however, this is not the case. The definition of convergent stability misses the essential ingredient contained in Maynard Smith's definition, the presence of a mutant strategy. Convergent stability examines the stability of the system only in terms of the extant populations and strategies. It need not produce evolutionary stability in the sense of Maynard Smith's ESS concept. In fact it is possible for a minimum point on the adaptive landscape

---

[1] Under non-equilibrium Darwinian dynamics an ecological cycle (see Definition 6.8.1) is used in place of an equilibrium point.

to be convergent stable. Such a point is not an ESS. This is not to say that a convergent stable minimum point could not occur in nature as the outcome of natural selection. For example, in Subsection 10.2.2 a cell model is proposed that has a convergent stable minimum on the adaptive landscape. In order to sustain this equilibrium solution, the genetic structure of the cells must be such that perfect clonal reproduction of the cells is maintained.

A goal of this chapter is to develop a precise mathematical ESS definition useful for the study of evolutionary stability. We arrive at a definition which extends Maynard Smith's original ESS concept. The reason for the extension is threefold. First a mathematical definition is needed if we are going to use mathematics to solve for an ESS. Second, a definition is needed that addresses issues of convergence stability and resistance to invasion by mutant strategies. Third, a definition is needed that applies to the broad classes of evolutionary games and $G$-functions.

## 6.1 Evolution of evolutionary stability

In his pioneering definition of an ESS, Maynard Smith showed how to model and characterize the outcomes of frequency-dependent selection. Natural selection, without frequency dependence, results in an engineering perspective of adaptations where $FF\&F$ is optimized (Vincent and Vincent, 2000). This viewpoint holds that adaptations must be optimal in the sense of maximizing fitness. This optimization perspective has appeal. The unusually long loop of Henle within the kidney of the desert-dwelling kangaroo rats permits it to produce extremely concentrated urine. They can survive solely on the water contained in their food. Tree trunks allow plants access to light. Zebras run fast to escape their cursorial predators. At first glance, common sense seems sufficient to model natural selection in this way.

This common sense approach to explaining adaptations crumbles under close inspection. For example, the adaptations for speed in the zebra challenge our sense of sensible adaptations. Given that the zebra's chief predators attack following a short (in the case of ambushing lions) or long (in the case of hyenas and hunting dogs) chase, it seems sensible to be able to accelerate and run fast, and to have sufficient endurance. What is fast enough and at what cost to the zebra? What if zebra predators did not run at all but used other tactics for ensnaring their prey (witness the pitfall traps of humans and ant-lions, or the webs of spiders)? What if zebras did not flee but fought back, hunkered down, or relied on superior camouflage? In short, what is ideally fast for a zebra depends on what strategies the predators exhibit as well as the strategies of other zebras (it will often suffice to be a wee bit faster than your neighbor). This means that evolution will generally be frequency dependent, which, in turn, means that we need frequency-dependent models to understand them. An optimal *FF&F* approach is not going to be valid for the zebra predator–prey system or most other biological systems.

Prior to the formal definition given by Maynard Smith and Price (1973), several links between natural selection and adaptation had been forged. From the great population geneticists of the 1920s and 1930s, it was known that density-dependent selection, in which an individual's fitness is influenced by its strategy and the population's size, resulted in strategies that maximize the population's growth rate. If an equilibrium exists, density-dependent selection results in strategies that maximize population size.

Levins's (1962) fitness set approach (see also Levins, 1968) provided an early departure from the traditional models of population genetics, and took a step on a path leading to the ESS concept. Levins examined evolution within an ecological context (usually one of habitat heterogeneity, or resource patchiness). Levins imagined a set of evolutionarily feasible strategies. Rather than consider them explicitly, he assumed that these strategies influenced an organism's performance in each of two habitats. A **fitness set** defined the mapping of all possible strategy values onto a fitness space composed of fitness in habitat $A$ versus fitness in habitat $B$ as illustrated in Figure 6.1. The only fitness values of interest in this set are those that lie on the upper right boundary known as the **active edge** shown by the solid line in the figure.[2] The active edge is that subset of fitness pairs that represent trade-offs in performance between habitats $A$ and $B$. To see this, consider any other point in the fitness set not on the active edge. For any such point, the fitness in both habitat $A$ and habitat $B$ can be improved upon by choosing a different strategy corresponding to some point on the active edge. Natural selection should always favor strategies that simultaneously increase performance in both habitats. Following this logic, we see a critical facet of adaptation. Natural selection should find just the right compromise between competing demands or opportunities, that is natural selection should choose among points on the active edge.

Given an environment with habitats $A$ and $B$, we may determine the fitness consequence to an individual using particular strategies in both habitats $A$ and $B$. Suppose that an individual is in habitat $A$ with probability $p$ and in habitat $B$ with probability $(1 - p)$. Its overall fitness, $F$, is given by

$$F = pF_a + (1 - p)F_b$$

---

[2] In game theory, the boundary shown by the solid line is usually called the **Pareto-optimal set** (Vincent and Grantham, 1981).

## 6.1 Evolution of evolutionary stability

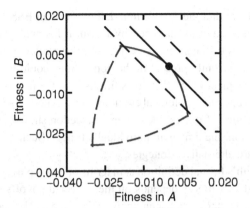

**Figure 6.1** The fitness set, depicted here by the interior of the top-shaped region, represents the fitness in habitats $A$ and $B$ as a result of using every possible stratgy.

where $F_a$ is its fitness in $A$ and $F_b$ is its fitness in $B$. For a given $p$, lines of equal fitness can be constructed in the fitness space as illustrated by the straight dashed lines in Figure 6.1. Lines of higher fitness require successively higher performances in $A$ and $B$. When these fitness isoclines have points that lie inside the fitness set, there exists more than one strategy that could evolve within the population to attain that level of fitness. Levins reasoned that natural selection should evolve to produce a strategy that lies on the highest possible fitness isocline. This strategy must lie on the active edge as indicated by the large dot[3] in Figure 6.1. This solution occurs where the fitness isocline is just tangent to the fitness set. Because the slope of an isocline depends on $p$, it follows that any point on the active edge is a possible solution.[4]

The solution that Levins proposes with fitness sets anticipates Maynard Smith's ESS concept. Strategies that lie on a fitness isocline that slices through the fitness set are not evolutionarily stable because they can be invaded by strategies that produce higher fitness in both habitats. Points lying on the active edge have two properties that make them attractive solutions for natural selection. They are feasible and they cannot be replaced by alternative strategies yielding higher fitness in both habitats. Solutions identified through fitness set analysis are adaptive in the sense of maximizing an individual's fitness given ecological circumstances that involve the strategies of others. That is, when the model of natural selection is frequency dependent.

---

[3] There can be more than one solution if the active edge is non-convex.
[4] Levins's approach has a parallel in microeconomics. The optimal consumption of two goods by a consumer occurs where a utility isopleth – analogous to a line of equal fitness – is just tangent to the person's budget constraint – analogous to the fitness set.

The ESS concept and the first formal definition of an ESS apply a similar logic to natural selection to that used by Levin in his analysis. At the heart of Maynard Smith's ESS concept is the premise that strategies offering higher fitness than those currently present in the population should be able to invade and increase in frequency. Like Levins's fitness sets, the fitness advantages of strategies depend upon the ecological circumstances. When these circumstances involve the strategies of others, then natural selection should favor strategies that, when common, maximize an individual's fitness. Such strategies cannot be invaded by rare alternative strategies.

Maynard Smith's formal definition is in terms of a two-player matrix game[5] where an individual's fitness depends on the expected payoffs obtained from using a particular strategy against an opponent's strategy. He draws from classical game theory the notion of matrix game with two discrete strategies: $A, B$. Under his formal definition, $A$ is an ESS if either

$$E(A, A) > E(B, A)$$

or

$$E(A, A) = E(B, A) \text{ and } E(A, B) > E(B, B)$$

where $E(A, B)$ is the expected payoff to an individual that plays strategy $A$ against an opponent using strategy $B$. The first argument of $E$ is the individual's strategy and the second argument is the opponent's strategy.

The next 10 years following the paper by Maynard Smith and Price (1973) saw great progress in applying this definition to many situations, some of which lay outside the scope of the original formulation. Application of the formal definition revealed difficulties (Hines, 1987). The formal definition becomes unsatisfactory with the introduction of mixed strategies, strategy dynamics, and fitness formulations with explicit population sizes and population dynamics.

The idea of mixed strategies will be discussed at length in Chapter 9, but the issue of mixed strategies arises when the following condition occurs

$$E(A, B) > E(B, B)$$

and

$$E(B, A) > E(A, A).$$

In this case, strategy $A$ can invade a population of $B$, and strategy $B$ can invade a population of strategy $A$. In this case, natural selection favors the coexistence of the two strategies within the population. Coexistence, though, can come

---

[5] Matrix games are discussed in detail in Chapter 9.

## 6.1 Evolution of evolutionary stability

about in two ways. The population may represent a mixture of two distinct sets of individuals, some that are $A$ and some that are $B$. Or there may be mixed strategies in which a strategy $u$ represents an individual's probability of exhibiting strategy $A$ or exhibiting strategy $B$. Superficially, the two interpretations of a mixture of discrete strategies and a mixed strategy seem the same. However, when formalized they represent very different games. In the first, the number of strategies remains small and discrete. Evolution simply alters the frequencies of the two types of individuals within the population; whereas a mixed strategy involves a strategy set that is continuous. Strategy dynamics can change the value of $u$ used by the population.

Maynard Smith's definition applied to mixed strategies results in an ESS where each pure strategy $A$ and $B$ has equal fitness in the population. At the ESS, all strategies $0 \leq u \leq 1$ have equal fitness. In other words, the adaptive landscape is completely flat (Vincent and Cressman, 2000). The following conclusions have emerged from exhaustive analyses of mixed-strategy matrix games (Taylor and Jonker, 1978; Zeeman, 1981). Maynard Smith's formal definition identifies likely outcomes of natural selection so long as rare alternative strategies occur in the population only one at a time. If more than one rare alternative strategy can be present in the population simultaneously then natural selection will not return a population to the mixed strategy ESS. Thus a mixed-strategy matrix game ESS, while resistant to invasion by a single mutant strategy, is not resistant to invasion by multiple mutant strategies. A stronger ESS definition is needed.

Eshel (1983) exposed other weaknesses in the above Maynard Smith definition. He extended the class of games from matrix games with a finite number of discrete strategies to continuous strategies where an individual's trait represents a quantitative character such as physical size. Eshel imagined that an individual's fitness was influenced by its own strategy and by the mean strategy value of the population. He used a technique borrowed from classical game theory called **rational reaction sets** (Simaan and Cruz, 1973) to find a strategy value that has the properties of an ESS. The rational reaction set graphs an individual's best strategy response (on the $y$-axis) to the mean strategy of the population (on the $x$-axis) as illustrated in Figure 6.2. Consider those points corresponding to the intersection of the rational reaction set with the $y = x$ line. At these points, an individual's fitness maximizing strategy is the same as the population's mean strategy. Hence, when everyone uses a strategy corresponding to one of these points, no individual can increase fitness by unilaterally altering strategy (similar to the no-regret strategy of Nash). This implies that such a strategy cannot be invaded by any number of rare alternative strategies, an ESS property. But will such strategies evolve from natural selection? Will a population with a

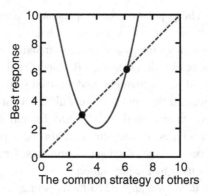

**Figure 6.2** The rational reaction set is given by the solid line.

strategy near but not at the "ESS" converge by natural selection on the "ESS"? Not necessarily, as Eshel showed.

Eshel's work and much work since have shown that evolutionary convergence depends upon the slope of the rational reaction set at the intersection point. Suppose that the rational reaction set intersects the 45-degree line from above (slope < 1). This means that when a population has a strategy just below the intersection point an individual will do better with a strategy greater than the population's mean. When the population's mean strategy is greater than the intersection value natural selection will favor individuals with a strategy less than the population's. In this way, natural selection favors strategies that converge on the ESS. But we lose convergence when the rational reaction set intersects the $y = x$ line from below (slope > 1). If the population's mean strategy is above the intersection, then natural selection favors even greater values for the individual. If the population's mean is less than the intersection value then natural selection favors even smaller strategy values. Instead of convergence to the ESS candidate, natural selection actually favors divergence from this strategy.

Eshel's work and the work of others have shown that the ESS concept has two somewhat distinct properties. The first is **resistance to invasion**, where natural selection favors strategies that cannot be invaded by rare alternative strategies. The second is **convergence stability** where natural selection favors strategies that will maintain an equilibrium subject to perturbations in **x** and **u**. Authors have independently derived these properties of evolutionary stability and have used their own terms. Eshel refers to resistance to invasion as an evolutionarily unbeatable strategy, and has referred to convergence stability as continuous stability. Taylor (1997) uses $\delta$-stability and $m$-stability and Apaloo (1997) uses the concept of neighborhood invader strategy to deal with these same properties.

## 6.1 Evolution of evolutionary stability

Maynard Smith's formal ESS definition ensures only resistance to invasion. Strategies that satisfy his definition may not be convergent stable. One could attempt to define evolutionary stability using only the property of convergence stability. That is, define an evolutionary dynamic where strategies close to the population mean either succeed or fail depending upon whether they have higher fitness than the population's mean. However, convergence stability (as happens on a two-dimensional rational reaction set when it intersects the 45-degree line with a slope less than 1) does not ensure resistance to invasion. Abrams *et al.* (1993*b*) showed that requiring only convergence stability can result in strategies that actually minimize fitness. They referred to these as **evolutionarily stable minima**. Frequency dependence results in a flexible adaptive landscape so that it is possible for a strategy to be constantly evolving up a gradient only to arrive at a convergent stable point that is a fitness minimum. Even though strategies on either side of a convergent stable minimum point offer higher fitness, as soon as a population's strategy shifts away from the minimum, the fitness gradient changes and favors strategies closer to the minimum (Brown and Pavlovic, 1992).

It remains to be seen in nature whether natural selection can actually leave the strategies of populations at a fitness minimum (Abrams, 2001a; Cohen *et al.*, 2001) without protection from invasion by rare alternative strategies. If a strategy is introduced into the population on either side of the minimum, it can invade and increase in frequency. Such minimum points invite speciation, where the single strategy splits into two distinct strategies. This diversification of strategies from evolutionarily stable minima has been viewed either as **competitive speciation** (Cohen *et al.*, 1999) or as **evolutionary branching** (Metz *et al.*, 1996; Geritz *et al.*, 1997, 1998). We will explore this possible mechanism of speciation in Chapter 8.

Roughgarden (1983) experimented with a modeling approach to multi-species evolutionary games. He formulated an individual's fitness as influenced by the strategy of its own species and the strategies present in other species. No special consideration was given to the individual's own strategy independent of its population size. The absence of the individual's own strategy was justified on the grounds that the model may exhibit frequency dependence among species but not within a species. A **coevolutionary stable strategy** was defined as the strategies for each species that maximized the fitness of a particular species given the strategies of the other species (Rummel and Roughgarden, 1985). The problem with this approach is that frequency dependence among species sharing both the same strategy sets and fitness consequences of using those strategies requires frequency dependence within species. The coevolutionary stable strategies identified by Roughgarden's theory need not be resistant to invasion or convergent stable (Brown and Vincent, 1987c; Taper and Case, 1992; Abrams *et al.*, 1993a).

The ESS concept requires a game theory formulation that explicitly considers frequency-dependent effects within and among evolving and/or coevolving species (Brown and Vincent, 1987a). Other similar fitness formulations for evolutionary games achieve a distinction between a focal individual's strategy and the strategy of its population by specifying a variable for a rare mutant strategy within the existing community of strategies (Abrams *et al.*, 1993a; Dieckmann and Law, 1996) and they consider multiple species by establishing a new fitness function each time a new distinct strategy or species is added to the community (Geritz *et al.*, 1997).

The $G$-function approach to evolutionary games makes transparent the sources of frequency-dependent selection. Frequency dependence enters through the influence of **u** on $G$. When there are several extant species having the same $G$-function, frequency dependence occurs simultaneously within and among species. When more than one $G$-function is required, frequency dependence can occur within and/or between $G$-functions. Interspecific frequency dependence can occur in the absence of intraspecific frequency dependence.

The above discussion points out the need to generalize Maynard Smith's ESS concept so that it includes all aspects of frequency-dependent selection as well as convergence stability (lacking in most prior ESS definitions). Natural selection favors strategies that are both resistant to invasion and convergent stable. We feel that both of these properties should be an integral part of the ESS definition. Defining the ESS with these properties provides a direct way for evaluating likely outcomes of natural selection, and it avoids the proliferation of terms and definitions that result when one does not work from Maynard Smith's original ESS definition.

Using the same categories of $G$-functions introduced in Chapter 4, we start by defining an ESS in terms of scalar populations $x_i$ and scalar strategies $u_i$. We then show that this ESS definition remains unchanged for the remaining categories provided that we correctly interpret **x**, **u**, $n_s$, $n_{s^*}$ used in the various $G$-functions. We also need to introduce **y** when resource dynamics is a part of the model.

## 6.2 $G$-functions with scalar strategies

For this category of $G$-functions we have an ecological community composed of $n_s$ species. All species have a similar life history. Species are identified by their respective population size, $x_i$, and scalar strategy, $u_i$.

## 6.2.1 Population dynamics

We have

$$\mathbf{x} = \begin{bmatrix} x_1 & \cdots & x_{n_s} \end{bmatrix}$$
$$\mathbf{u} = \begin{bmatrix} u_1 & \cdots & u_{n_s} \end{bmatrix}$$

as the densities and strategies associated with the $n_s$ species in a community. The strategy used by species $i$ is given by $u_i$ and must satisfy constraints specified in Section 4.3. In terms of this notation, the dynamics of the $i$th species's population density is given by the following

Population dynamics in terms of fitness functions

| | |
|---|---|
| Difference: | $x_i(t+1) = x_i \left[ 1 + H_i(\mathbf{u}, \mathbf{x}) \right]$ |
| Exp. Difference: | $x_i(t+1) = x_i \exp H_i(\mathbf{u}, \mathbf{x})$ |
| Differential: | $\dot{x}_i = x_i H_i(\mathbf{u}, \mathbf{x})$ |

(6.1)

where $i = 1, \cdots, n_s$ and the dot denotes differentiation with respect to time.

## 6.2.2 Ecological stability

Population sizes must remain non-negative and finite. Even in isolation, limits to growth ensure that no population can grow to infinite size. On the other hand, generally only a few species in an initial community are able to persist and coexist at positive population sizes. In the following, we characterize issues of non-zero population sizes, the persistence of a species, and the coexistence of several species. Persistence of a population simply requires population dynamics bounded from zero (including the possibility of limit cycles and chaotic dynamics).

Because population density is a non-negative quantity, the dynamical model given by (6.1) must produce trajectories of population density that satisfy $\mathbf{x}(t) \geq 0$ for all $t \geq 0$. For a given strategy vector $\mathbf{u}$ and non-negative initial population densities $\mathbf{x}(0)$, the population trajectory, $\mathbf{x}(t)$, generated by (6.1) has $n_s$ dimensions. It must remain in the non-negative space, $\mathcal{O}$, defined as the subset of points in the $n_s$ dimensional real state space, $\mathcal{R}^{n_s}$ satisfying

$$\mathcal{O} = \left\{ \mathbf{x} \in \mathcal{R}^{n_s} \mid \mathbf{x} \geq 0 \right\}.$$

The set of points $\mathcal{O}$ is also referred to as the **non-negative orthant**.

Any realistic model will not generate solutions with arbitrarily large components of $\mathbf{x}$, so that the system trajectories must remain in a bounded subset of $\mathcal{O}$. Many types of motion are possible, including stable motion to an equilibrium point, periodic orbits, and chaos (May, 1973, 1976). For all of the $G$-function categories, the main focus is on motion with stable **fixed points** (for difference equations) or stable **equilibrium points** (for differential equations). Recall that the meanings of a fixed point and an equilibrium point are the same: no change in state with time. When referring to this condition for both types of systems, we will use the single term "equilibrium point" as a replacement for "fixed point" commonly used with difference equations.

We need a workable definition for an equilibrium point. It follows from the definition of $\mathcal{O}$ that an equilibrium solution $\mathbf{x}^*$ may have some components positive and some components zero. Roberts (1974) made the important point that, in order for a non-linear population model to represent a viable ecosystem, it must have at least one positive equilibrium. Hence we need to distinguish between species (one or more) whose equilibrium populations are positive, $x_i^* > 0$, and those with equilibrium populations of zero, $x_i^* = 0$. For convenience we assign those species with positive equilibrium populations the first $n_{s^*} \leq n_s$ indices. By doing this, we are now able to give the following formal definition of an **ecological equilibrium** for $G$-functions with scalar strategies.

**Definition 6.2.1 (ecological equilibrium)** *Given a strategy vector* $\mathbf{u} \in \mathcal{U}$, *a point* $\mathbf{x}^* \in \mathcal{O}$ *is said to be an ecological equilibrium point for (6.1) provided that there exists an index* $n_{s^*}$ *with* $1 \leq n_{s^*} \leq n_s$ *such that*

$$H_i(\mathbf{u}, \mathbf{x}^*) = 0 \text{ and } x_i^* > 0 \quad \text{for} \quad i = 1, \cdots, n_{s^*}$$
$$x_i^* = 0 \quad \quad \quad \quad \quad \quad \quad \text{for} \quad i = n_{s^*}+1, \cdots, n_s.$$

In order to examine the stability of an ecological equilibrium point, points in the neighborhood of $\mathbf{x}^*$ need to be defined. The concept of a **ball** is useful in this regard. A ball $\mathcal{B}$ centered at $x^*$ is the set of points in the $n_s$-dimensional state space, with a Euclidean norm, $E^{n_s}$, satisfying

$$\mathcal{B} = \left\{ \mathbf{x} \in E^{n_s} \mid \|\mathbf{x} - \mathbf{x}^*\| < \epsilon \right\}.$$

We assume that, for every strategy vector $\mathbf{u} \in \mathcal{U}$, an ecological equilibrium solution $\mathbf{x}^*$ exists. For an ecological equilibrium point to be stable, we require that every trajectory which begins in a non-negative neighborhood of the point remains in that neighborhood for all $t$, and converges to the equilibrium as $t \to \infty$. If an ecological equilibrium point has this property, we will refer to it as an **ecologically stable equilibrium** (ESE). Goh (1980) was the first to introduce this concept and referred to it as **sector stability**.

**Definition 6.2.2 (ESE)** *Given a strategy vector* $\mathbf{u} \in \mathcal{U}$, *an ecological equilibrium point* $\mathbf{x}^* \in \mathcal{O}$ *is said to be an ecologically stable equilibrium (ESE) if there exists a ball* $\mathcal{B}$ *such that for any* $\mathbf{x}(0) \in \mathcal{O} \cap \mathcal{B}$ *the solution generated*

## 6.2 G-functions with scalar strategies

by (6.1) satisfies $\mathbf{x}(t) \in \mathcal{O}$ for all $t > 0$ and asymptotically approaches $\mathbf{x}^*$ as $t \to \infty$. If the radius of the ball can be made arbitrarily large, the ecological equilibrium point is said to be a **global ESE**, otherwise it is said to be a **local ESE**.

There can be only one global ESE, but there can be one or more local ESE solutions for $\mathbf{x}^*$ with a corresponding **domain of attraction** (see Subsection 2.5.2). While Definition 6.2.2 requires that $x^*(t)$ asymptotically approaches $x^*$ it is not equivalent to the definition of asymptotic stability, since the latter is defined in terms of open neighborhoods (an open neighborhood does not exist about a point on the boundary of the non-negative orthant).

**Lemma 6.2.1 (ESE)** *Given* $\mathbf{u} \in \mathcal{U}$, *if an ecological equilibrium* $\mathbf{x}^*$ *is an ESE then*

$$H_i(\mathbf{u}, \mathbf{x}^*) = 0 \quad \text{for} \quad i = 1, \ldots, n_{s^*}$$
$$H_i(\mathbf{u}, \mathbf{x}^*) \leq 0 \quad \text{for} \quad i = n_{s^*} + 1, \ldots, n_s.$$

*Proof.* The first condition follows from the definition of an ecological equilibrium. Suppose that $H_i(\mathbf{u}, \mathbf{x}^*) > 0$ for some $i = n_s^* + 1, \ldots, n_s$. By continuity, we can imply that there exists a ball of non-zero radius centered at $\mathbf{x}^*$ such that $H_i(\mathbf{u}, \mathbf{x}^*) > 0, i = n_s^* + 1, \ldots, n_s$ in the neighborhood $\mathcal{O} \cap \mathcal{B}$. This means that for any initial condition located in this neighborhood, the system dynamics will produce an increase in $x_i, i = n_s^* + 1, \ldots, n_s$. Because $\mathbf{x}^*$ is non-negative, it is not possible for the zero components of $\mathbf{x}^*$ to be reached by moving positively from negative values; hence, the dynamics cannot return the system to $\mathbf{x}^*$. This contradicts the assumption that $\mathbf{x}^*$ is an ESE, hence the second condition follows. ∎

The conditions provided by the ESE lemma are necessary conditions for stability and are by no means complete. For example, if $n_s^* = n_s$ then we have only the equality conditions that provide no information regarding the stability of $\mathbf{x}^*$. Additional stability conditions are available using other methods (e.g., eigenvalue analysis). However, our interest is usually with the $n_s^* > n_s$ case and the conditions provided by this lemma are useful for proving an important theorem in Chapter 7.

### 6.2.3 Evolutionary stability

To define evolutionary stability for a species, we need a way to characterize those species that survive as the biological system evolves to its evolutionary equilibrium.

**Definition 6.2.3 (coalition vector)** *If for the system (6.1) there exists an ecological equilibrium point* $\mathbf{x}^*$ *(implying* $x_i^* > 0$ *for the indices* $1, \cdots, n_{s^*} \geq 1$ *and* $x_i^* = 0$ *for the indices* $n_{s^*} + 1, \cdots, n_s$*) corresponding to the strategy vector* $\mathbf{u} \in \mathcal{U}$*, then the composite of the strategies for the first group of indices is called a* **coalition vector**, $\mathbf{u}_c = [u_1, \cdots, u_{n_{s^*}}]$, *and the composite strategies of the second group of indices (with equilibrium population of sizes of zero) is designated by the vector* $\mathbf{u}_m = [u_{n_{s^*}+1}, \cdots, u_{n_s}]$.

A coalition vector describes the strategies of an inclusive set of individuals that can persist together. The strategies need not be evolutionarily stable or exclusive. By adding individuals with different strategies to the initial vector of strategies, the resulting coalition vector may change in size (more or fewer strategies) and/or composition. While it is not necessary to distinguish between phenotypes and species in the following definition (as we did with strategy dynamics) it is useful to think in these terms. In particular, the coalition vector may be thought of as being made up of species's strategies and the vector $\mathbf{u}_m$ composed of species's strategies and/or phenotypic strategies within a species.

**Definition 6.2.4 (ESS)** *A coalition vector* $\mathbf{u}_c \in \mathcal{U}$ *is said to be an evolutionarily stable strategy (ESS) for the equilibrium point* $\mathbf{x}^*$ *if, for all* $n_s > n_{s^*}$ *and all strategies* $\mathbf{u}_m \in \mathcal{U}$*, the equilibrium point* $\mathbf{x}^*$ *is an ecological stable equilibrium (ESE).*

The ESS will be a **local ESS** if $\mathbf{x}^*$ is a local ESE and the ESS will be a **global ESS** if $x^*$ is a global ESE. As with the ESE, there can be only one global ESS, but there can be one or more local ESS solutions, each with its own $\mathbf{x}^*$ and domain of attraction. Note that under this definition an ESS must be an ESE, but given $\mathbf{u}_c$, with an ecologically stable equilibrium $\mathbf{x}^*$, an ESE does not imply that $\mathbf{u}_c$ will be an evolutionarily stable strategy.

The ESS is made up of the same number of species as there are strategies contained in the coalition vector $\mathbf{u}_c$. However, the vector $\mathbf{u}_m$ may be made up of either strategies corresponding to species different from those in the coalition vector or strategies corresponding to phenotypes of the species in the coalition vector or some combination of both. Thus the definition requires stability with respect to both intraspecific and interspecific competition. The definition requires that the ESS be resistant to invasion by rare alternative strategies. In addition, we show below that the definition implies convergent stability. This definition is useful because it is compatible with known properties of evolutionary equilibria and reduces the proliferation of terms regarding the different stability properties inherent in the outcomes of natural selection under frequency-dependent selection (Eshel, 1996).

If we go back to the original definition of an ESS given by Maynard Smith (Maynard Smith, 1982) "An ESS is a strategy such that, if all members of a population adopt it, then no mutant strategy could invade the population under the influence of natural selection" we see that his definition is equivalent to the local ESS definition given here by simply letting $n_s = 2$, and $n_{s^*} = 1$. In this case, the ESS definition states that the ESS strategy $u_1$ cannot be invaded by the mutant strategy $u_2$. That is, for any $u_2$ the system will always return to the equilibrium point $\begin{bmatrix} x_1^* & x_2^* \end{bmatrix}$ from any point in the neighborhood of the equilibrium point (with $x_1$ common, $x_2$ rare). Thus the ESS Definition includes Maynard Smith's definition but it has been expanded so that the ESS may now be a coalition of strategies which must satisfy the ecological stability property against multiple mutants.

### 6.2.4 Convergent stability

The ESS definition does not explicitly contain a convergent stability requirement. However, the condition in the ESS definition, that $\mathbf{x}^*$ remain an ecologically stable equilibrium, does imply that $\mathbf{x}^*$, $\mathbf{u}_c$ are convergent stable. Recall from Chapter 5, there must be some distribution of strategies (by whatever mechanism) for evolution to take place. We noted in Chapter 5 that $\mathbf{x}_i$ and $\mathbf{u}_i$ are mean values. That is, the $i$th population has a mean of $x_i$ with a distribution given by $x_{ij}$ and the strategy used by this population has a mean $\mathbf{u}_i$ with a distribution given by $\mathbf{u}_{ij}$. Suppose that $x^*$ and $\mathbf{u}_c$ satisfy the ESS definition for a coalition of one, $n_s^* = 1$ (the argument which follows is easily modified for coalitions greater than one). Choose $n_s \gg 1$ and let the additional strategies represent a distribution about some mean strategy $\bar{\mathbf{u}}$ at a mean population $\bar{x}$. Since $x^*$ and $\mathbf{u}_c$ are included in the distributions that define the means and since $\mathbf{u}_c$ is an ESS at $x^*$, it follows that all the populations in the distribution will die out with time except for $x^*$. Hence

$$\bar{x} \to x^*$$
$$\bar{\mathbf{u}} \to \mathbf{u}_c$$

that, in turn, implies convergent stability.

We may also visualize convergent stability geometrically, in terms of the adaptive landscape. We have seen in Chapter 5 that Darwinian dynamics on the adaptive landscape is one of hill climbing. However, due to the flexible nature of the landscape, an equilibrium $\mathbf{x}^*$ and $\mathbf{u}_c$ may correspond to a local maximum, minimum, or saddle point on the adaptive landscape. Such a point, by definition, is a **convergent stable point** on the adaptive landscape since in the process of reaching equilibrium

$$\mathbf{x} \to \mathbf{x}^*$$
$$\mathbf{u} \to \mathbf{u}_c.$$

We will show in Chapter 7 that, if $\mathbf{x}^*$ is the equilibrium population corresponding to the ESS strategy $\mathbf{u}_c$, then each strategy in the coalition vector $\mathbf{u}_c$ must correspond to a maximum point of $G(v, \mathbf{u}, \mathbf{x}^*)$ as given by the requirement

$$\max_{v \in \mathcal{U}} G(v, \mathbf{u}, \mathbf{x}^*) = G(v, \mathbf{u}, \mathbf{x}^*)\big|_{v=u_i} = 0 \qquad (6.2)$$

for $i = 1, \ldots, n_s^*$. While an ESS is convergent stable, not all convergent stable equilibrium points need correspond to evolutionarily stable strategies (since they may correspond to minima or saddlepoints). Since this maximum condition is a necessary condition, a solution obtained using (6.2) does not, by itself, imply convergent stability. Such a solution must be tested by some other method to show that the solution obtained is indeed convergent stable (such as using Darwinian dynamics to obtain the same result) before claiming the solution is an ESS. These issues will be discussed in more detail in Chapter 7.

### 6.2.5 Using $G$-functions with scalar strategies

In the following examples we obtain equilibrium solutions using Darwinian dynamics (population dynamics plus strategy dynamics). Most often we will use first-order strategy dynamics in these calculations. In terms of the $G$-functions with scalar strategies, the population dynamics and first-order strategy dynamics from Sections 4.3 and 5.4 are given by the following

Population dynamics

$$\begin{aligned}
\text{Difference:} \quad & x_i(t+1) = x_i \left[ 1 + G(v, \mathbf{u}, \mathbf{x})\big|_{v=u_i} \right] \\
\text{Exp. Difference:} \quad & x_i(t+1) = x_i \exp G(v, \mathbf{u}, \mathbf{x})\big|_{v=u_i} \\
\text{Differential:} \quad & \dot{x}_i = x_i G(v, \mathbf{u}, \mathbf{x})\big|_{v=u_i}
\end{aligned} \qquad (6.3)$$

First-order strategy dynamics

$$\begin{aligned}
\text{Difference:} \quad & \Delta u_i = \frac{\sigma_i^2}{1+G|_{u_i}} \frac{\partial G}{\partial v}\bigg|_{u_i} \\
\text{Exp. Difference:} \quad & \Delta u_i = \sigma_i^2 \frac{\partial G}{\partial v}\bigg|_{u_i} \\
\text{Differential:} \quad & \dot{u}_i = \sigma_i^2 \frac{\partial G}{\partial v}\bigg|_{u_i}
\end{aligned} \qquad (6.4)$$

where $\Delta u_i = u_i(t+1) - u_i$.

**Example 6.2.1 (L–V competition game)** *In Section 4.3 we introduced the L–V competition game with the G-function given by*

$$G_i(v, \mathbf{u}, \mathbf{x}) = r - \frac{r}{K(v)} \sum_{j=1}^{n_s} a(v, u_j) x_j,$$

*and with a symmetric distribution for the carrying capacity*

$$K(v) = K_m \exp\left\{-\frac{v^2}{2\sigma_k^2}\right\}$$

*and a non-symmetric distribution function*

$$a(v, u_j) = 1 + \exp\left\{-\frac{(v - u_j + \beta)^2}{2\sigma_a^2}\right\} - \exp\left\{-\frac{\beta^2}{2\sigma_a^2}\right\}.$$

*Consider now the specific case with*

$$r = 0.25$$
$$K_m = 100$$
$$\sigma_a^2 = \sigma_k^2 = \beta^2 = 4.$$

*In Section 5.4, using Darwinian dynamics, we obtain the following one species solution for the equilibrium strategy and density*

$$u_c = u_1 = 1.213$$
$$x^* = x_1^* = 83.20.$$

*By changing to $\sigma_k^2 = 12.5$, we obtained a coalition of two species with strategies and densities given by*

$$\mathbf{u}_c = \begin{bmatrix} u_1 & u_2 \end{bmatrix} = \begin{bmatrix} 3.1294 & -0.2397 \end{bmatrix}$$
$$\mathbf{x}^* = \begin{bmatrix} x_1^* & x_1^* \end{bmatrix} = \begin{bmatrix} 51.062 & 39.283 \end{bmatrix}.$$

*Coalitions greater than two species are also possible for this game. These solutions must satisfy the ESS definition to qualify as ESS strategies. One way to verify this is to test the strategies using the definition. For example, one could test the ESS coalition of two by introducing two other fixed strategies at non-zero population numbers*

$$\mathbf{u}_c = \begin{bmatrix} 3.1294 & -0.2397 \end{bmatrix}$$
$$\mathbf{u}_m = \begin{bmatrix} 2 & 3 \end{bmatrix}$$

*with some initial population distribution such as*

$$\mathbf{x}(0) = \begin{bmatrix} 25 & 25 & 50 & 50 \end{bmatrix}$$

and then integrating (6.3) until equilibrium is reached. The equilibrium solution obtained for $\mathbf{x}^*$

$$\mathbf{x}^* = \begin{bmatrix} 51.062 & 39.283 & 0 & 0 \end{bmatrix}$$

demonstrates that the candidate solution satisfies the definition for this case. The problem with this approach is that an ESS candidate solution must be tested for all possible vectors $\mathbf{u}_m$. In Chapter 7, we develop a tool that allows us to test solutions in a much more direct way.

In the above example, all three dynamical systems as given by (6.3) and (6.4) produce the same ESS candidate.

## 6.3 $G$-functions with vector strategies

Since the $G$-function approach was originally developed for the case of vector-valued strategies (Vincent, 1985; Brown and Vincent, 1987c) only a small change in notation is needed to extend the above results for the ESE and ESS definitions. In this case, the density vector $\mathbf{x}$ has the same meaning as in the scalar strategy case, but the strategies used by species $i$ are now vectors as defined in Section 4.4 where

$$\mathbf{u}_i = \begin{bmatrix} u_{i1} & \cdots & u_{in_u} \end{bmatrix}.$$

Collectively all of the strategies in the population can be expressed in terms of a single vector $\mathbf{u}$ by forming a concatenation of all the vectors $\mathbf{u}_i$. This results in a vector partitioned according to

$$\mathbf{u} = \begin{bmatrix} \mathbf{u}_1 \mid \cdots \mid \mathbf{u}_{n_s} \end{bmatrix}.$$

Thus we are able to use the same boldface notation for $\mathbf{u}$ (as in the scalar strategy case), but now the components of $\mathbf{u}$ may be scalars or vectors. Definition 6.2.1 for an ecological equilibrium, the ESE definition 6.2.2, and the ESE lemma 6.2.1 all remain unchanged. Definition 6.2.3 for a coalition vector remains unchanged if we replace scalar $u_i$ with vector $\mathbf{u}_i$ with the understanding that the vectors $\mathbf{u}_c$ and $\mathbf{u}_m$ are concatenations of vectors formed by

$$\mathbf{u}_c = \begin{bmatrix} \mathbf{u}_1 \mid \cdots \mid \mathbf{u}_{n_{s^*}} \end{bmatrix}$$
$$\mathbf{u}_m = \begin{bmatrix} \mathbf{u}_{n_{s^*}+1} \mid \cdots \mid \mathbf{u}_{n_s} \end{bmatrix}.$$

With this understanding, the ESS definition 6.2.4 also remains unchanged, for $G$-functions with vector strategies.

### 6.3.1 Using *G*-functions with vector strategies

Darwinian dynamics for the vector case as obtained in Sections 4.4 and 5.5 is similar in form to the scalar case

Population dynamics

$$\begin{aligned}
\text{Difference:} \quad & x_i(t+1) = x_i \left[1 + G(\mathbf{v}, \mathbf{u}, \mathbf{x})|_{\mathbf{v}=\mathbf{u}_i}\right] \\
\text{Exp. Difference:} \quad & x_i(t+1) = x_i \exp G(\mathbf{v}, \mathbf{u}, \mathbf{x})|_{\mathbf{v}=\mathbf{u}_i} \\
\text{Differential:} \quad & \dot{x}_i = x_i\, G(\mathbf{v}, \mathbf{u}, \mathbf{x})|_{\mathbf{v}=\mathbf{u}_i}
\end{aligned} \tag{6.5}$$

First-order strategy dynamics

$$\begin{aligned}
\text{Difference:} \quad & \Delta \mathbf{u}_i = \frac{\mathcal{D}_i}{1+G|_{\mathbf{u}_i}} \frac{\partial G}{\partial \mathbf{v}}\bigg|_{\mathbf{u}_i} \\
\text{Exp. Difference:} \quad & \Delta \mathbf{u}_i = \mathcal{D}_i \frac{\partial G}{\partial \mathbf{v}}\bigg|_{\mathbf{u}_i} \\
\text{Differential:} \quad & \dot{\mathbf{u}}_i = \mathcal{D}_i \frac{\partial G}{\partial \mathbf{v}}\bigg|_{\mathbf{u}_i}
\end{aligned}$$

where $\Delta \mathbf{u}_i = \mathbf{u}_i(t+1) - \mathbf{u}_i$.

**Example 6.3.1 (L–V big bully game)** *In Section 4.4 we introduced*

$$G(\mathbf{v}, \mathbf{u}, \mathbf{x}) = r - \frac{r}{K(\mathbf{v})} \sum_{j=1}^{r} a(\mathbf{v}, \mathbf{u}_j) x_j$$

*with a vector-valued strategy that has two components. The first component influences both the carrying capacity*

$$K(\mathbf{v}) = \left(1 - v_2^2\right) K_{\max} \exp\left(-\frac{v_1^2}{2\sigma_k^2}\right)$$

*and the competition coefficient*

$$a(\mathbf{v}, \mathbf{u}_j) = 1 + B_j \exp\left[-\frac{(v_1 - u_{j1} + \beta)^2}{2\sigma_a^2}\right] - \exp\left[-\frac{\beta^2}{2\sigma_a^2}\right].$$

*The second component, $v_2$, influences both the carrying capacity and the competition coefficients via a "bully" function*

$$B_j = 1 + B_{\max}\left(u_{j2} - v_2\right).$$

*Given the parameters*

$$K_{\max} = 100$$
$$R = 0.25$$
$$\sigma_\alpha = 2$$

$$\sigma_k = \sqrt{2}$$
$$\beta = 2$$
$$B_{\max} = 1$$

in Section 5.5 we discovered that Darwinian dynamics resulted in the following solution:

$$x^* = 84.08$$
$$\mathbf{u}^* = [0.6065 \quad 0.2796].$$

Hence convergent stability has been demonstrated (but not proven). The definition of an ESS can be further tested by introducing a second species. One of two things can happen. If both species are allowed to evolve they can coevolve to the same solution for **u**, dividing $x^*$ between them, or if they are not allowed to evolve, the introduced species will die out. For example integrating (6.5) using

$$\mathbf{u} = [\,0.6065 \quad 0.2796 \,|\, 0.5 \quad -0.2\,]$$
$$\mathbf{x}(0) = [50 \quad 50]$$

and setting (no evolution)

$$\mathcal{D}_1 = \mathcal{D}_2 = \begin{bmatrix} 0 & 0 \\ 0 & 0 \end{bmatrix}$$

the expected equilibrium solution $\mathbf{x}^* = [84.08 \quad 0]$ is obtained.

## 6.4 G-functions with resources

There are many situations of interest that explicitly examine the role resources play in determining the ESS. Here, we introduce resources in a game with vector-valued strategies. The population dynamics in terms of fitness functions is given by

Population dynamics with resouces

| | |
|---|---|
| Difference: | $x_i(t+1) = x_i[1 + H_i(\mathbf{u}, \mathbf{x}, \mathbf{y})]$ |
| Exp. Difference: | $x_i(t+1) = x_i \exp H_i(\mathbf{u}, \mathbf{x}, \mathbf{y})$ |
| Differential: | $\dot{x}_i = x_i H_i(\mathbf{u}, \mathbf{x}, \mathbf{y})$ |

(6.6)

with the addition of the resource equations of the form

Resource dynamics

| | |
|---|---|
| Difference: | $\mathbf{y}(t+1) = \mathbf{y} + \mathbf{N}(\mathbf{u}, \mathbf{x}, \mathbf{y})$ |
| Differential: | $\dot{\mathbf{y}} = \mathbf{N}(\mathbf{u}, \mathbf{x}, \mathbf{y}).$ |

(6.7)

## 6.4 G-functions with resources

The resource equations are of a more general form than the population dynamics equations in that they need not have an identifiable fitness function. Note however the difference equation and differential equation forms have been structured so that each has equilibrium solutions when $N = 0$. We do not include an exponential difference form with resource dynamics since, in general, these equations are not derivable from a population dynamic form (with a $y_i$ multiplier on the left-hand side). This feature is required for the exponential form that we have been using (see Subsection 2.3.3).

For a given fixed $\mathbf{u} \in \mathcal{U}$ and $\mathbf{x} \in \mathcal{O}$ we assume that there exists an equilibrium solution for $\mathbf{y}^* \overset{>}{\sim} \mathbf{0}$ such that[6]

$$\mathbf{N}\left(\mathbf{u}, \mathbf{x}, \mathbf{y}^*\right) = 0.$$

Furthermore, we assume that this equilibrium point is locally asymptotically stable. With $\mathbf{u}$ and $\mathbf{x}$ having the same meaning as in the previous section, we need to restate the definitions and theorem of Section 6.2 so that they include $\mathbf{y}^*$. Definitions 6.2.1, 6.2.2, and Lemma 6.2.1 are modified as follows.

**Definition 6.4.1 (ecological equilibrium – resource)** *Given a strategy vector $\mathbf{u} \in \mathcal{U}$ and a resource vector $\mathbf{y}^*$ a point $\mathbf{x}^* \in \mathcal{O}$ is said to be an ecological equilibrium point for (6.6) provided that there exists an index $n_{s^*}$ with $1 \leq n_{s^*} \leq n_s$ such that*

$$H_i\left(\mathbf{u}, \mathbf{x}^*, \mathbf{y}^*\right) = 0, \; x_i^* > 0 \quad \text{for} \quad i = 1, \ldots, n_{s^*}$$
$$x_i^* = 0 \quad \text{for} \quad i = n_{s^*} + 1, \ldots, n_s.$$

**Definition 6.4.2 (ecologically stable equilibrium – resource)** *Given a strategy vector $\mathbf{u} \in \mathcal{U}$ and a resource vector $\mathbf{y}^*$, an ecological equilibrium point $\mathbf{x}^* \in \mathcal{O}$ is said to be an ecologically stable equilibrium (ESE) if there exists a ball $\mathcal{B}$ such that for any $\mathbf{x}(0) \in \mathcal{O} \cap \mathcal{B}$ the solution generated by (6.6) and (6.7) satisfies $\mathbf{x}(t) \in \mathcal{O}$ for all $t > 0$ and asymptotically approaches $x^*$ as $t \to \infty$. If the radius of the ball can be made arbitrarily large, the ecological equilibrium point is said to be a **global ESE**, otherwise it is said to be a **local ESE**.*

**Lemma 6.4.1 (ESE – resourse)** *Given $\mathbf{u} \in \mathcal{U}$ and a resource vector $\mathbf{y}^*$, if an ecological equilibrium point $\mathbf{x}^*$ is an ESE then*

$$H_i\left(\mathbf{u}, \mathbf{x}^*, \mathbf{y}^*\right) = 0 \quad \text{for} \quad i = 1, \ldots, n_s^*$$
$$H_i\left(\mathbf{u}, \mathbf{x}^*, \mathbf{y}^*\right) \leq 0 \quad \text{for} \quad i = n_s^* + 1, \ldots, n_s.$$

*Proof.* The first condition follows from the definition of an ecological equilibrium. Suppose that $H_i(\mathbf{u}, \mathbf{x}^*, \mathbf{y}^*) > 0$ for some $i = n_s^* + 1, \ldots, n_s$. By

---

[6] The notation $\overset{>}{\sim}$ is used to imply that every component of the vector is $\geq 0$ with at least one component $> 0$.

continuity, we know that there exists a ball of non-zero radius centered at $\mathbf{x}^*$ such that $H_i(\mathbf{u}, \mathbf{x}^*, y^*) > 0$, $i = n_s^* + 1, \ldots, n_s$ in the neighborhood $\mathcal{O} \cap \mathcal{B}$. This means that for any initial condition located in this neighborhood, the system dynamics will produce an increase in $x_i$, $i = n_s^* + 1, \ldots, n_s$. Because $\mathbf{x}^*$ is non-negative, it is not possible for the zero components of $\mathbf{x}^*$ to be reached by moving positively from negative values, hence the dynamics cannot return the system to $\mathbf{x}^*$. This contradicts that assumption that $\mathbf{x}^*$ is an ESE, hence the second condition follows. ∎

Definition 6.2.3 for a coalition vector remains unchanged if we replace scalar $u_i$ with vector $\mathbf{u}_i$ with the understanding that the vectors $\mathbf{u}_c$ and $\mathbf{u}_m$ are concatenations of vectors formed by

$$\mathbf{u}_c = \begin{bmatrix} \mathbf{u}_1 \mid \cdots \mid \mathbf{u}_{n_{s*}} \end{bmatrix}$$
$$\mathbf{u}_m = \begin{bmatrix} \mathbf{u}_{n_{s*}+1} \mid \cdots \mid \mathbf{u}_{n_s} \end{bmatrix}.$$

With this understanding, the ESS definition 6.2.4 also remains unchanged.

### 6.4.1 Using $G$-functions with resources

Darwinian dynamics for this category of $G$-functions was developed in Sections 4.5 and 5.6.

Population dynamics

| | |
|---|---|
| Difference: | $x_i(t+1) = x_i \left[ 1 + G(\mathbf{v}, \mathbf{u}, \mathbf{x}, \mathbf{y}) \vert_{\mathbf{v}=\mathbf{u}_i} \right]$ |
| Exp. Difference: | $x_i(t+1) = x_i \exp G(\mathbf{v}, \mathbf{u}, \mathbf{x}, \mathbf{y}) \vert_{\mathbf{v}=\mathbf{u}_i}$ |
| Differential: | $\dot{x}_i = x_i G(\mathbf{v}, \mathbf{u}, \mathbf{x}, \mathbf{y}) \vert_{\mathbf{v}=\mathbf{u}_i}$ |

Resource dynamics

| | |
|---|---|
| Difference: | $\mathbf{y}(t+1) = \mathbf{y} + \mathbf{N}(\mathbf{u}, \mathbf{x}, \mathbf{y})$ |
| Differential: | $\dot{\mathbf{y}} = \mathbf{N}(\mathbf{u}, \mathbf{x}, \mathbf{y})$ |

First-order strategy dynamics

| | |
|---|---|
| Difference: | $\Delta \mathbf{u}_i = \frac{\mathcal{D}_i}{1 + G(\mathbf{v},\mathbf{u},\mathbf{x})\vert_{\mathbf{v}=\mathbf{u}_i}} \frac{\partial G(\mathbf{v},\mathbf{u},\mathbf{x},\mathbf{y})}{\partial \mathbf{v}} \bigg\vert_{\mathbf{v}=\mathbf{u}_i}$ |
| Exp. Difference: | $\Delta \mathbf{u}_i = \mathcal{D}_i \frac{\partial G(\mathbf{v},\mathbf{u},\mathbf{x},\mathbf{y})}{\partial \mathbf{v}} \bigg\vert_{\mathbf{v}=\mathbf{u}_i}$ |
| Differential: | $\dot{\mathbf{u}}_i = \mathcal{D}_i \frac{\partial G(\mathbf{v},\mathbf{u},\mathbf{x},\mathbf{y})}{\partial \mathbf{v}} \bigg\vert_{\mathbf{v}=\mathbf{u}_i}$ |

where $\Delta \mathbf{u}_i = \mathbf{u}_i(t+1) - \mathbf{u}_i$.

## 6.4 G-functions with resources

**Example 6.4.1 (Bergmann's rule)** *In Section 4.5 we introduced this consumer resource game with*

$$G(v, \mathbf{u}, \mathbf{x}) = \frac{Av^\alpha y}{1 + AHv^{(\alpha-\beta)}y} - Cv^\gamma$$

$$\dot{y} = r(K - y) - \sum_{i=1}^{n_s} \frac{Au_i^\alpha y x_i}{1 + AHu_i^{(\alpha-\beta)}y}$$

*In Section 5.6, with the parameter values*

$$r = 0.25$$
$$K = 100$$
$$A = 0.05$$
$$C = 1.5$$
$$H = 0.3$$
$$\alpha = 0.5$$
$$\beta = 1$$
$$\gamma = 0.75$$

*we obtained, using Darwinian dynamics, the equilibrium solution*

$$x^* = 10.517, \quad y^* = 54.000, \quad u^* = 0.6561$$

*for both the differential and exponential difference equation models. We may now do a trial test to see whether this solution satisfies the ESS definition. We do so by adding a second species and checking to see if we obtain the same solution. If we allow the second species to evolve at the same rate as the first one, with the initial conditions*

$$x_1(0) = 10.517, \; x_2(0) = 50, \; y^* = 54.000, \; u_1(0) = 0.6561, \; u_2(0) = 0.2$$

*we obtain the following solution*

$$x_1^* = 2.9962, \quad x_2^* = 7.5205, \quad y^* = 54.000, \quad u_1^* = u_2^* = 0.6561.$$

*Both species coevolve to the same ESS strategy, with a combined population density the same as before. However, if we do not let the second species evolve, we obtain*

$$x_1^* = 10.517, \quad x_2^* = 0.0000, \quad y^* = 54.000, \quad u_1^* = 0.6561, \quad u_2^* = 0.2.$$

*Both results satisfy the ESS definition.*

## 6.5 Multiple G-functions

With multiple $G$-functions, there are different strategies within and between $G$-functions. Groups of species that can co-exist at positive population sizes (ESE) can now include members from different $G$-functions. And, the ESS definition must now apply across as well as within $G$-functions. Extending the stability conditions across $G$-functions requires some additional notational complexity. From Section 4.6, the population dynamics for all the species

$$\mathbf{x} = [\, \mathbf{x}_1 \mid \cdots \mid \mathbf{x}_{n_g} \,] \tag{6.8}$$

using strategies

$$\mathbf{u} = [\, \mathbf{u}_1 \mid \cdots \mid \mathbf{u}_{n_g} \,] \tag{6.9}$$

are given by the following

Population dynamics

$$\begin{array}{ll} \text{Difference:} & x_i(t+1) = x_i \left[1 + G_j(\mathbf{v}, \mathbf{u}, \mathbf{x})\big|_{\mathbf{v}=\mathbf{u}_i}\right] \\ \text{Exp. Difference:} & x_i(t+1) = x_i \exp G_j(\mathbf{v}, \mathbf{u}, \mathbf{x})\big|_{\mathbf{v}=\mathbf{u}_i} \\ \text{Differential:} & \dot{x}_i = x_i \, G_j(\mathbf{v}, \mathbf{u}, \mathbf{x})\big|_{\mathbf{v}=\mathbf{u}_i} \end{array}$$

(6.10)

where $i$ and $j$ are determined using the rising number r

$$i = r_{j-1} + 1, \ldots, r_j \quad \text{for } j = 1, \ldots, n_g \tag{6.11}$$

where

$$r_0 = 0$$
$$r_i = \sum_{j=1}^{i} n_{s_j} \quad \text{for } i = 1, \ldots, n_g.$$

The definition of the non-negative orthant $\mathcal{O}$ remains the same as before. We can now provide a definition of an ecological equilibrium similar to Definition (6.2.1) paying attention to the definition of $n_{s^*}$.

**Definition 6.5.1 (ecological equilibrium – multiple)** *Given a strategy vector $\mathbf{u} \in \mathcal{U}$, a point $\mathbf{x}^* \in \mathcal{O}$ is said to be an ecological equilibrium point for (6.10) provided that there exists an index $n_{s^*}$ with $1 \le n_{s^*} \le n_s$ such that*

$$\begin{array}{ll} H_i(\mathbf{u}, \mathbf{x}^*) = 0 \text{ and } x_i^* > 0 & \text{for } i = 1, \ldots, n_{s^*} \\ x_i^* = 0 & \text{for } i = n_{s^*} + 1, \ldots, n_s \end{array}$$

*where the order in which $H_i$ is numbered is determined from the functions $G_j$*

## 6.5 Multiple G-functions

*according to*

$$H_i(\mathbf{u}, \mathbf{x}) = G_j(\mathbf{v}, \mathbf{u}, \mathbf{x})\big|_{\mathbf{v}=\mathbf{u}_i}$$

*and $i$ and $j$ are determined from the rising number $r$*

$$i = r_{j-1} + 1, \ldots, r_j \quad \text{for } j = 1, \ldots, n_g$$

*with*

$$r_0 = 0$$
$$r_i = \sum_{j=1}^{i} n_{s_j^*} \quad \text{for } i = 1, \ldots, n_g$$

*where $n_{s_j^*}$ is the number of strategies within each G-function that have non-zero equilibrium densities.*

As in our previous definitions, $n_{s^*}$ is used to distinguish between species (one or more) whose equilibrium populations are positive, $x_i^* > 0$, and those with equilibrium populations of zero, $x_i^* = 0$. Those species with positive equilibrium populations are assigned the first $n_{s^*} \leq n_s$ indices. The same is done here. The assignment process occurs one G-function at a time. For example, suppose that we have three G-functions with three species in each, and, at equilibrium, the first G-function has only one species $\left(n_{s_1^*} = 1\right)$ with positive density, the second G-function none $\left(n_{s_2^*} = 0\right)$ with positive density and the third G-function has two species with positive density $\left(n_{s_3^*} = 2\right)$. In this case, there are three species with positive densities. The indices are ordered by using one from G-function 1 and the remaining two from the third G-function ($n_{s^*} = 3$).

With this understanding, the ESE definition remains exactly the same as in Subsection 6.2.2.

**Definition 6.5.2 (ESE – multiple)** *Given a strategy vector $\mathbf{u} \in \mathcal{U}$, an ecological equilibrium point $\mathbf{x}^* \in \mathcal{O}$ is said to be an ecologically stable equilibrium (ESE) if there exists a ball $\mathcal{B}$ such that for any $\mathbf{x}(0) \in \mathcal{O} \cap \mathcal{B}$ the solution generated by (6.10) satisfies $\mathbf{x}(t) \in \mathcal{O}$ for all $t > 0$ and asymptotically approaches $x^*$ as $t \to \infty$. If the radius of the ball can be made arbitrarily large, the ecological equilibrium point is said to be a **global ESE**, otherwise it is said to be a **local ESE**.*

Likewise we obtain the same ESE lemma. The proof is the same as in 6.2.2.

**Lemma 6.5.1 (ESE – multiple)** *Given $\mathbf{u} \in \mathcal{U}$, if an ecological equilibrium $\mathbf{x}^*$is an ESE then*

$$H_i(\mathbf{u}, \mathbf{x}^*) = 0 \quad \text{for} \quad i = 1, \ldots, n_{s^*}$$
$$H_i(\mathbf{u}, \mathbf{x}^*) \leq 0 \quad \text{for} \quad i = n_{s^*} + 1, \ldots, n_s.$$

To define an ESS, we need a definition that applies to a coalition vector that includes species from within and among multiple $G$-functions. This coalition vector contains the strategies of the non-zero equilibrium populations contained in each of the $G$-functions. Let $\mathbf{u}$ be given ($\mathbf{u}$ as in (6.9)) and let $\mathbf{x}^*$ ($\mathbf{x}$ as in (6.8)) be the equilibrium solution to (6.10). We now reorder the indices by moving all those with zero density to the end (leaving the others in the original order). As in the above example we would start with the species in the order

$$\mathbf{x} = \begin{bmatrix} x_1 & x_2 & | & x_3 & | & x_4 & x_5 & x_6 \end{bmatrix}.$$

Suppose that the strategies are scalars given by

$$\mathbf{u} = \begin{bmatrix} 1 & 2 & 3 & 4 & 5 & 6 \end{bmatrix}$$

and at equilibrium suppose the densities are given by

$$\mathbf{x}^* = \begin{bmatrix} 0 & 10 & | & 0 & | & 5 & 0 & 3 \end{bmatrix}$$

We would then reorder according to $x_2 \to x_1$, $x_4 \to x_2$, $x_6 \to x_3$ with the others reordered in any fashion. In this case we have a coalition of one in the first $G$-function, a coalition of zero in the second $G$-function and a coalition of two in the third $G$-function. Thus the densities in the coalition are defined by only two ($n_{g^*} = 2$) of the original three $G$-functions ($n_g = 3$)

$$\mathbf{x}_c^* = \begin{bmatrix} 10 & | & 5 & 3 \end{bmatrix}.$$

Likewise the coalition vector is composed of three strategies ($n_{s^*} = 3$), one from the first $G$-function and two from the third $G$-function

$$\mathbf{u}_c = \begin{bmatrix} 2 & | & 4 & 6 \end{bmatrix}.$$

In other words, at equilibrium, we look to see which $G$-functions have species at non-zero population numbers and then define a coalition vector composed of only these. The remaining species are assigned to $\mathbf{x}_m^*$ with strategies $\mathbf{u}_m$.

We keep track of individuals in the coalition in the same way as before by using the rising number for the species in the coalition as given by

$$r_0 = 0$$
$$r_i = \sum_{j=1}^{i} n_{s_j^*} \quad \text{for } i = 1, \ldots, n_{g^*}$$

with $i$ and $j$ determined from

$$i = r_{j-1} + 1, \ldots, r_j \quad \text{for } j = 1, \ldots, n_{g^*}$$

where $n_{s_j^*}$ is the number of species in the $j$th $G$-function of the coalition and

$n_{g*}$ is the number of $G$-functions in the coalition. Thus

$$\mathbf{u}_c = \begin{bmatrix} \mathbf{u}_1 & \cdots & \mathbf{u}_{n_{s*}} \end{bmatrix}$$

where each species in the coalition satisfies constraints according to

$$\mathbf{u}_i \in \mathcal{U}_j$$

for $j = 1, \ldots, n_{g*}$. When the constraints are satisfied in each of the bauplans we also use the notation

$$\mathbf{u} \in \mathcal{U}.$$

In the above example, it follows that, for the coalition, $n_{s*} = 3$, $n_{g*} = 2$, $r_1 = 1$, $r_2 = 3$. With this interpretation of $\mathbf{u}_c$, $\mathbf{u}_m$, and $n_{s*}$, our previous definition 6.2.3 for a coalition vector and definition 6.2.4 for an ESS apply to this case.

### 6.5.1 Using multiple $G$-functions

The following equations for Darwinian dynamics were obtained in Sections 4.6 and 5.7.

Population dynamics

| | |
|---|---|
| Difference: | $x_i(t+1) = x_i \left[1 + G_j(\mathbf{v}, \mathbf{u}, \mathbf{x})\big|_{v=u_i}\right]$ |
| Exp. Difference: | $x_i(t+1) = x_i \exp G_j(\mathbf{v}, \mathbf{u}, \mathbf{x})\big|$ |
| Differential: | $\dot{x}_i = x_i G_j(\mathbf{v}, \mathbf{u}, \mathbf{x})\big|_{v=u_i}$ |

First-order strategy dynamics

| | |
|---|---|
| Difference: | $\Delta \mathbf{u}_i = \dfrac{\mathcal{D}_i}{1 + G(\mathbf{v},\mathbf{u},\mathbf{x})\big|_{v=u_i}} \dfrac{\partial G_j(\mathbf{v},\mathbf{u},\mathbf{x})}{\partial v}\bigg|_{v=u_i}$ |
| Exp. Difference: | $\Delta \mathbf{u}_i = \mathcal{D}_i \dfrac{\partial G_j(\mathbf{v},\mathbf{u},\mathbf{x})}{\partial v}\bigg|_{v=u_i}$ |
| Differential: | $\dot{\mathbf{u}}_i = \mathcal{D}_i \dfrac{\partial G_j(\mathbf{v},\mathbf{u},\mathbf{x})}{\partial v}\bigg|_{v=u_i}$ |

where $\Delta \mathbf{u}_i = \mathbf{u}_i(t+1) - \mathbf{u}_i$ and $i$ and $j$ are determined from

$$i = r_{j-1} + 1, \ldots, r_j \quad \text{for } j = 1, \ldots, n_g$$

where

$$r_0 = 0$$
$$r_i = \sum_{j=1}^{i} n_{s_j} \quad \text{for } i = 1, \ldots, n_g.$$

**Example 6.5.1 (predator–prey coevolution)** *In this game all the strategies are scalars, allowing us to avoid the double subscript notation. Since there are* *only two G-function, $n_g = 2$ ($\mathbf{x}_1 =$ prey, $\mathbf{x}_2 =$ predators), we have $r_1 = n_{s_1}$ and $r_2 = n_{s_1} + n_{s_2} = n_s$. The G-functions for this game as given in Section 4.6 are*

$$G_1(v, \mathbf{u}, \mathbf{x}) = \frac{r_1}{K(v)} \left[ K(v) - \sum_{j=1}^{n_{s_1}} a(v, u_j) x_j \right] - \sum_{j=n_{s_1}+1}^{n_s} b(v, u_j) x_j$$

$$G_2(v, \mathbf{u}, \mathbf{x}) = r_2 \left[ 1 - \frac{\sum_{j=n_{s_1}+1}^{n_s} x_j}{c \sum_{j=1}^{n_{s_1}} b(v, u_j) x_j} \right]$$

*with the following assumed functional forms*

$$K(v) = K_{\max} \exp\left[-\frac{v^2}{\sigma_k^2}\right]$$

$$a(v, u_j) = \exp\left[-\frac{(v - u_j)^2}{\sigma_a^2}\right]$$

$$b(v, u_j) = b_{\max} \exp\left[-\frac{(v - u_j)^2}{\sigma_b^2}\right].$$

*In Section 5.7, using the following set of parameters*

$$r_1 = r_2 = c = 0.25$$
$$K_{\max} = 100$$
$$b_{\max} = 0.15$$
$$\sigma_k^2 = 2$$
$$\sigma_a^2 = 4$$
$$\sigma_b^2 = 10$$

*we obtained, using Darwinian dynamics, the equilibrium solution*

$$u_1^* = u_2^* = 0, \quad x_1^* = 30.77, \quad x_2^* = 1.154.$$

*This represents a multiple G-functions ESS coalition of two strategies candidates with $n_{s_1^*} = 1$, $n_{s_2^*} = 1$ (one prey and one predator). However, by changing $\sigma_b$ other solutions are possible. For example, when $\sigma_b^2 = 4$, the ESS candidate is a multiple G-functions coalition of three with $n_{s_1^*} = 2$, and $n_{s_2^*} = 1$ (two prey and one predator)*

$$u_1^* = 0.90, \quad u_2^* = -0.90, \quad u_3^* = 0, \quad x_1^* = x_2^* = 19.35, \quad x_3^* = 1.19.$$

## 6.5 Multiple G-functions

When $\sigma_b^2 = 1$, the ESS candidate is a multiple G-functions coalition of four with $n_{s_1^*} = 2$, and $n_{s_2^*} = 2$ (two prey and two predators)

$$u_1^* = 0.79, \; u_2^* = -0.79, \; u_3^* = 0.56, \; u_4^* = -0.56 \quad x_1^* = x_2^* = 28.71,$$
$$x_3^* = x_4^* = 0.60.$$

*Other solutions are also possible.*

Any of the solutions in this example can be tested using the ESS definition. We do so by introducing additional prey and predators to the candidate solutions obtained, but without any strategy dynamics. Consider the $\sigma_b^2 = 4$ case. Suppose we introduced one additional prey so that $s_1 = 3$ and two additional predators so that $s_2 = 3$. We need to re-number the old strategies ($u_1 \Rightarrow u_1, u_2 \Rightarrow u_2, u_3 \Rightarrow u_4$) when introducing the new ones ($u_3 = 0.5$, $u_5 = 0.6, u_6 = 0.8$)

$$u_1 = 0.90, \; u_2 = -0.90, \; u_3 = 0.5, \; u_4 = 0, \; u_5 = 0.6, \; u_6 = 0.8.$$

Setting non-zero initial conditions

$$x_1(0) = x_2(0) = x_3(0) = x_4(0) = x_5(0) = x_6(0) = 10$$

and solving (6.10) for the differential and exponential difference equations results in the equilibrium solution

$$x_1^* = x_2^* = 19.35, \; x_3^* = 0, \; x_4^* = 1.19, \; x_5^* = x_6^* = 0. \quad (6.12)$$

Because the domain of attraction for the difference equation model is small, in order to get the same result for this case, we must start with initial conditions much closer to the ESS candidate solution. For example, using

$$x_1(0) = 19.35, \; x_2(0) = 19.35, \; x_3(0) = 0.2,$$
$$x_4(0) = 1.19, \; x_5(0) = 0.2, \; x_6(0) = 0.2$$

we again obtain the equilibrium solution given by (6.12). Clearly, for the difference equation case, we have a local ESS candidate. As a notation review, we have

$$n_g = 2$$
$$n_{s_1} = 3, \; n_{s_2} = 3$$
$$r_1 = 3, \; r_2 = 6$$
$$n_{g^*} = 2$$
$$n_{s_1^*} = 2, \; n_{s_2^*} = 1$$
$$r_1^* = 2, \; r_2^* = 3.$$

## 6.6 G-functions in terms of population frequency

In Section 4.7, we wrote the population dynamics equations in terms of frequency. This formulation is useful for making frequency dependence explicit and is the preferred notation in matrix games. Recall that the exponential difference equations are not included in the frequency case.

Using the definition introduced in Section 4.7

$$p_i = \frac{x_i}{N} \quad \text{where} \quad N = \sum_{i=1}^{n_s} x_i$$

we found that the corresponding frequency dynamics are given by the following

Frequency dynamics

$$\text{Difference:} \quad p_i(t+1) = p_i \frac{1 + H_i(\mathbf{u}, \mathbf{p}, N)}{1 + \bar{H}}$$
$$\text{Differential:} \quad \dot{p}_i = p_i \left[ H_i(\mathbf{u}, \mathbf{p}, N) - \bar{H} \right] \quad (6.13)$$

where $\bar{H}$ is the average fitness of the population as a whole

$$\bar{H} = \sum_{i=1}^{n_s} p_i H_i(\mathbf{u}, \mathbf{p}, N)$$

and

Total population size dynamics

$$\text{Difference:} \quad N(t+1) = N\left(1 + \bar{H}\right)$$
$$\text{Differential:} \quad \dot{N} = N\bar{H}. \quad (6.14)$$

Working with this notation, we follow the same general development as in the scalar $G$-function case, except now the strategies are vectors and, instead of the non-negative orthant, the dynamics of $\mathbf{p}$ lies in the **frequency space** defined by

$$\Delta^{n_s} = \{\mathbf{p} \in R^{n_s} \mid \sum_{i=1}^{n_s} p_i = 1, \; p_i \geq 0\}.$$

The various definitions given Section 6.2 are reformulated as follows:

**Definition 6.6.1 (ecological equilibrium – frequency)** *Given a strategy vector* $\mathbf{u} \in \mathcal{U}$, *the frequency* $\mathbf{p}^* \in \Delta^{n_s}$ *and population* $N^*$ *are said to be an ecological equilibrium for (6.13) and (6.14) provided that there exists an index* $n_{s^*}$ *with* $1 \leq n_{s^*} \leq n_s$ *such that*

$$H_i(\mathbf{u}, \mathbf{p}^*, N^*) = 0 \text{ and } p_i^* > 0 \quad \text{for} \quad i = 1, \ldots, n_{s^*}$$
$$p_i^* = 0 \quad \text{for} \quad i = n_{s^*}+1, \ldots, n_s.$$

**Definition 6.6.2 (ESE – frequency)** *Given a strategy vector* $\mathbf{u} \in \mathcal{U}$, *the ecological equilibrium* $\mathbf{p}^* \in \Delta^{n_s}, N^* > 0$ *is said to be an ecologically stable equilibrium (ESE) if there exists a ball* $\mathcal{B}$ *centered at* $\mathbf{p}^*$ *such that, for any*

## 6.6 G-functions in terms of population frequency

$\mathbf{p}(0) \in \Delta^{n_s} \cap \mathcal{B}$ and $N(0) = N^* + \delta N > 0$, the solutions generated by (6.13) and (6.14) satisfy $\mathbf{p}(t) \in \Delta^{n_s}$ and $N(t) > 0$ for all $t > 0$ and asymptotically approach $p^*$ and $N^*$ as $t \to \infty$. If the radius of the ball can be made arbitrarily large and for any $\delta N$ satisfying $N^* + \delta N > 0$, the ecological equilibrium point is said to be a **global ESE**, otherwise it is said to be a **local ESE**.

**Lemma 6.6.1 (ESS – frequency)** *Given* $\mathbf{u} \in \mathcal{U}$, *if an ecological equilibrium* $\mathbf{p}^*, N^*$ *is an ESE then*

$$H_i(\mathbf{u}, \mathbf{p}^*, N^*) = \bar{H} \quad \text{for} \quad i = 1, \ldots, n_s^*$$
$$H_i(\mathbf{u}, \mathbf{p}^*, N^*) \leq \bar{H} \quad \text{for} \quad i = n_s^* + 1, \ldots, n_s.$$

*Proof.* The first condition follows from the definition of an ecological equilibrium. Suppose that $H_i(\mathbf{u}, \mathbf{p}^*, N^*) > \bar{H}$ for some $i = n_s^* + 1, \ldots, n_s$. By continuity, we know that there exists a ball of non-zero radius centered at $\mathbf{p}^*$ such that $H_i(\mathbf{u}, \mathbf{p}^*, N^*) > \bar{H}$ in the neighborhood $\Delta^{n_s} \cap \mathcal{B}$. This means that, for any initial condition located in this neighborhood, the system dynamics will produce an increase in $p_i$, $i = n_s^* + 1, \ldots, n_s$. Because $\mathbf{p}^*$ is non-negative, it is not possible for the zero components of $\mathbf{p}^*$ to be reached by moving positively from negative values, hence the dynamics cannot return the system to $\mathbf{p}^*$. This contradicts the assumption that $\mathbf{p}^*$ is an ESE, hence the second condition follows. ∎

**Definition 6.6.3 (coalition vector – frequency)** *If for the system (6.13) and (6.14) there exists an ecological equilibrium* $\mathbf{p}^*, N^*$ *corresponding to the strategy vector* $\mathbf{u} \in \mathcal{U}$, *then the composite of the strategies for the first group of indices* $\left(i = 1, \ldots, n_s^*\right)$ *is called a* **coalition vector**, $\mathbf{u}_c = [u_1 \ldots u_{n_{s*}}]$, *and the composite strategies of the second group of indices* $\left(j = n_s^* + 1, \ldots, n_s\right)$ *are designated by the vector* $\mathbf{u}_m = [u_{n_{s*}+1} \ldots u_{n_s}]$.

**Definition 6.6.4 (ESS – frequency)** *A coalition vector* $\mathbf{u}_c \in \mathcal{U}$ *is said to be an evolutionarily stable strategy (ESS) for the ecological equilibrium* $\mathbf{p}^*, N^*$ *if, for all* $n_s > n_{s*}$ *and all strategies* $\mathbf{u}_m \in \mathcal{U}$, *the ecological equilibrium* $\mathbf{p}^*, N^*$ *is an ecologically stable equilibrium (ESE)*.

### 6.6.1 Using G-functions in terms of population frequency

From Section 4.7 and 5.8 the population dynamics and first-order strategy dynamics are given by the following

Frequency dynamics

| | |
|---|---|
| Difference: | $p_i(t+1) = p_i \dfrac{1 + G(\mathbf{v}, \mathbf{u}, \mathbf{p}, N)\vert_{\mathbf{v}=\mathbf{u}_i}}{1 + \bar{G}}$ |
| Differential: | $\dot{p}_i = p_i \left[ G(\mathbf{v}, \mathbf{u}, \mathbf{p}, N)\vert_{\mathbf{v}=\mathbf{u}_i} - \bar{G} \right]$ |

where

$$\bar{G} = \sum_{i=1}^{n_s} p_i \, G(\mathbf{v}, \mathbf{u}, \mathbf{p}, N)|_{\mathbf{v}=\mathbf{u}_i}$$

Total population size dynamics

| | |
|---|---|
| Difference: | $N(t+1) = N\left(1+\bar{G}\right)$ |
| Differential: | $\dot{N} = N\bar{G}$ |

First-order strategy dynamics

| | |
|---|---|
| Difference: | $\Delta \mathbf{u}_i = \dfrac{\mathcal{D}_i}{1+G(\mathbf{v},\mathbf{u},\mathbf{p},N)|_{\mathbf{v}=\mathbf{u}_i}} \left.\dfrac{\partial G(\mathbf{v},\mathbf{u},\mathbf{p},N)}{\partial \mathbf{v}}\right|_{\mathbf{v}=\mathbf{u}_i}$ |
| Differential: | $\dot{\mathbf{u}}_i = \mathcal{D}_i \left.\dfrac{\partial G(\mathbf{v},\mathbf{u},\mathbf{p},N)}{\partial \mathbf{v}}\right|_{\mathbf{v}=\mathbf{u}_i}$ |

where $\Delta \mathbf{u}_i = \mathbf{u}_i(t+1) - \mathbf{u}_i$.

**Example 6.6.1 (L–V competition game in terms of frequency)** *This game (see Example 4.7.1) formulated in terms of frequency is defined by*

$$G(v, \mathbf{u}, \mathbf{p}, N) = \frac{r}{K(v)}\left(K(v) - N\sum_{j=1}^{n_s} a(v, u_j)\, p_j\right),$$

with

$$K(v) = K_m \exp\left\{-\frac{v^2}{2\sigma_k^2}\right\}$$

$$\alpha(v, u_i) = 1 + \exp\left\{-\frac{(v-u_i+\beta)^2}{2\sigma_\alpha^2}\right\} - \exp\left\{-\frac{\beta^2}{2\sigma_\alpha^2}\right\}.$$

*As in Example 6.2.1 using*

$$r = 0.25$$
$$K_m = 100$$
$$\sigma_\alpha^2 = \sigma_k^2 = \beta^2 = 4$$

*we obtain, using Darwinian dynamics, the equivalent coalition of one solution*

$$u_c = u_1 = 1.213$$
$$p^* = p_1^* = 1$$
$$N^* = 83.20.$$

*Likewise, by changing to $\sigma_k^2 = 12.5$, we obtained the equivalent coalition of* two solution

$$\mathbf{u}_c = \begin{bmatrix} u_1 & u_2 \end{bmatrix} = \begin{bmatrix} 3.1294 & -0.2397 \end{bmatrix}$$
$$\mathbf{p}^* = \begin{bmatrix} 0.5652 & 0.4348 \end{bmatrix}$$
$$N^* = 90.347.$$

Note that when comparing with Example 6.2.1 $x_1^* = p_1 N^*$ and $x_2^* = p_2 N^*$.

The differential equation and difference equation population dynamics models have the same solutions.

## 6.7 Multistage $G$-functions

A single multistage **G**-matrix, with scalar strategies, is used to model species with $n_h$ life-history stages. Recall from Section 4.8, the population dynamics for the multistage case is expressed in terms of a population matrix

$$\begin{array}{ll} \text{Difference:} & \mathbf{x}_i(t+1) = \mathbf{x}_i \left[ \mathcal{I} + \mathbf{H}_i^T(\mathbf{u}, \mathbf{x}) \right] \\ \text{Differential:} & \dot{\mathbf{x}}_i = \mathbf{x}_i \mathbf{H}_i^T(\mathbf{u}, \mathbf{x}). \end{array} \quad (6.15)$$

As with the frequency bauplan, no exponential difference formulation is considered. Once again we need to re-examine the basic definitions. We must be able to distinguish between species (one or more) whose equilibrium populations have at least one stage positive, $\mathbf{x}_i^* \,\bar{>}\, 0$ (the notation $\bar{>}$ is used when every component of the vector is $\geq 0$ with at least one component $> 0$), and those whose equilibrium populations have every component zero, $\mathbf{x}_i^* = 0$. As before, we refer to the first group using the indices $1 \cdots n_{s^*}$ and the second group using the indices $n_{s^*} + 1 \cdots n_s$.

**Definition 6.7.1 (ecological equilibrium – multistage)** *Given a strategy vector* $\mathbf{u} \in \mathcal{U}$, *a point* $\mathbf{x}^* \in \mathcal{O}$ *is said to be an ecological equilibrium point for* (6.15) *provided that there exists an index* $n_{s^*}$ *with* $1 \leq n_{s^*} \leq n_s$ *such that*

$$\begin{array}{ll} \det[\mathbf{H}_i(\mathbf{u}, \mathbf{x}^*)] = 0 \text{ and } \mathbf{x}_i^* \,\bar{>}\, 0 & \text{for} \quad i = 1, \cdots, n_{s^*} \\ \mathbf{x}_i^* = 0 & \text{for} \quad i = n_{s^*} + 1, \cdots, n_s. \end{array}$$

A change in the inequality notation distinguishes this definition from the previous scalar ecological equilibrium definition 6.2.1. Definition 6.7.1 is more general in that it contains the other (when $n_h = 1$, $\det[\mathbf{H}_i(\mathbf{u}, \mathbf{x}^*)] = 0$ is the same as $H_i(\mathbf{u}, \mathbf{x}^*) = 0$ and $\mathbf{x}_i^* \,\bar{>}\, 0$ is the same as $x_i^* > 0$). By using this more general definition of an ecological equilibrium, the previous definition 6.2.2 for an ESE remains unchanged.

However, we do need a new ESE lemma that requires some preliminary work. Following the methods of Goh (1980), we determine conditions for an ecological equilibrium point of (6.15) to be an ESE.

**Lemma 6.7.1 (multistage eigenvalues)** *Given* $\mathbf{u} \in \mathcal{U}$, *if an ecological equilibrium point* $\mathbf{x}^*$ *is an ESE then all the eigenvalues of the matrices*

$$\begin{array}{ll} \text{Difference:} & [\mathcal{I}+\mathcal{D}] \text{ and } [\mathcal{I}+\mathcal{H}] \\ \text{Differential:} & \mathcal{D} \text{ and } \mathcal{H} \end{array} \quad (6.16)$$

*must satisfy* $|\lambda_i| \leq 1$ *(difference equations) or have non-positive real parts (differential equations). Furthermore, if the eigenvalues satisfy* $|\lambda_i| < 1$ *(difference equations) or have negative real parts (differential equations), then* $\mathbf{x}^*$ *is a local ESE. The matrix* $\mathcal{D}$ *is defined by*

$$\mathcal{D} = \begin{bmatrix} x_1 \frac{\partial H_1}{\partial x_1} & \cdots & x_1 \frac{\partial H_1}{\partial x_{n_{s*}}} \\ \vdots & \ddots & \vdots \\ x_{n_{s*}} \frac{\partial H_{n_{s*}}}{\partial x_1} & \cdots & x_{n_{s*}} \frac{\partial H_{n_{s*}}}{\partial x_{n_{s*}}} \end{bmatrix}_{(\mathbf{u},\mathbf{x}^*)}$$

*and the diagonal matrix* $\mathcal{H}$ *is defined by*

$$\mathcal{H} = \begin{bmatrix} H_{n_{s*}+1} & \cdots & 0 \\ \vdots & \ddots & \vdots \\ 0 & \cdots & H_{n_s} \end{bmatrix}_{(\mathbf{u},\mathbf{x}^*)}$$

*where* $\mathcal{I}$ *is an identity matrix of the same dimension corresponding to* $\mathcal{D}$ *or* $\mathcal{H}$.

*Proof.* Given $\mathbf{u} \in \mathcal{U}$, a first-order Taylor series expansion of (6.15) about $\mathbf{x}^*$ yields the following perturbation equations

$$\begin{array}{ll} \text{Difference:} & \delta x_i(t+1) = \delta x_i \left[1 + H_i|_{(\mathbf{u},\mathbf{x}^*)}\right] + x_i \sum_{j=1}^{n_s} \left.\frac{\partial H_i}{\partial x_j}\right|_{(\mathbf{u},\mathbf{x}^*)} \delta x_j \\ \text{Differential:} & \delta \dot{x}_i = \delta x_i \, H_i|_{(\mathbf{u},\mathbf{x}^*)} + x_i \sum_{j=1}^{n_s} \left.\frac{\partial H_i}{\partial x_j}\right|_{(\mathbf{u},\mathbf{x}^*)} \delta x_j \end{array}$$

where $\delta x_i$ is the perturbation in $x_i$ from the ecological equilibrium solution. These equations may be written as

$$\begin{array}{ll} \text{Difference:} & \delta \mathbf{x}_i(t+1) = \begin{bmatrix} \mathcal{I}+\mathcal{D} & \mathcal{I}+\mathcal{D}_m \\ 0 & \mathcal{I}+\mathcal{H} \end{bmatrix} \delta \mathbf{x} \\ \text{Differential:} & \delta \dot{\mathbf{x}}_i = \begin{bmatrix} \mathcal{D} & \mathcal{D}_m \\ 0 & \mathcal{H} \end{bmatrix} \delta \mathbf{x} \end{array} \quad (6.17)$$

where $\delta \mathbf{x}$ and $\delta \dot{\mathbf{x}}$ are the column vectors

$$\delta \mathbf{x} = \begin{bmatrix} \delta x_1 \\ \vdots \\ \delta x_{n_{s^*}} \\ \delta x_{n_{s^*}+1} \\ \vdots \\ \delta x_{n_s} \end{bmatrix}, \quad \delta \dot{\mathbf{x}} = \begin{bmatrix} \delta \dot{x}_1 \\ \vdots \\ \delta \dot{x}_{n_{s^*}} \\ \delta \dot{x}_{n_{s^*}+1} \\ \vdots \\ \delta \dot{x}_{n_s} \end{bmatrix},$$

and

$$\mathcal{D}_m = \begin{bmatrix} x_1 \frac{\partial \mathbf{H}_1}{\partial x_{n_{s^*}+1}} & \cdots & x_1 \frac{\partial \mathbf{H}_1}{\partial x_{n_s}} \\ \vdots & \ddots & \vdots \\ x_{n_{s^*}} \frac{\partial \mathbf{H}_{n_c}}{\partial x_{n_{s^*}+1}} & \cdots & x_{n_{s^*}} \frac{\partial \mathbf{H}_{n_c}}{\partial x_{n_s}} \end{bmatrix}_{(\mathbf{u},\mathbf{x}^*)}.$$

If $\mathbf{x}^*$ is an ESE then by definition, in the limit, as $t \to \infty$, $\mathbf{x}(t)$ must asymptotically approach $\mathbf{x}^*$. This means that the eigenvalues of the matrix in (6.17) must lie on the unit disk (difference equations) or have non-positive real parts (differential equations). Note that eigenvalues on the border between stability and instability are allowed since higher-order terms in the expansion can provide for stability. ∎

Because the matrix in (6.17) is in block upper triangular form, this condition on the eigenvalues implies that the matrices given in (6.16) must lie on the unit disk (difference equations) or have non-positive real parts (differential equations). Furthermore if the matrices in (6.16) have eigenvalues that are inside the unit circle (difference equations) or have negative real parts (differential equations) then $\mathbf{x}^*$ will be locally asymptotically stable.

The ESE theorem is stated in terms of the **critical value** definitions introduced in Section 4.8. Recall that if $\lambda_i = \begin{bmatrix} \lambda_{i1} \cdots \lambda_{in_s} \end{bmatrix}$ is the vector of eigenvalues corresponding to $\mathbf{H}_i|_{(\mathbf{u},\mathbf{x}^*)}$ then max $(\text{abs}(\lambda_i))$ is the maximum component of the vector of absolute values and max $(\text{Re}(\lambda_i))$ is the maximum component of the vector of real parts. The definition of the critical value depends on whether difference or differential equations are used to model the system

$$\boxed{\begin{array}{ll} \text{Difference:} & \text{crit}\left[\mathbf{H}_i|_{(\mathbf{u},\mathbf{x}^*)}\right] = \max(\text{abs}(\lambda_i)) \\ \text{Differential:} & \text{crit}\left[\mathbf{H}_i|_{(\mathbf{u},\mathbf{x}^*)}\right] = \max(\text{Re}(\lambda_i)). \end{array}}$$

**Lemma 6.7.2 (ESE – multistage)** *Given $\mathbf{u} \in \mathcal{U}$, if an ecological equilibrium point $\mathbf{x}^*$ is an ESE then*

$$\text{crit}\left[\mathbf{H}_i|_{(\mathbf{u},\mathbf{x}^*)}\right] = 0$$

for $i = 1, \ldots, n_{s^*}$ and

$$\text{crit}\left[\mathbf{H}_i|_{(\mathbf{u},\mathbf{x}^*)}\right] \leq 0$$

for $i = n_{s^*} + 1, \ldots n_s$.

*Proof.* At an ecological equilibrium point $\mathbf{x}_i(t+1) = \mathbf{x}_i$ (difference) or $\dot{\mathbf{x}}_i = \mathbf{0}$ (differential) the following equilibrium condition is obtained from (6.15)

$$\mathbf{x}_i \mathbf{H}_i^T(\mathbf{u}, \mathbf{x}) = 0.$$

For a non-trivial solution for $\mathbf{x}_i^*$ to exist the following must hold

$$\det[\mathbf{H}_i(\mathbf{u}, \mathbf{x}^*)] = 0 \quad \text{for} \quad i = 1, \ldots, n_{s^*}. \tag{6.18}$$

It follows from (6.18) that at least one eigenvalue of $\mathbf{H}_i|_{(\mathbf{u},\mathbf{x}^*)}$ must be zero ($i = 1, \ldots, n_{s^*}$). Consider the differential equation case first. Stability for the $n_h$ life-history stages (see Section 4.8) requires that the eigenvalues of $\mathbf{H}_i|_{(\mathbf{u},\mathbf{x}^*)}$ have non-positive real parts for $i = 1, \ldots, n_{s^*}$. The equilibrium condition implies that the critical values must be zero as well, implying the first condition of Lemma 6.7.2. Lemma 6.7.1 requires that all of the eigenvalues of $\mathcal{H}$ must have non-positive real parts. The second condition follows from the fact that the eigenvalues of a diagonal block form are equal to the eigenvalues of each block. Hence the critical value of $\mathbf{H}_i|_{(\mathbf{u},\mathbf{x}^*)}$ must be less than or equal to zero for $i = n_{s^*} + 1, \cdots n_s$. A similar argument holds for the difference equation case. Stability for the $n_h$ life history stages requires that the eigenvalues of $\mathcal{I} + \mathbf{H}_i|_{(\mathbf{u},\mathbf{x}^*)}$ have absolute values $\leq 1$ for $i = 1, \ldots, n_{s^*}$. Because $\text{eig}\left[\mathcal{I} + \mathbf{H}_i|_{(\mathbf{u},\mathbf{x}^*)}\right] = 1 + \lambda_j$ (where $\lambda_j$ are the eigenvalues of $\mathbf{H}_i|_{(\mathbf{u},\mathbf{x}^*)}$) the equilibrium condition implies that the critical values must equal zero and the first condition follows. The second condition follows from the fact that the eigenvalues of a diagonal block form are equal to the eigenvalues of each block. Hence the critical value of $\left[\mathcal{I} + \mathbf{H}_i|_{(\mathbf{u},\mathbf{x}^*)}\right]$ must be less than or equal to zero for $i = n_{s^*} + 1, \cdots n_s$. ∎

**Definition 6.7.2 (coalition vector – multistage)** *If for the system (6.15) there exists an ecological equilibrium point $\mathbf{x}^*$ corresponding to the strategy vector $\mathbf{u} \in \mathcal{U}$, then the composite of the strategies for the first group of indices is called a **coalition vector**, $\mathbf{u}_c = [\mathbf{u}_1 \cdots \mathbf{u}_{n_c}]$, and the composite strategies of the second group of indices are designated by the vector $\mathbf{u}_m = [\mathbf{u}_{n_c+1} \cdots \mathbf{u}_{n_x}]$.*

**Definition 6.7.3 (ESS – multistage)** *A coalition vector $\mathbf{u}_c \in \mathcal{U}$ is said to be an evolutionarily stable strategy (ESS) for the equilibrium point $\mathbf{x}^*$ if, for all $n_s > n_{s^*}$ and all strategies $\mathbf{u}_m \in \mathcal{U}$, the equilibrium point $\mathbf{x}^*$ is an ecologically stable equilibrium (ESE).*

### 6.7.1 Using multistage $G$-functions

The **G**-matrix is a generating matrix for the **H**-matrices. If the **G**-matrix is a scalar, then it is identical to a scalar $G$-function. However, if **G** is a true matrix, then there is an important difference, namely (from Section 4.8) the $G$-function for this case is given by

$$G(v, \mathbf{u}, \mathbf{x}) = \text{crit}\,[\mathbf{G}(v, \mathbf{u}, \mathbf{x})]. \tag{6.19}$$

The **G**-matrix is used with the population dynamics equations, while the $G$-function (as determined from the **G**-matrix using (6.19)) is used with the first-order strategy dynamics equations.

The population dynamics and first-order strategy dynamics as derived in terms of the **G**-matrix and/or $G$-function from Sections 4.8 and 5.9 are given by the following

Population dynamics

$$\begin{aligned}
\text{Difference:} \quad & \mathbf{x}_i(t+1) = \mathbf{x}_i \left[\mathcal{I} + \mathbf{G}^\mathrm{T}(v, \mathbf{u}, \mathbf{p})\big|_{v=\mathbf{u}_i}\right] \\
\text{Differential:} \quad & \dot{\mathbf{x}}_i = \mathbf{x}_i\, \mathbf{G}^\mathrm{T}(v, \mathbf{u}, \mathbf{p})\big|_{v=\mathbf{u}_i}
\end{aligned}$$

First-order strategy dynamics

$$\begin{aligned}
\text{Difference:} \quad & \Delta u_i = \tfrac{\sigma_i^2}{1+G|_{u_i}} \tfrac{\partial G(v,\mathbf{u},\mathbf{x})}{\partial v}\Big|_{v=u_i} \\
\text{Differential:} \quad & \dot{u}_i = \sigma_i^2 \tfrac{\partial G(v,\mathbf{u},\mathbf{x})}{\partial v}\Big|_{v=u_i}
\end{aligned}$$

where $\Delta u_i = u_i(t+1) - u_i$.

**Example 6.7.1 (multistage tutorial game)** *This game was introduced in Section 4.8. It has a two-stage life history with scalar strategies (only single subscripts are required) with the **G**-matrix*

$$\mathbf{G}(v, \mathbf{u}, \mathbf{x}) = \begin{bmatrix} -1 + 4v - v^2 - \sum_{j=1}^{n_x} u_j x_{j1} & v \\ v & -\left(\sum_{j=1}^{n_x} x_{j1}\right) \end{bmatrix}.$$

*In Section 5.9 we obtained, for the differential equation case, the equilibrium solution*

$$x_{11}^* = 1.781,\ x_{12}^* = 4.562,\ u_1^* = 4.562.$$

*This solution may be tested by introducing mutant strategies at non-zero populations. Let two mutant strategies be added to the population so that*

$$u_1 = 4.562,\ u_2 = 2,\ u_3 = 6$$

*and reset initial conditions*

$$x_{11}(0) = x_{12}(0) = x_{21}(0) = x_{22}(0) = x_{31}(0) = x_{32}(0) = 2.$$

*Integrating (6.15) results in the equilibrium solution*

$$x_{11}^* = 1.781, \ x_{12}^* = 4.562, \ x_{21}^* = x_{22}^* = x_{31}^* = x_{32}^* = 0$$

*as required by the ESS definition.*

## 6.8 Non-equilibrium Darwinian dynamics

We limit our analysis of non-equilibrium Darwinian dynamics to the category of $G$-functions with vector strategies.[7] The population dynamics in this case are given by the following

Population dynamics

| | | |
|---|---|---|
| Difference: | $x_i(t+1) = x_i \left[ 1 + G(\mathbf{v}, \mathbf{u}, \mathbf{x})\|_{\mathbf{v}=\mathbf{u}_i} \right]$ | |
| Exp. Difference: | $x_i(t+1) = x_i \exp G(\mathbf{v}, \mathbf{u}, \mathbf{x})\|_{\mathbf{v}=\mathbf{u}_i}$ | (6.20) |
| Differential: | $\dot{x}_i = x_i G(\mathbf{v}, \mathbf{u}, \mathbf{x})\|_{\mathbf{v}=\mathbf{u}_i}.$ | |

Here we relax the assumption that an equilibrium exists, but require that solutions generated by the dynamics remain bounded. This allows for periodic, quasi-periodic, or chaotic motion with trajectories that lie on periodic orbits, limit cycles (continuous) or $n$-cycles (discrete), and strange attractors (see Subsection 2.5.7). In order to arrive at an ESS definition under non-equilibrium dynamics, we need to restate a number of previous definitions. We use the notation $\mathbf{x}^\circ$ to denote all the points in a periodic orbit, limit cycle, $n$-cycle, or strange attractor.

**Definition 6.8.1 (ecological cycle)** *Given a strategy vector* $\mathbf{u} \in \mathcal{U}$, *the vector* $\mathbf{x}^\circ \in \mathcal{O}$ *is said to be an* **ecological cycle** *(periodic orbit, limit cycle, n-cycle, or strange attractor) for (6.20) provided that there exists an index* $n_{s^\circ}$ *with* $1 \leq n_{s^\circ} \leq n_s$ *such that*

$$\mathbf{x}_i^\circ > 0 \quad \text{for} \quad i = 1, \ldots, n_{s^\circ}$$
$$\mathbf{x}_i^\circ = 0 \quad \text{for} \quad i = n_{s^\circ} + 1, \ldots, n_s.$$

A modified vector notation applies (e.g., $\mathbf{x}_i^\circ > 0$ means that for every point in the ecological cycle, $x_i > 0$). The following definitions are almost identical to those of Section 6.2 with $\mathbf{x}^\circ$ replacing $x^*$. The ecologically stable equilibrium

---

[7] A similar approach can be used with any of the other categories with appropriate notational changes.

(ESE) definition is replaced by an ecologically stable cycle (ESC) definition, but the definitions of a coalition vector and ESS are unchanged except for $\mathbf{x}°$ replacing $x^*$.

**Definition 6.8.2 (ESC)** *Given a strategy vector* $\mathbf{u} \in \mathcal{U}$, *an ecological cycle* $\mathbf{x}° \in \mathcal{O}$ *is said to be an* **ecologically stable cycle** *(ESC) if there exists a ball* $\mathcal{B}$ *such that for any* $\mathbf{x}(0) \in \mathcal{O} \cap \mathcal{B}$ *the solution generated by (6.20) satisfies* $\mathbf{x}(t) \in \mathcal{O}$ *for all* $t > 0$ *and asymptotically approaches* $\mathbf{x}°$ *as* $t \to \infty$. *If the radius of the ball can be made arbitrarily large, the ecological equilibrium point is said to be a* **global ESC**, *otherwise it is said to be a* **local ESC**.

In order to arrive at necessary conditions for the stability of an ecological cycle, the concept of compound fitness is introduced. A system is said to complete an ecological cycle when it arrives back to where it started.[8] We will show that, while fitness at each point of the cycle may change, the system would not return to the starting point unless the "accumulation" of fitness functions along the way added up to zero. The compound fitness function defines this accumulation.

**Definition 6.8.3 (compound fitness function)** $H_i(\mathbf{u}, \mathbf{x}°)$ *is a* **compound fitness function** *defined by*

| | |
|---|---|
| Difference: | $1 + H_i(\mathbf{u}, \mathbf{x}°) = \prod [1 + H_i(\mathbf{u}, \mathbf{x})]$ |
| Exp. Difference: | $H_i(\mathbf{u}, \mathbf{x}°) = \sum H_i(\mathbf{u}, \mathbf{x})$ |
| Differential: | $H_i(\mathbf{u}, \mathbf{x}°) = \int H_i(\mathbf{u}, \mathbf{x}) \, dt$ |

*where* $\mathbf{u}, \mathbf{x}$ *in the product, summation, and integral are the strategy and current state at each time step. For an n-cycle the product and sum would be over n. For a limit cycle the integral is over the time required to complete the cycle, and, for quasi-periodic or chaotic motion, the product, sum, and integral are infinite (however, a finite approximation is used in practice).*

**Example 6.8.1 (compound fitness)** *Suppose that each difference equation model produces a 3-cycle and the differential equation model takes three time units to complete a limit cycle. From (6.20) it follows that from any starting point on the ecological cycle*

| | |
|---|---|
| Difference: | $x_i(t+3) = x_i(t+2)\{1 + H_i[\mathbf{u}(t+2), \mathbf{x}(t+2)]\}$ |
| Exp. Difference: | $x_i(t+3) = x_i(t+2) \exp\{H_i[\mathbf{u}(t+2), \mathbf{x}(t+2)]\}$ |
| Differential: | $x_i(t_0+3) = x_i(t_0) + \int_{t_0}^{3} H_i[\mathbf{u}(t), \mathbf{x}(t)] \, dt$ |

*where t is the starting time for the difference equations and $t_0$ is the starting*

---

[8] For quasi-periodic or chaotic motion the system never returns to its starting point. In these cases, the system is said to complete an ecological cycle if it returns to a point very near to where it started.

time for the differential equations. Using the shorthand notation
$$H_i[\mathbf{u}(t+n), \mathbf{x}(t+n)] = H_i^{(n)}$$
$$x_i(t+n) = x_i^{(n)}$$
from the first difference equation we have
$$\begin{aligned}x_i^{(3)} &= x_i^{(2)}\left[1+H_i^{(2)}\right]\\&= x_i^{(1)}\left[1+H_i^{(1)}\right]\left[1+H_i^{(2)}\right]\\&= x_i^{(0)}\left[1+H_i^{(0)}\right]\left[1+H_i^{(1)}\right]\left[1+H_i^{(2)}\right]\\&= x_i^{(0)}\prod_{j=0}^{2}\left[1+H_i^{(j)}\right]\end{aligned}$$
and from the second difference equation
$$\begin{aligned}x_i^{(3)} &= x_i^{(2)}\exp\left[H_i^{(2)}\right]\\&= x_i^{(1)}\exp\left[H_i^{(1)}\right]\exp\left[H_i^{(2)}\right]\\&= x_i^{(0)}\exp\left[H_i^{(0)}\right]\exp\left[H_i^{(1)}\right]\exp\left[H_i^{(2)}\right]\\&= x_i^{(0)}\exp\left[\sum_{j=0}^{2}H_i^{(j)}\right].\end{aligned}$$

From the definition of the compound fitness function and the above example, it follows that the value of $x_i$ obtained after completing one ecological cycle is given by

| | |
|---|---|
| Difference: | $x_i(t+n) = x_i(t)[1+H_i(\mathbf{u},\mathbf{x}°)]$ |
| Exp. Difference: | $x_i(t+n) = x_i(t)\exp H_i(\mathbf{u},\mathbf{x}°)$ |
| Differential: | $x_i(t_0+t_f) = x_i(t_0) + H_i(\mathbf{u},\mathbf{x}°)$ |

Since the system returns to where it started, it follows that
$$H_i(\mathbf{u},\mathbf{x}°) = 0$$
represents an **equilibrium requirement** for an ecological cycle.

**Lemma 6.8.1 (ESC)** *Given* $\mathbf{u} \in \mathcal{U}$, *if an ecological cycle* $\mathbf{x}°$ *is an ESC then*
$$H_i(\mathbf{u},\mathbf{x}°) = 0 \quad \text{for} \quad i = 1,\ldots,n_{s°}$$
$$H_j(\mathbf{u},\mathbf{x}°) \le 0 \quad \text{for} \quad j = n_{s°}+1,\ldots,n_s.$$

*Proof.* Pick any point $\mathbf{x}(t_0)$ on an ecological cycle and iterate or integrate (6.20) until the system returns to $\mathbf{x}(t_0)$ (or an arbitrarily small neighborhood of $\mathbf{x}(t_0)$ for the quasi-periodic or chaotic approximation). Since, after completing the cycle, $x_i = x_i(t_0)$ for $i = 1, \ldots, n_s^\circ$ the compound fitness function must be zero. Suppose that, after completing the cycle, $H_j(\mathbf{u}, \mathbf{x}^\circ) > 0$ for some $j \in [n_{s^\circ} + 1 \cdots n_s]$. By continuity, this implies that, if we start at a neighboring point close to $\mathbf{x}(t_0)$ (lying in the positive orthant with $x_j > 0$) and iterate or integrate (6.1) over the same time interval, the ending values for $x_j$, will be greater than the starting values for $x_j$. This contradicts the assumption that the ecological cycle $\mathbf{x}^\circ$ is an ESC, since this ending value should be approaching $\mathbf{x}(t_0)$ (recall that $x_j = 0$ is on the ecological cycle; hence such a point can only be approached by decreasing values of $x_j$). The second condition thus follows. ∎

**Definition 6.8.4 (coalition vector – non-equilibrium)** *If for the system (6.1) there exists an ecological cycle $\mathbf{x}^\circ$ corresponding to the strategy vector $\mathbf{u} \in \mathcal{U}$, then the composite of the strategies for the indices $1, \ldots, n_{s^\circ}$ is called a* **coalition vector**, $\mathbf{u}_c = [u_1 \ldots u_{n_{s^\circ}}]$, *and the composite strategies for the indices $n_{s^\circ} + 1, \ldots, n_s$ is designated by the vector $\mathbf{u}_m = [u_{n_{s^\circ}+1} \ldots u_{n_s}]$.*

**Definition 6.8.5 (ESS – non-equilibrium)** *A coalition vector $\mathbf{u}_c \in \mathcal{U}$ is said to be an evolutionarily stable strategy (ESS) for the ecological cycle $\mathbf{x}^\circ$ if, for all $n_s > n_{s^\circ}$ and all strategies $\mathbf{u}_m \in \mathcal{U}$, the ecological cycle $\mathbf{x}^\circ$ is an ecological stable cycle (ESC).*

This definition is identical to the ESS definition for equilibrium dynamics when the ecological cycle is an equilibrium point.

### 6.8.1 Using $G$-functions with non-equilibrium dynamics

In this case, the Darwinian dynamics is the same as those for $G$-functions with vector strategies as given in Subsection 6.3.1.

Population dynamics

| | |
|---|---|
| Difference: | $x_i(t+1) = x_i \left[ 1 + G(\mathbf{v}, \mathbf{u}, \mathbf{x}) \vert_{\mathbf{v}=\mathbf{u}_i} \right]$ |
| Exp. Difference: | $x_i(t+1) = x_i \exp G(\mathbf{v}, \mathbf{u}, \mathbf{x}) \vert_{\mathbf{v}=\mathbf{u}_i}$ |
| Differential: | $\dot{x}_i = x_i G(\mathbf{v}, \mathbf{u}, \mathbf{x}) \vert_{\mathbf{v}=\mathbf{u}_i}$ |

First-order strategy dynamics

| | |
|---|---|
| Difference: | $\Delta \mathbf{u}_i = \frac{\mathcal{D}_i}{1+G\vert_{\mathbf{u}_i}} \frac{\partial G}{\partial \mathbf{v}} \vert_{\mathbf{u}_i}$ |
| Exp. Difference: | $\Delta \mathbf{u}_i = \mathcal{D}_i \frac{\partial G}{\partial \mathbf{v}} \vert_{\mathbf{u}_i}$ |
| Differential: | $\dot{\mathbf{u}}_i = \mathcal{D}_i \frac{\partial G}{\partial \mathbf{v}} \vert_{\mathbf{u}_i}$ |

where $\Delta \mathbf{u}_i = \mathbf{u}_i(t+1) - \mathbf{u}_i$.

**Example 6.8.2 (non-equilibrium L–V game)** *The G-function is defined by*

$$G(v, \mathbf{u}, \mathbf{x}) = \frac{r}{K(v)} \left( K(v) - \sum_{j=1}^{n_s} a(v, u_j) x_j \right)$$

*where*

$$K(v) = K_m \exp\left(-\frac{v^2}{2\sigma_k^2}\right)$$

$$a(v, u_i) = 1 + \exp\left\{-\frac{(v - u_i + \beta)^2}{2\sigma_a^2}\right\} - \exp\left\{-\frac{\beta^2}{2\sigma_a^2}\right\}.$$

*We will consider the same coalition of two case as in Example 6.2.1, except that we now use the difference equation model and increase r sufficiently high to produce chaos. Using*

$$r = 2.8$$
$$K_m = 100$$
$$\sigma_\alpha^2 = \beta^2 = 4, \ \sigma_k^2 = 12.5$$

*with $\sigma_1^2 = \sigma_2^2 = 0.5$ yields the results illustrated in Figure 6.3. Once again we*

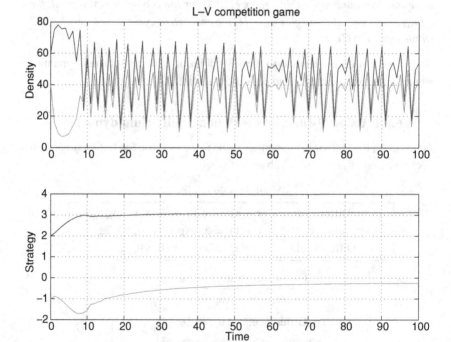

**Figure 6.3** An ESS coalition of two under chaotic density dynamics.

*obtain the equilibrium solution of*

$$\mathbf{u}^* = \begin{bmatrix} 3.1294 & -0.2397 \end{bmatrix}.$$

In the above example, the equilibrium solution obtained is actually an ESS coalition of two. In spite of the fact that the population fluctuates chaotically, the strategies evolve to the ESS in a nice asymptotic fashion. There are two reasons why this can happen. First, the time scale chosen (via the speed term) for the first-order strategy dynamics equation is set sufficiently smaller than the time scale of the population dynamics equation that the strategies cannot track rapid population changes. This would be the normal situation in the real world. If the time scale of the strategy dynamics equations is increased significantly, oscillations may be observed in the strategy dynamics and may even cause the entire system to become unstable. A second reason why the strategies can asymptotically approach equilibrium is that, at the ESS, the $G$-function is either weakly dependent or independent of density (it is easy to show that for a coalition of one, in the above model, the $G$-function is indeed independent of $\mathbf{x}$). This need not always be the case as illustrated in the next example.

**Example 6.8.3 (non-equilibrium L–V game with $\beta$ a function of x)** *In this case we have the same model as above, except that, instead of $\beta$ constant, we introduce $\mathbf{x}$ dependence according to*

$$\beta = \frac{1}{50} \sum_{i=1}^{n_s} x_i.$$

*If we let the other parameters be given by*

$$r = 2.6$$
$$K_m = 100$$
$$\sigma_\alpha^2 = \sigma_k^2 = 4$$

*and choose a moderate time scale for the first-order strategy dynamics (with the speed term $\sigma_1^2 = 0.5$) we obtain, using the exponential difference equations model, the results shown in Figure 6.4. If $\sigma_1^2$ is increased much above 0.5, the system dynamics becomes unstable. However, if we decrease $\sigma_1^2$ ($\sigma_1^2 = 0.02$) then the strategy dynamics does not track the population dynamics as well and a smoother strategy dynamics is obtained as illustrated in Figure 6.5. Of course it takes much longer for an ESS strategy to evolve and the strategy is still in a 4-cycle, but now the oscillations are sufficiently small that we can approximate a candidate ESS strategy to be $u = 0.845$. Figure 6.6*

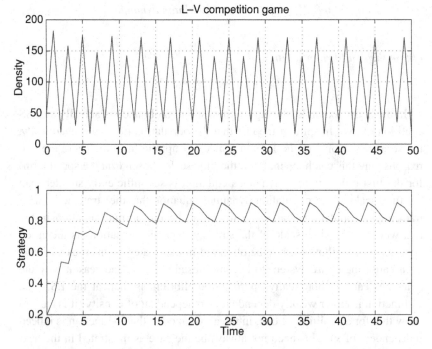

**Figure 6.4** When the ESS is strongly dependent on **x**, the strategy dynamics will also cycle.

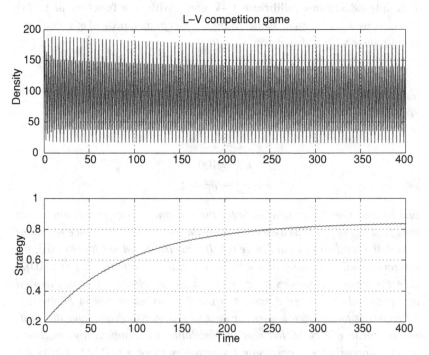

**Figure 6.5** At a slower rate of evolution, the strategy dynamics becomes smoother.

## 6.8 Non-equilibrium Darwinian dynamics

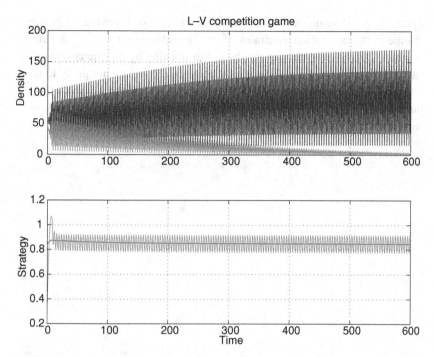

**Figure 6.6** An ESS under non-equilibrium dynamics.

*illustrates that $u = 0.845$ cannot be invaded by an evolving 4-cycle strategy. This figure was generated by using the above model with two identical species with the same parameters and same initial conditions as above except that the first species has $\sigma_1^2 = 0$ and the second species has $\sigma_2^2 = 0.5$. The first species uses $u = 0.845$; the second species dies out with time as it evolves to its 4-cycle strategy. As a final demonstration that $u = 0.845$ is a likely ESS candidate, consider using this strategy against the four strategies in the 4-cycle obtained when $\sigma_1^2 = 0.5$ (as in Figure 6.4). The four strategies obtained in that case are $\mathbf{u}_m = \begin{bmatrix} 0.828 & 0.798 & 0.920 & 0.887 \end{bmatrix}$. Choosing non-zero initial conditions for all the species, setting $\sigma_2^2 = 0$ in all the strategy dynamics equations and iterating the population dynamic equations, we find that, after many iterations, the 4-cycle population fluctuations associated with the species using the strategies of $\mathbf{u}_m$ approach zero. It takes many iterations for this to happen since the strategies of $\mathbf{u}_m$ are close in value to $u$.*

Our ESS definition for non-equilibrium dynamics is applicable only for strategies that are constant on the evolutionary cycle. The above example illustrates the fact that this is a reasonable assumption provided that the strategy dynamics is not so quick that the strategies end up **fast tracking** every rapid change in the population dynamics. If, in the real world, there are situations in which Darwinian dynamics results in fast tracking by the strategy, then an alternative definition for an ESS would be required. We are not going to take that route except to note that if a fast-tracking strategy were to maximize the $G$-function at every point on the ecological cycle then such a sequence of strategies would have the desired ESS properties.

# 7
# The ESS maximum principle

The definition of an ESS in Chapter 6 requires that an ESS be convergent stable and resistant to invasion by alternative strategies. The Darwinian dynamics discussed in Chapter 5 provides convergent stability. In this chapter, an **ESS maximum principle** is obtained that characterizes the property of resistance to invasion. We show that, when **u** and **x**$^*$ have values corresponding to an ESS, the $G$-function must take on a maximum with respect to the focal strategy, **v**, when **v** is set equal to one of the strategies of the ESS. Like the Nash equilibrium, an ESS is a no-regret strategy for the $G$-function in the sense that at an ESS no individual can gain a fitness advantage by unilaterally changing strategy. This property of the ESS, as expressed in terms of the $G$-function, can be used as a necessary condition for solving for candidate ESS solutions. This necessary condition is formalized in this chapter as the ESS maximum principle. This principle also describes the property of adaptation, the true sense of *FF&F*. An adaptation is a strategy that maximizes individual fitness as determined from the $G$-function given the circumstances, and these circumstances include the strategies and population sizes of others.

We use the term **ESS candidate** to refer to any solutions obtained using the ESS maximum principle. While an ESS candidate will be resistant to invasion, it need not satisfy the convergent stability property required by the ESS definition. Thus, convergence stability must be checked by some other method. If it turns out that an ESS candidate is convergent stable, then it is indeed an ESS. One method for checking the convergence stability of an ESS candidate is to see whether the candidate can be obtained using Darwinian dynamics.

In this chapter, we use the $G$-function concept to develop the ESS maximum principle. It turns out that there are many versions of the ESS maximum principle, depending upon the nature of the game and the dynamical system under consideration. As in the previous chapter we will first develop the principle for the simplest case of scalar strategies for an ecological system that can be described by a single $G$-function. We then generalize this result for the other categories of $G$-functions. The one unifying theme is that the $G$-function(s) must take on a maximum with respect to the virtual variable contained in the $G$-function.

We present a maximum principle for each category of $G$-functions including non-equilibrium dynamics as described in Section 6.8. Each maximum principle will have a title. For example, the ESS maximum principle for $G$-functions with scalar strategies is designated by "ESS-scalar." The Darwinian dynamics used to determine ESS strategies for this chapter's examples is summarized in Chapter 6 in the "using $G$-functions" sections. In each case, $n_s$ is the number of species in the community. We assume that the number of species in the coalition vector is less than the initial number of species in the community: $n_s^* < n_s$. This assumption guarantees that the species in a coalition vector corresponding to an ESS are subject to invasion by at least one mutant strategy, an essential requirement in the ESS definitions. It is assumed that both the virtual variable and all the strategies in the community must lie within a constraint set $\mathcal{U}$ as defined in Chapter 4. The reader just interested in applications may skip most of the details and simply make note of the various versions of the maximum principle for use in the following chapters. All important results are stated as theorems.

## 7.1 Maximum principle for $G$-functions with scalar strategies

**Theorem 7.1.1 (ESS-scalar)** *Let $G(v, \mathbf{u}, \mathbf{x})$ be the fitness generating function for a community defined by a $G$-function with scalar strategies. For a given $\mathbf{u} \in \mathcal{U}$ assume that there exists an ESE designated by $\mathbf{x}^*$ and let $\mathbf{u}$ be partitioned in such a way that the first $n_{s^*} < n_s$ components of $\mathbf{u}$ make up the coalition vector $\mathbf{u}_c$*

$$\mathbf{u} = \begin{bmatrix} u_1 & \cdots & u_{n_s} \end{bmatrix} = \begin{bmatrix} \mathbf{u}_c \mid \mathbf{u}_m \end{bmatrix}.$$

*If the coalition vector $\mathbf{u}_c$ is an ESS for $\mathbf{x}^*$ then*

$$\max_{v \in \mathcal{U}} G(v, \mathbf{u}, \mathbf{x}^*) = G(v, \mathbf{u}, \mathbf{x}^*)\big|_{v=u_i} = 0 \qquad (7.1)$$

*for $i = 1, \ldots, n_{s^*}$.*

*Proof.* From the ESE lemma 6.2.1, a necessary condition for an ecological equilibrium is given by

$$H_i(\mathbf{u}, \mathbf{x}^*) = 0 \quad \text{for} \quad i = 1, \ldots, n_{s^*}$$
$$H_i(\mathbf{u}, \mathbf{x}^*) \le 0 \quad \text{for} \quad i = n_{s^*}+1, \ldots, n_s.$$

## 7.1 Maximum principle for G-functions with scalar strategies

It follows that for any $i = 1, \ldots, n_{s^*}$ and any $j = n_{s^*} + 1, \ldots, n_s$

$$H_j\left(\mathbf{u}, \mathbf{x}^*\right) \leq H_i\left(\mathbf{u}, \mathbf{x}^*\right) = 0$$

or in terms of the $G$-function

$$G(v, \mathbf{u}, \mathbf{x}^*)\big|_{v=u_j} \leq G(v, \mathbf{u}, \mathbf{x}^*)\big|_{v=u_i} = 0.$$

The ESS definition requires that $\mathbf{x}^*$ remain an ESE for any $\mathbf{u}_m \in \mathcal{U}$. Condition (7.1) satisfies this requirement. ∎

Equation (7.1) is a very compact way of stating the following. Let $\mathbf{u}_c$ be an ESS. If one substitutes the vector $\mathbf{u}$ and the ecological stable equilibrium point $\mathbf{x}^*$ into the $G$-function[1] then this function must take on a maximum value of zero with respect to $v$ evaluated over the set $\mathcal{U}$ for every strategy in the coalition vector $\mathbf{u}_c$. The necessary conditions as provided by the ESS maximum principle are constructive in the sense that we can use them to find ESS candidates. For example, if the strategies are unconstrained, then we have the following necessary conditions

$$\left.\frac{\partial G(v, \mathbf{u}, \mathbf{x}^*)}{\partial v}\right|_{v=u_i} = 0, \quad i = 1, \ldots, n_c \tag{7.2}$$

along with

$$G(v, \mathbf{u}, \mathbf{x}^*)\big|_{v=u_i} = 0, \quad i = 1, \ldots, n_c$$

to solve for an ESS candidate solution $\mathbf{u}_c$ and equilibrium density $\mathbf{x}^*$ (only $n_c$ equilibrium equations are needed since $x_i^* = 0$ for $i = n_c + 1, \ldots, n_s$). Note that (7.2) is also an equilibrium condition for the first-order strategy dynamics. Thus, equilibrium in the first-order strategy dynamics will be assured. Furthermore, since the ESS maximum principle applies to all three dynamical models, any solution obtained using these conditions will be an ESS candidate for all three models. However, there is no assurance that if the candidate solution is an ESS for one model it will be an ESS for all three models. Convergent stability must be checked for each model independently. For example, it is possible for an ESS candidate to be convergent stable for the differential equation model (and hence an ESS) but not convergent stable for one of the difference equation models.

The ESS maximum principle requires that the strategies of an ESS must correspond to global-maximum points of the adaptive landscape obtained by plotting of $G^*(v)$ versus $v$ where by definition

$$G^*(v) = G(v, \mathbf{u}_c, \mathbf{x}^*).$$

---

[1] $G$ is now thought of as a function of $v$ only.

Since we have imposed equilibrium population dynamics, these peaks occur at a fitness value of zero.

**Example 7.1.1 (L–V competition game – coalition of one)** *Example 4.3.1 introduced the L–V competition game using the Lotka–Volterra G-function from Example 4.1.1*

$$G_i(v, \mathbf{u}, \mathbf{x}) = \frac{r}{K(v)} \left[ K(v) - \sum_{j=1}^{n_s} a(v, u_j) x_j \right]$$

*with a symmetric distribution for the carrying capacity*

$$K(v) = K_m \exp\left(-\frac{v^2}{2\sigma_k^2}\right)$$

*and a non-symmetric distribution function*

$$a(v, u_j) = 1 + \exp\left[-\frac{(v - u_j + \beta)^2}{2\sigma_a^2}\right] - \exp\left[-\frac{\beta^2}{2\sigma_a^2}\right].$$

*Let us now seek an ESS coalition of one using Theorem 7.1.1. We first set*

$$G(u_1, u_1, x_1^*) = 0$$

*and use it to determine $x_1^*$ as a function of $u_1$. We obtain*

$$x_1^* = K_m \exp\left(\frac{u_1^2}{2\sigma_k^2}\right).$$

*We then set*

$$\left.\frac{\partial G(v, u_1, x_1^*)}{\partial v}\right|_{v=u_1} = 0$$

*to obtain*

$$u_1 = \beta \left(\frac{\sigma_k}{\sigma_\alpha}\right)^2 \exp\left(-\frac{\beta^2}{2\sigma_a^2}\right).$$

*As in Example 6.2.1, consider again the specific case with*

$$r = 0.25$$
$$K_m = 100$$
$$\sigma_\alpha^2 = \sigma_k^2 = \beta^2 = 4.$$

*For these parameters, the above equations yield the same results as we obtained in Example 6.2.1 using Darwinian dynamics*

$$u_c = u_1 = 1.213$$
$$x^* = x_1^* = 83.19.$$

## 7.1 Maximum principle for G-functions with scalar strategies

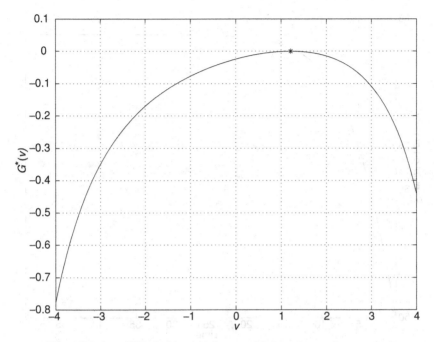

**Figure 7.1** At an ESS, $G^*(v)$ must take on a global maximum when $v = u_1$.

We must now check to see whether this solution yields a global maximum to $G(v, u_1, x_1^*)$. This is most easily done by simply plotting the $G^*$-function vs. $v$ as illustrated in Figure 7.1. Since the solution yields a global maximum at $u_1 = 1.213$, it follows that Theorem 7.1.1 is satisfied by this solution. However, we cannot claim this is an ESS until we demonstrate that the solution is at least a local ESE (implying, it is convergent stable). The ESS maximum principle is a necessary condition and, this being so, solutions that satisfy the ESS maximum principle are only candidate ESS solutions. We must rely on the ESS definition for completeness. An easy way to demonstrate convergent stability is by using Darwinian dynamics. We have already done this in Example 6.2.1 and we know that strategy dynamics will result in $u_1 = 1.213$. One can use the population dynamics (Section 6.2.5) to show that the ecological dynamics converges on the equilibrium population size. For example, choosing $u_1 = 1.213$ and $x_1 = 50$ we see from Figure 7.2 that the differential equation system quickly returns to $x_1^*$. Strictly speaking one must check all points in the neighborhood of $x_1^*$. An alternative eigenvalue analysis may also be used (Section 5.11); however, this analysis is difficult to apply to solutions involving an ESS coalition greater than one.

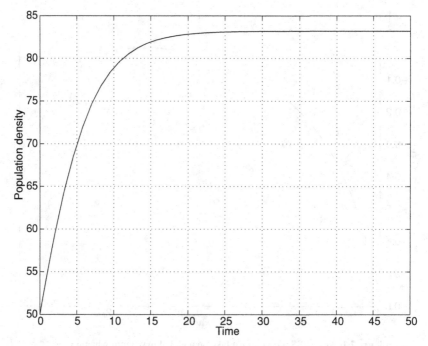

**Figure 7.2** A convergent stable system will return to $\mathbf{x}^*$ when $\mathbf{u} = \mathbf{u}_c$.

**Example 7.1.2 (L–V competition game – coalition of two)** *By changing to $\sigma_k^2 = 12.5$, the above game produces a more interesting ESS candidate solution. In this case, the conditions used to obtain a coalition of one yield*

$$u_1 = 3.639$$
$$x_1^* = 57.58.$$

*This time, the plot of $G^*(v) = G\left(v, u_1, x_1^*\right)$ vs. $v$ as illustrated in Figure 7.3 does not produce a global maximum at $u_1$. Instead $u_1$ resides at a local minimum – almost a saddle – of the adaptive landscape. The shape of the curve suggests an ESS coalition of two. Since the ESS maximum principle cannot be satisfied by an ESS coalition of one, we must seek an ESS coalition of two. This requires the simultaneous solution of the following system of equations*

$$G\left(u_1, \mathbf{u}, \mathbf{x}^*\right) = 0$$
$$G\left(u_2, \mathbf{u}, \mathbf{x}^*\right) = 0$$
$$\left.\frac{\partial G(v, \mathbf{u}, \mathbf{x}^*)}{\partial v}\right|_{v=u_1} = 0$$
$$\left.\frac{\partial G(v, \mathbf{u}, \mathbf{x}^*)}{\partial v}\right|_{v=u_2} = 0$$

## 7.1 Maximum principle for G-functions with scalar strategies

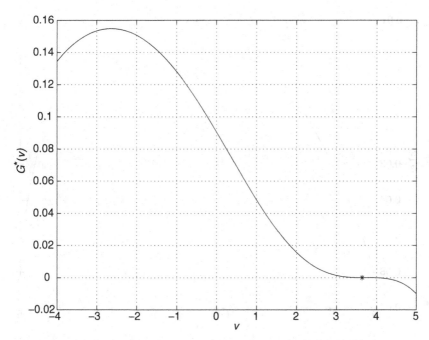

**Figure 7.3** The solution obtained does not satisfy the ESS maximum principle.

where $\mathbf{u} = \begin{bmatrix} u_1 & u_2 \end{bmatrix}$ and $\mathbf{x} = \begin{bmatrix} x_1 & x_2 \end{bmatrix}$. These equations are more complicated. For example the first equilibrium condition is given by

$$r - \frac{r}{K_m \exp\left(-\frac{u_1^2}{2\sigma_k^2}\right)} \left[ x_1 + \left\{ 1 + \exp\left[-\frac{(u_1 - u_2 + \beta)^2}{2\sigma_a^2}\right] \right. \right.$$
$$\left. \left. - \exp\left[-\frac{\beta^2}{2\sigma_a^2}\right] \right\} x_2 \right] = 0.$$

*It follows that an analytical solution is very difficult without the help of symbolic manipulation software.* One obtains

$$\mathbf{u}_c = \begin{bmatrix} u_1 & u_2 \end{bmatrix} = \begin{bmatrix} 3.036 & -0.3320 \end{bmatrix}$$
$$\mathbf{x}^* = \begin{bmatrix} x_1^* & x_1^* \end{bmatrix} = \begin{bmatrix} 52.27 & 37.58 \end{bmatrix}.$$

*However, Darwinian dynamics may also be used to find this equilibrium solution. Once a solution is obtained (by any method), it should be checked to see whether (7.1) is satisfied. One advantage with using Darwinian dynamics to obtain a solution is that this trajectory is convergent stable. While it does not prove that the solution point is an ESE (since not all possible trajectories have been checked) it is at least a good indication that the solution is convergent*

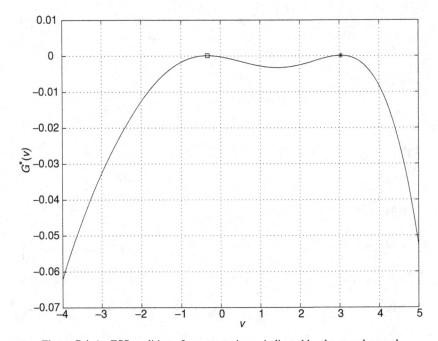

**Figure 7.4** An ESS coalition of two strategies as indicated by the open box and asterisk.

stable. We illustrate the procedure by starting with two species using the same parameters as above, but with initial conditions

$$x_1(0) = x_2(0) = 10$$

and strategies

$$u_1(0) = 0, \quad u_2(0) = -1.$$

Integrating the population and strategy dynamics equations for the differential equation case (Subsection 6.2.5) with $\sigma_1^2 = \sigma_2^2 = 0.1$ results in the equilibrium solution

$$x_1^* = 52.27, \quad x_2^* = 37.58$$

and the ESS coalition of two candidate

$$u_1 = 3.036 \quad u_2 = -0.3320.$$

We now check this candidate by plotting $G^*(v) = G\left(v, u_1, u_2, x_1^*, x_2^*\right)$ as illustrated in Figure 7.4. We see that this solution satisfies the ESS maximum principle and is most likely convergent stable (thus a true ESS).

## 7.2 Maximum principle for $G$-functions with vector strategies

**Theorem 7.2.1 (ESS – vector)** *Let $G(\mathbf{v}, \mathbf{u}, \mathbf{x})$ be the fitness generating function for a community defined by a $G$-function with vector strategies. For a given $\mathbf{u} \in \mathcal{U}$ assume that there exists an ESE designated by $\mathbf{x}^*$ and let $\mathbf{u}$ be partitioned in such a way that the first $n_c \leq n_s$ components of $\mathbf{u}$ make up the coalition vector $\mathbf{u}_c$*

$$\mathbf{u} = [\,\mathbf{u}_1 \mid \cdots \mid \mathbf{u}_{n_s}\,] = [\,\mathbf{u}_c \mid \mathbf{u}_m\,].$$

*If the coalition vector $\mathbf{u}_c$ is an ESS for $\mathbf{x}^*$ then*

$$\max_{\mathbf{v} \in \mathcal{U}} G(\mathbf{v}, \mathbf{u}, \mathbf{x}^*) = G(\mathbf{v}, \mathbf{u}, \mathbf{x}^*)\big|_{\mathbf{v}=\mathbf{u}_i} = 0 \qquad (7.3)$$

*for $i = 1, \ldots, n_c$.*

*Proof.* From Lemma 6.2.1 (this lemma is valid for both scalar and vector strategies), a necessary condition for an ecological equilibrium is given by

$$H_i(\mathbf{u}, \mathbf{x}^*) = 0 \quad \text{for} \quad i = 1, \ldots, n_{s^*}$$
$$H_i(\mathbf{u}, \mathbf{x}^*) \leq 0 \quad \text{for} \quad i = n_{s^*}+1, \ldots, n_s.$$

It follows that for any $i = 1, \ldots, n_{s^*}$ and any $j = n_{s^*}+1, \ldots, n_s$

$$H_j(\mathbf{u}, \mathbf{x}^*) \leq H_i(\mathbf{u}, \mathbf{x}^*) = 0$$

or in terms of the $G$-function

$$G(\mathbf{v}, \mathbf{u}, \mathbf{x}^*)\big|_{\mathbf{v}=\mathbf{u}_j} \leq G(\mathbf{v}, \mathbf{u}, \mathbf{x}^*)\big|_{\mathbf{v}=\mathbf{u}_i} = 0.$$

The ESS definition requires that $\mathbf{x}^*$ remain an ESE for any $\mathbf{u}_m \in \mathcal{U}$. Condition (7.3) satisfies this requirement. ∎

For the case of $n_u = 1$ or 2 we have a nice graphical interpretation of the maximum principle in terms of the adaptive landscape when we plot $G(\mathbf{v}, \mathbf{u}, \mathbf{x}^*)$ vs. $\mathbf{v}$ since we can see the hills and valleys of the landscape on such a plot. The maximum principle requires that an ESS coalition solution be located on the highest peaks (all of which must have the same height). If the strategies are unconstrained, it follows that we may use the requirement that the gradient of the $G$-function with respect to $v$ must be zero at an ESS

$$\left[\frac{\partial G(\mathbf{v}, \mathbf{u}, \mathbf{x}^*)}{\partial \mathbf{v}}\right]_{\mathbf{v}=\mathbf{u}_i} = \mathbf{0}, \quad i = 1, \ldots, n_c.$$

Setting the gradient equal to zero produces $n_c \times n_u$ equations. These equations along with the $n_c$ equilibrium equations

$$G(\mathbf{v}, \mathbf{u}, \mathbf{x}^*)\big|_{\mathbf{v}=\mathbf{u}_i} = 0, \quad i = 1, \ldots, n_c$$

give $n_c \times (n_u + 1)$ equations. These equations can be solved for the $n_c \times (n_u + 1)$ unknowns contained in $\mathbf{u}$ and $\mathbf{x}^*$. However, depending on the complexity of the problem under consideration, using these equations may not be the best way to solve for ESS candidates. As the following examples illustrate, it is often easier to use Darwinian dynamics to find a solution and then verify that the solution obtained satisfies the ESS maximum principle.

**Example 7.2.1 (L–V big bully game – coalition of one)** *This game, introduced in Example 4.4.1, uses the Lotka–Volterra G-function*

$$G(\mathbf{v}, \mathbf{u}, \mathbf{x}) = \frac{r}{K(\mathbf{v})} \left[ K(\mathbf{v}) - \sum_{j=1}^{r} a(\mathbf{v}, \mathbf{u}_j) x_j \right]$$

*with a vector-valued strategy that has two components. The first component influences both the carrying capacity*

$$K(\mathbf{v}) = \left(1 - v_2^2\right) K_{\max} \exp\left(-\frac{v_1^2}{2\sigma_k^2}\right)$$

*and the competition coefficient*

$$a(\mathbf{v}, \mathbf{u}_j) = 1 + B_j \exp\left[-\frac{(v_1 - u_{j1} + \beta)^2}{2\sigma_a^2}\right] - \exp\left[-\frac{\beta^2}{2\sigma_a^2}\right].$$

*The second component of an individual's strategy, $v_2$, influences both the carrying capacity and the competition coefficients via a "bully" function*

$$B_j = 1 + B_{\max}(u_{j2} - v_2).$$

*We set the model's parameters equal to the following values.*

$$K_{\max} = 100$$
$$R = 0.25$$
$$\sigma_\alpha^2 = 4$$
$$\sigma_k^2 = 2$$
$$\beta = 2$$
$$B_{\max} = 1.$$

*Due to the complexity of the problem we use Darwinian dynamics to look for*

## 7.2 Maximum principle for G-functions with vector strategies

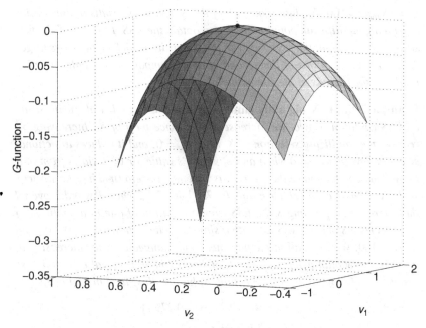

**Figure 7.5** An ESS coalition of one strategy. Regardless of the number of starting species or their initial strategy values, adaptive Darwinian results in the single-strategy ESS.

*candidate solutions. We start by assuming an ESS coalition of one. For example, choose* $\mathbf{u}_1(0) = [0 \quad 0]$ *and* $x_1(0) = 100$. *Then integrate Darwinian dynamics (both strategy and population dynamics together) for the differential equation case. This integration continues until the following equilibrium solution results*

$$\mathbf{u}_c = \mathbf{u}_1 = [\,0.6065 \quad 0.2796\,]$$
$$x^* = x_1^* = 84.08.$$

*Figure 7.5 illustrates that this solution satisfies the ESS maximum principle. The candidate ESS appears to be global with respect to convergence stability and with respect to the number of species. Regardless of the starting value for the species's strategy, strategy dynamics evolve to* $\mathbf{u}_c$. *If one begins with several species, the strategies of all of the species eventually evolves to the same peak. We conclude that the differential equation model has an ESS coalition of one species. At the ESS, the species has a population size that is less than it would have had at* $\mathbf{u} = [0 \quad 0]$. *When both strategy components equal zero,*

$x^* = K_{\max} = 100$. *Evolution in each of the components results in the species sacrificing equilibrium population size to obtain the ESS. In terms of the first strategy component, the asymmetry in the competition coefficient favors larger values for* $v_1$. *In terms of the second strategy component, the bully function produces the tragedy of the commons:* $v_2^* > 0$.

**Example 7.2.2 (L–V big bully game – coalition of two)** *In order to illustrate an ESS coalition of two, we increase the variance term of K. Increasing* $\sigma_k^2$ *reduces the stabilizing selection of K when* $v_1 = 0$, *and it reduces directional selection towards* $v_1 = 0$ *when an individual's value for* $v_1$ *deviates from 0. If the second component of the strategy is set equal to zero (making the strategy scalar valued), we again have Examples 7.1.1 and 7.1.2 where it was discovered that increasing* $\sigma_k^2$ *changes the ESS from a coalition of one to a coalition of two species. As before, we start with a single species with* $\mathbf{u}(0) = \begin{bmatrix} 0 & 0 \end{bmatrix}$ *and* $x(0) = 100$. *We leave all of the parameters unchanged save for increasing to* $\sigma_k^2 = 8$. *Darwinian dynamics corresponding to the differential equation case result in the following convergent stable solution (illustrated in Figure 7.6)*

$$\mathbf{u}_c = \mathbf{u}_1 = \begin{bmatrix} 2.426 & 0.2796 \end{bmatrix}$$
$$x^* = x_1 = 63.82.$$

*This solution is at a local maximum point of the adaptive landscape. But it is not an ESS. The G-function is greater than zero elsewhere, so that the solution is susceptible to invasion. The convergent stability of this solution can be verified by starting the system with two species,* $n_s = 2$, *in the neighborhood of this local maximum and noting that the system returns to the same peak with the two species using the same* $\mathbf{u}_c$ *strategies. However, if we start the system with two species,* $n_s = 2$, *with strategy values far enough away from the local maximum*

$$\mathbf{u}_1(0) = \begin{bmatrix} -1.2 & 0.3 \end{bmatrix}$$
$$\mathbf{u}_2(0) = \begin{bmatrix} 2.3 & 0.26 \end{bmatrix}$$
$$\mathbf{x}(0) = \begin{bmatrix} 14.49 & 60.99 \end{bmatrix}$$

*then under Darwinian dynamics, we obtain*

$$\mathbf{u}_c = \begin{bmatrix} \mathbf{u}_1 & \mathbf{u}_2 \end{bmatrix}$$
$$\mathbf{u}_1 = \begin{bmatrix} -1.204 & 0.3025 \end{bmatrix}$$
$$\mathbf{u}_2 = \begin{bmatrix} 2.275 & 0.2609 \end{bmatrix}$$
$$\mathbf{x}^* = \mathbf{x}_1 = \begin{bmatrix} 13.69 & 61.72 \end{bmatrix}.$$

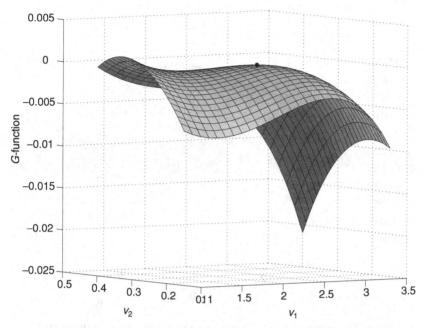

**Figure 7.6** Decreasing the prey's niche breadth from that of Figure 7.5 changes the outcome. When the system is constrained to have a single species, then, regardless of initial conditions, it evolves to a local maximum. This single-species strategy is not an ESS.

*Figure 7.7 illustrates that this solution is an ESS coalition of two. Each species strategy satisfies the ESS maximum principle. This example has two convergent stable solutions. If we start the system with just a single species, it always evolves to the non-ESS local maximum shown in Figure 7.6. If we start with two or more species, either strategy values continue to evolve towards the non-ESS solution of one species, or they evolve to the ESS of two species. In other words, the ESS is locally convergent stable but not globally convergent stable. The strategy values for at least two of the initial species must be sufficiently far apart for the system to evolve to the ESS. It is noteworthy that the adaptive landscape at the ESS provides no insights or clues into the non-ESS convergent stable solution, and the adaptive landscape at the non-ESS convergent stable solution reveals little regarding the actual ESS.*

In Brown *et al.* (2005) a larger value of $\sigma_k^2 = 15$ was investigated for this example. Single species, under adaptive dynamics, now evolve to a non-ESS convergent stable saddlepoint solution. At this point, the first component of the

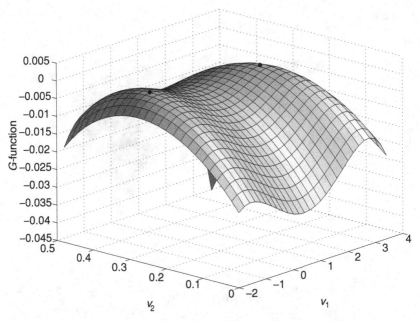

**Figure 7.7** Darwinian dynamics results in ESS when the system starts with two or more species with sufficiently distinct initial strategy values. However, not all starting conditions need produce this result. For some starting conditions (with two or more species) the system will converge on the single, local, non-ESS peak of Figure 7.6.

strategy is subject to disruptive selection (it is at a minimum) and the second component experiences stabilizing selection (it is at a maximum). We can permit **adaptive speciation** by introducing a new species at a strategy value close to the saddle point. With two species, the system evolves to an ESS of two species similar to Figure 7.7.

Under adaptive speciation, with a convergent-stable saddle point it is easy for evolution to produce an ESS coalition of two species (this process is discussed in more detail in Chapter 8). The two species' strategies will diverge from the saddle point no matter how close their starting values. This speciation is being driven by the disruptive selection on the first strategy component. Nonetheless, two species that differ only with respect to their second components will still diverge as initial differences in the second component will drive coadapted changes in the first component.

Both $\sigma_k^2 = 8$ and $\sigma_k^2 = 15$ environments have a convergent stable solution that contains just one species, the first at a local maximum and the second at a saddle point. Thus, changes in a parameter can produce significant changes in the character of non-ESS convergent-stable solutions, yet produce no change in the character of the ESS. This behavior can also apply to scalar strategies (Cohen et al., 1999).

## 7.3 Maximum principle for $G$-functions with resources

Recall from Section 4.5 that this case includes resource dynamics of the form

$$\text{Difference:} \quad \mathbf{y}(t+1) = \mathbf{y} + \mathbf{N}(\mathbf{u}, \mathbf{x}, \mathbf{y})$$
$$\text{Differential:} \quad \dot{\mathbf{y}} = \mathbf{N}(\mathbf{u}, \mathbf{x}, \mathbf{y})$$

where the resource vector $\mathbf{y} = \begin{bmatrix} y_1 & \cdots & y_{n_y} \end{bmatrix}$ and the vector function $\mathbf{N} = \begin{bmatrix} N_1 & \cdots & N_{n_y} \end{bmatrix}$ both have $n_y$ components. In general, the $G$-function will include $\mathbf{y}$ dependence.

**Theorem 7.3.1 (ESS – resource)** *Let $G(\mathbf{v}, \mathbf{u}, \mathbf{x}, \mathbf{y})$ be the fitness generating function for a community defined by a $G$-function with resources. For a given $\mathbf{u} \in \mathcal{U}$ assume that there exists an ESE designated by $\mathbf{x}^*$ and $\mathbf{y}^*$ satisfying*

$$N_i\left(\mathbf{u}, \mathbf{x}^*, \mathbf{y}^*\right) = 0 \quad \text{for} \quad i = 1, \ldots, n_y.$$

*Let $\mathbf{u}$ be partitioned in such a way that the first $n_{s^*} \leq n_s$ components of $\mathbf{u}$ make up the coalition vector $\mathbf{u}_c$*

$$\mathbf{u} = \begin{bmatrix} \mathbf{u}_1 \mid \cdots \mid \mathbf{u}_{n_s} \end{bmatrix} = \begin{bmatrix} \mathbf{u}_c \mid \mathbf{u}_m \end{bmatrix}.$$

*If the coalition vector $\mathbf{u}_c$ is an ESS for $\mathbf{x}^*$ and $\mathbf{y}^*$ then*

$$\max_{v \in \mathcal{U}} G(\mathbf{v}, \mathbf{u}, \mathbf{x}^*, \mathbf{y}^*) = G(\mathbf{v}, \mathbf{u}, \mathbf{x}^*, \mathbf{y}^*)\big|_{v=u_i} = 0 \quad (7.4)$$

*for $i = 1, \ldots, n_c$.*

*Proof.* From Lemma 6.4.1, a necessary condition for an ecological equilibrium is given by

$$H_i\left(\mathbf{u}, \mathbf{x}^*, \mathbf{y}^*\right) = 0 \quad \text{for} \quad i = 1, \ldots, n_{s^*}$$
$$H_i\left(\mathbf{u}, \mathbf{x}^*, \mathbf{y}^*\right) \leq 0 \quad \text{for} \quad i = n_{s^*}+1, \ldots, n_s.$$

It follows that for any $i = 1, \ldots, n_{s^*}$ and any $j = n_{s^*}+1, \ldots, n_s$

$$H_j\left(\mathbf{u}, \mathbf{x}^*, \mathbf{y}^*\right) \leq H_i\left(\mathbf{u}, \mathbf{x}^*, \mathbf{y}^*\right) = 0$$

or in terms of the $G$-function

$$G(\mathbf{v}, \mathbf{u}, \mathbf{x}^*, \mathbf{y}^*)\big|_{v=u_j} \leq G(\mathbf{v}, \mathbf{u}, \mathbf{x}^*, \mathbf{y}^*)\big|_{v=u_i} = 0.$$

The ESS definition requires that $\mathbf{x}^*$ remain an ESE for any $\mathbf{u}_m \in \mathcal{U}$. Condition (7.4) satisfies this requirement. ∎

Recall from Section 6.4 that $\mathbf{y}^*$ is assumed to be locally asymptotically stable. If the strategies are unconstrained, we have the requirement that the gradient of the $G$-function must be zero at an ESS

$$\left[\frac{\partial G(\mathbf{v}, \mathbf{u}, \mathbf{x}^*, \mathbf{y}^*)}{\partial \mathbf{v}}\right]_{\mathbf{v}=\mathbf{u}_i} = \mathbf{0} \quad i = 1, \ldots, n_c. \tag{7.5}$$

Setting the gradient equal to zero produces $n_c \times n_u$ equations. These equations along with the $n_c + n_y$ equilibrium equations

$$G(\mathbf{v}, \mathbf{u}, \mathbf{x}^*)|_{\mathbf{v}=\mathbf{u}_i} = 0 \quad \text{for } i = 1, \ldots, n_c$$
$$N_i(\mathbf{u}, \mathbf{x}^*, \mathbf{y}^*) = \mathbf{0} \quad \text{for } i = 1, \ldots, n_y \tag{7.6}$$

give $n_c \times n_u + n_c + n_y$ equations to solve for the $n_c \times n_u + n_c + n_y$ unknowns contained in the vectors $\mathbf{u}$, $\mathbf{x}^*$, and $\mathbf{y}^*$.

**Example 7.3.1 (Bergmann's rule)** *This game was introduced in Example 4.5.1 with the following G-function and resource dynamics*

$$G(v, \mathbf{u}, \mathbf{x}) = \frac{Av^\alpha y}{1 + AHv^{(\alpha-\beta)}y} - Cv^\gamma$$

$$\dot{y} = r(K - y) - \sum_{i=1}^{n_s} \frac{Au_i^\alpha y x_i}{1 + AHu_i^{(\alpha-\beta)}y}.$$

*Assuming an ESS coalition of one and applying (7.5) and (7.6) to this problem results in the following analytical solution for body size*

$$u = \left[\frac{HC(\alpha - \beta)}{(\alpha - \gamma)}\right]^{\beta-\gamma}. \tag{7.7}$$

*This result satisfies Bergmann's rule: optimal body size should go up when the temperature goes down. This happens because C, the energetic cost of foraging, increases with a decrease in temperature. And u increases with increasing C. We see that this rule results because body size is an ESS! If we now apply (7.7) with the same parameters as used in Examples 5.6.1 and 6.4.1 we obtain*

$$u = 0.6561.$$

*This is the same answer as obtained in Example 5.6.1 using Darwinian dynamics. Plotting the adaptive landscape using the Example 5.6.1 equilibrium solution*

$$x^* = 10.5167, \quad y^* = 54.000, \quad u^* = 0.6561$$

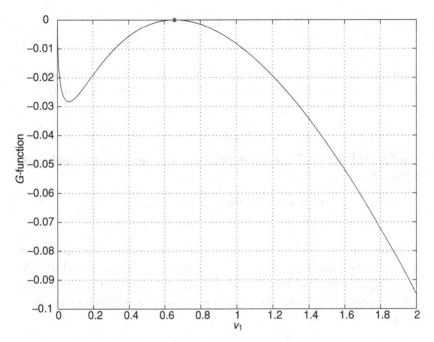

**Figure 7.8** Adaptive landscape for Bergmann's rule $G$-function. Because only a positive body size is allowed $G(v, \mathbf{u}^*, \mathbf{x}^*, y^*)$ has a unique maximum.

we see that this solution satisfies the ESS maximum principle, as illustrated in Figure 7.8. Since $u$ must satisfy the constraint $u > 0$, there is one unique maximum.

## 7.4 Maximum principle for multiple $G$-functions

In the multiple $G$-function case, $n_{s^*}$ and $n_{g^*}$ have a special meaning in the definitions of the coalition vector and the ESS (see Section 6.5).

**Theorem 7.4.1 (ESS – multiple)** *Let $G_j(\mathbf{v}, \mathbf{u}, \mathbf{x})$, $j = 1, \ldots, n_g$ be the fitness generating functions for a community defined by multiple $G$-functions. For a given $\mathbf{u} \in \mathcal{U}$ (constraints are satisfied in all of the bauplans) assume that there exists an ESE designated by $\mathbf{x}^*$ and let $\mathbf{u}$ be partitioned in such a way that the first $n_{s^*} \leq n_s$ components of $\mathbf{u}$ make up the coalition vector $\mathbf{u}_c$*

$$\mathbf{u} = \begin{bmatrix} \mathbf{u}_1 & \cdots & \mathbf{u}_{n_s} \end{bmatrix} = \begin{bmatrix} \mathbf{u}_c & \mathbf{u}_m \end{bmatrix}.$$

*If the coalition vector $\mathbf{u}_c$ is an ESS for $\mathbf{x}^*$ then*

$$\max_{\mathbf{v} \in \mathcal{U}_j} G_j(\mathbf{v}, \mathbf{u}, \mathbf{x}^*) = G_j(\mathbf{v}, \mathbf{u}, \mathbf{x}^*)\big|_{\mathbf{v}=\mathbf{u}_i} = 0 \qquad (7.8)$$

where $i$ and $j$ are determined according to

$$i = r_{j-1} + 1, \ldots, r_j \quad \text{for } j = 1, \ldots, n_{g^*} \tag{7.9}$$

where $n_{g^*}$ is the number of G-functions in the coalition, and $r_i$ is the rising number (see Sections 4.6 and 6.5) for species in the coalition defined by

$$r_0 = 0$$
$$r_i = \sum_{j=1}^{i} n_{s_j^*} \quad \text{for } i = 1, \ldots, n_{g^*}.$$

*Proof.* From Lemma 6.5.1, a necessary condition for an ecological equilibrium is given by

$$H_i(\mathbf{u}, \mathbf{x}^*) = 0 \quad \text{for } i = 1, \ldots, n_{s^*}$$
$$H_i(\mathbf{u}, \mathbf{x}^*) \leq 0 \quad \text{for } i = n_{s^*} + 1, \ldots, n_s.$$

We must now sort out which fitness functions belong to which G-functions. Let all of the fitness functions that belong to G-function $j$ be designated with the superscript $H^j$. Assume that $n_{s_j^*} < n_{s_j}$ for each of the G-functions as well. The above necessary conditions are restated as follows: at an ESE it is necessary that for $j = 1, \ldots, n_{g^*}$

$$H_i^j(\mathbf{u}, \mathbf{x}^*) = 0 \quad \text{for } i = 1, \ldots, n_{s_j^*}$$
$$H_i^j(\mathbf{u}, \mathbf{x}^*) \leq 0 \quad \text{for } i = n_{s_j^*} + 1, \ldots, n_{s_j}.$$

Thus for any $i = 1, \ldots, n_{s_j^*}$ and for any $k = n_{s_j^*} + 1, \ldots, n_{s_j}$

$$H_k^j(\mathbf{u}, \mathbf{x}^*) \leq H_i^j(\mathbf{u}, \mathbf{x}^*) = 0.$$

Using the definition of the G-function, the above condition implies

$$G_j(v, \mathbf{u}, \mathbf{x}^*)\big|_{v=u_k} \leq G_j(v, \mathbf{u}, \mathbf{x}^*)\big|_{v=u_i} = 0.$$

The ESS definition requires that $\mathbf{x}^*$ remain an ESE for any $\mathbf{u}_m \in \mathcal{U}$. Condition (7.8) satisfies this requirement. ∎

If the strategies are unconstrained, it follows that the gradient of the G-function must be zero at an ESS

$$\left[\frac{\partial G_j(v, \mathbf{u}, \mathbf{x}^*)}{\partial \mathbf{v}}\right]_{\mathbf{v}=\mathbf{u}_i} = 0 \quad i = 1, \ldots, n_{c^*}$$

where $i$ and $j$ are determined according to (7.9). Note that setting the gradient equal to zero produces $n_c \times n_u$ equations. These equations along with the $n_c$ equilibrium equations

$$G(v, \mathbf{u}, \mathbf{x}^*)\big|_{v=u_i} = 0, \quad i = 1, \ldots, n_c$$

## 7.4 Maximum principle for multiple G-functions

give $n_c \times (n_u + 1)$ equations to solve for the $n_c \times (n_u + 1)$ unknowns contained in **u** and **x**\*. However, depending on the complexity of the problem under consideration, using these equations may not be the best way to solve for ESS candidates. As the following examples illustrate, it is often easier to use Darwinian dynamics.

**Example 7.4.1 (predator–prey coevolution)** *The G-functions for the prey, $G_1$, and predator, $G_2$, introduced in Example 4.6.1 are given by*

$$G_1(v, \mathbf{u}, \mathbf{x}) = \frac{r_1}{K(v)} \left( K(v) - \sum_{j=1}^{n_{s_1}} a(v, u_j) x_j \right) - \sum_{j=n_{s_1}+1}^{n_s} b(v, u_j) x_j$$

$$G_2(v, \mathbf{u}, \mathbf{x}) = r_2 \left( 1 - \frac{\sum_{j=n_{s_1}+1}^{n_s} x_j}{c \sum_{j=1}^{n_{s_1}} b(v, u_j) x_j} \right)$$

*where*

$$K(v) = K_{\max} \exp\left[-\frac{v^2}{\sigma_k^2}\right]$$

$$a(v, u_j) = \exp\left[-\frac{(v - u_j)^2}{\sigma_a^2}\right]$$

$$b(v, u_j) = b_{\max} \exp\left[-\frac{(v - u_j)^2}{\sigma_b^2}\right].$$

*We have been using the following set of parameters*

$$r_1 = r_2 = c = 0.25$$
$$K_{\max} = 100$$
$$b_{\max} = 0.15$$
$$\sigma_k^2 = 2$$
$$\sigma_a^2 = 4.$$

*Using Darwinian dynamics, we obtained in Example 6.5.1 the following equilibrium solutions:*
$\sigma_b^2 = 10, \Rightarrow$ *one prey and one predator*

$$u_1 = u_2 = 0, \quad x_1^* = 30.77, \quad x_2^* = 1.154$$

$\sigma_b^2 = 4, \Rightarrow$ *two prey and one predator*

$$u_1 = 0.90, \, u_2 = -0.90, \, u_3 = 0, \quad x_1^* = x_2^* = 19.35, \quad x_3^* = 1.19$$

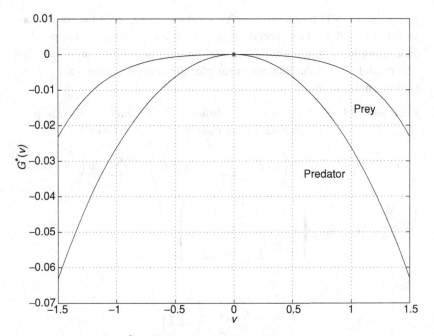

**Figure 7.9** Using $\sigma_b^2 = 10$ results in an ESS coalition with one prey and one predator. There is an illusion that the landscape for the prey dips. It is actually a true maximum as is the predator.

$\sigma_b^2 = 1, \Rightarrow$ two prey and two predators

$$u_1 = 0.79, \ u_2 = -0.79, \ u_3 = 0.56, \ u_4 = -0.56$$
$$x_1^* = x_2^* = 28.71, \ x_3^* = x_4^* = 0.60.$$

*In a similar fashion, we obtain using*
$\sigma_b^2 = 0.75, \Rightarrow$ three prey and two predators

$$u_1 = 0.83, \ u_2 = 0, \ u_3 = -0.83, \ u_4 = 0.55, \ u_5 = -0.55$$
$$x_1^* = 25.37, \ x_2^* = 9.72, \ x_3^* = 25.37, \ x_4^* = x_5^* = 0.59.$$

*Using the adaptive landscape, all of these solutions are shown to satisfy the ESS maximum principle. For clarity, we plot both the predator and prey landscapes in the same figure. Figure 7.9 illustrates the $\sigma_b^2 = 10$ case. Both the prey and the predator have the same strategy value that corresponds to a maximum of $G_1(v, \mathbf{u}, \mathbf{x}^*)$ and $G_2(v, \mathbf{u}, \mathbf{x}^*)$. The other cases demonstrate the effect of decreasing $\sigma_b^2$. Figure 7.10 illustrates the $\sigma_b^2 = 4$ case. The predator strategy is again $u_3 = 0$, with the prey strategies symmetric above and below the predator value. Figure 7.11 illustrates the $\sigma_b^2 = 1$ case. We now have two symmetric*

## 7.4 Maximum principle for multiple G-functions

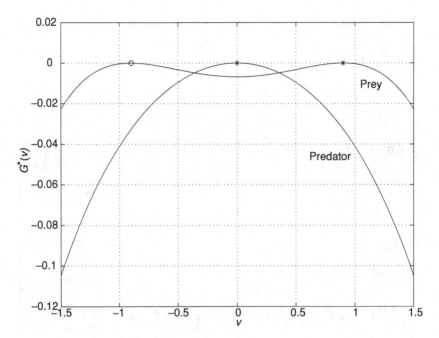

**Figure 7.10** Using $\sigma_b^2 = 4$ results in an ESS coalition with one prey and two predators.

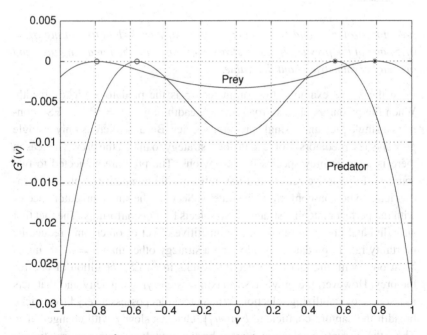

**Figure 7.11** Using $\sigma_b^2 = 1$ results in an ESS coalition with two prey and two predators.

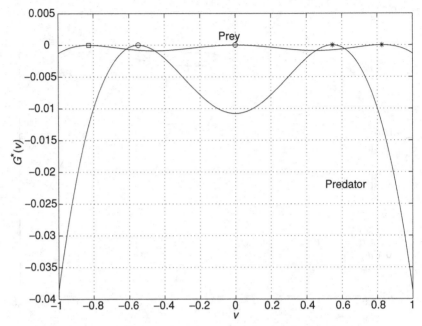

**Figure 7.12** Using $\sigma_b^2 = 0.75$ results in an ESS coalition with three prey and two predators.

*predator strategies and two symmetric prey strategies. Finally, setting $\sigma_b^2 = 0.75$, one of the prey again has a strategy of zero with the remaining prey and predators symmetric about this value.*

In the above example, the parameter $\sigma_b^2$ is the predator's niche breadth. When the predators have a broad niche breadth (e.g., $\sigma_b^2 = 10$) the ESS contains a single prey and a single predator species. Because there is only a single prey species, predators evolve to the best strategy to match the prey. The reason there is a single prey species is not obvious. The prey are subjected to two sources of disruptive selection. First, intraspecific competition selects for individuals to be non-conformist in strategy. Second, the single predator species with the perfect matching strategy also selects for nonconformity. The fact that an individual prey can reduce the competitive effect of others and reduce its mortality rate to predators by adopting a strategy other than $u = u_1$ seems to be at odds with the fact that selection results in an ESS coalition of one for the prey. However, the prey's ESS maximizes carrying capacity and that acts as a source of stabilizing selection. When predators possess a very broad niche breadth the capture coefficient, $b(v, u_j)$, changes slowly with changes in $v$. Thus, the disruptive selection exerted by the predators is small in relation to the stabilizing selection of a higher carrying capacity. The resultant ESS has a single prey species. When the predator has a narrower niche breadth, $\sigma_b^2 < 4.5$, the disruptive selection exerted by the predators on the single prey species is increased.

## 7.5 Maximum principle for $G$-functions in terms of population frequency

This category of $G$-functions is for those situations where it is useful to work in terms of the frequency of individuals using a particular strategy

$$p_i = \frac{x_i}{N}$$

and total number of individuals in a population $N$

$$N = \sum_{i=1}^{n_s} x_i$$

rather than density $x_i$.

**Theorem 7.5.1 (ESS – frequency)** *Let $G(\mathbf{v}, \mathbf{u}, \mathbf{p}, N)$ be the fitness generating function for a community defined by a G-function in terms of population frequency. For a given $\mathbf{u}$ assume that there exists an ESE designated by $\mathbf{p}^*$ and $N^*$. Let $\mathbf{u}$ be partitioned in such a way that the first $n_c \leq n_s$ components of $\mathbf{u}$ make up the coalition vector $\mathbf{u}_c$*

$$\mathbf{u} = [\, \mathbf{u}_1 \mid \cdots \mid \mathbf{u}_s \,] = [\, \mathbf{u}_c \mid \mathbf{u}_m \,].$$

*If the coalition vector $\mathbf{u}_c \in \mathcal{U}$ is an ESS for $\mathbf{p}^*$ and $N^*$ then*

$$\max_{\mathbf{v} \in \mathcal{U}} G(\mathbf{v}, \mathbf{u}, \mathbf{p}^*, N^*) = G(\mathbf{v}, \mathbf{u}, \mathbf{p}^*, N^*)\big|_{\mathbf{v}=\mathbf{u}_i} = 0 \quad (7.10)$$

*for $i = 1, \ldots, n_c$.*

*Proof.* From Lemma 6.6.1, a necessary condition for an ecological equilibrium is given by

$$H_i\left(\mathbf{u}, \mathbf{p}^*, N^*\right) = 0 \quad \text{for} \quad i = 1, \ldots, n_{s^*}$$
$$H_i\left(\mathbf{u}, \mathbf{p}^*, N^*\right) \leq 0 \quad \text{for} \quad i = n_{s^*}+1, \ldots, n_s.$$

It follows that for any $i = 1, \ldots, n_{s^*}$ and any $j = n_{s^*}+1, \ldots, n_s$

$$H_j\left(\mathbf{u}, \mathbf{p}^*, N^*\right) \leq H_i\left(\mathbf{u}, \mathbf{p}^*, N^*\right) = 0$$

or in terms of the $G$-function

$$G(\mathbf{v}, \mathbf{u}, \mathbf{p}^*, N^*)\big|_{\mathbf{v}=\mathbf{u}_j} \leq G(\mathbf{v}, \mathbf{u}, \mathbf{p}^*, N^*)\big|_{\mathbf{v}=\mathbf{u}_i} = 0.$$

The ESS definition requires that $\mathbf{p}^*$ remain an ESE for any $\mathbf{u}_m \in \mathcal{U}$. Condition (7.10) satisfies this requirement. ∎

We will see in Chapter 9 that the equilibrium assumption for $N$ is generally not required in matrix games.

**Example 7.5.1 (L–V competition game in terms of frequency)** *This game (scalar bauplan example) reformulated in terms of frequency is defined by*

$$G(v, \mathbf{u}, \mathbf{p}, N) = \frac{r}{K(v)} \left( K(v) - N \sum_{j=1}^{n_s} a(v, u_j) p_j \right),$$

with

$$K(v) = K_m \exp\left\{ -\frac{v^2}{2\sigma_k^2} \right\}$$

$$\alpha(v, u_i) = 1 + \exp\left\{ -\frac{(v - u_i + \beta)^2}{2\sigma_\alpha^2} \right\} - \exp\left\{ -\frac{\beta^2}{2\sigma_\alpha^2} \right\}.$$

*Here we can examine the solutions obtained in Examples 5.8.1 and 6.6.1 by plotting the adaptive landscape to see whether the ESS maximum principle is satisfied. With $\sigma_k^2 = 12.5$, we obtained a solution with a coalition of two strategies*

$$\mathbf{u}_c = \begin{bmatrix} u_1 & u_2 \end{bmatrix} = \begin{bmatrix} 3.1291 & -0.2397 \end{bmatrix}$$
$$\mathbf{p}^* = \begin{bmatrix} 0.5652 & 0.4348 \end{bmatrix}$$
$$N^* = 90.347.$$

*Figure 7.13 illustrates the corresponding adaptive landscape. Clearly the ESS maximum principle is satisfied by this solution. To generate this solution, Darwinian dynamics may be used by starting with two strategies with arbitrary initial conditions. It is worth noting that not all initial conditions need result in the same solution. For example using*

$$p_1(0) = 0.5, \quad p_2(0) = 0.5, \quad N(0) = 50, \quad u_1(0) = u_2(0) = 0.3$$

*we obtain the equilibrium solution*

$$p_1^* = 0.5, \quad p_2^* = 0.5, \quad N^* = 56.28, \quad u_1 = u_2 = 3.791.$$

*The corresponding adaptive landscape is illustrated in Figure 7.14. It is clear that this solution does not satisfy the ESS maximum principle and hence is not*

## 7.5 Maximum principle for G-functions

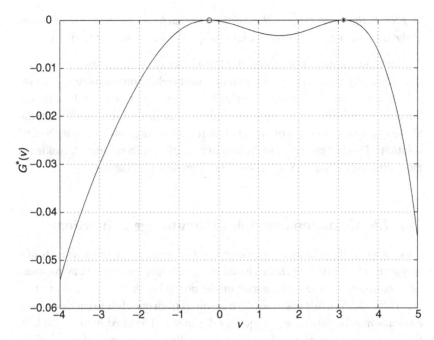

**Figure 7.13** An ESS coalition of two strategies as indicated by the circle and asterisk.

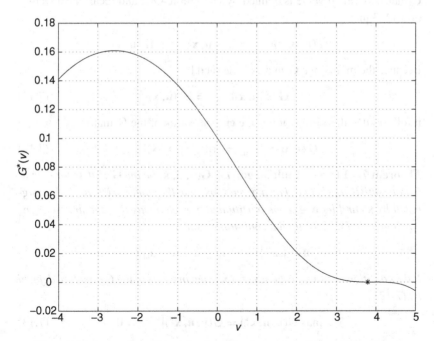

**Figure 7.14** A case where Darwinian dynamics does not result in an ESS solution.

an ESS. In this case, the initial conditions are such that the two species evolve to the same result as obtained if only one species were allowed to evolve.

Two requirements are necessary for Darwinian dynamics to drive a system to the ESS. In this case, the initial conditions must include one or more species for a coalition of one, two or more species for a coalition of two, etc. In addition, the initial conditions must lie in the region of attraction for the ESS. In the above example the region of attraction is relatively large and it is easy to find solutions. In other problems the region of attraction is more limited, making it more difficult to find ESS solutions using Darwinian dynamics.

## 7.6 Maximum principle for multistage $G$-functions

This category of evolutionary game applies to scenarios where individuals of a species reside in different classes based on ages, stages, genders or environmental circumstances. The evolutionary model now takes the form of a projection matrix (**G**-matrix), that describes the transitions from one class to another. These elements may be density and frequency dependent. The maximum principle in this case is in terms of a scalar $G$-function defined as the critical value of the **G**-matrix. Critical value is defined by Definition 4.8.2 and recall from Definition 4.8.3 that

$$G(v, \mathbf{u}, \mathbf{x})|_{v=u_i} = H_i(\mathbf{u}, \mathbf{x}) = \text{crit } \mathbf{H}_i(\mathbf{u}, \mathbf{x}).$$

Because the multistage **G**-matrix is defined by

$$\mathbf{G}(v, \mathbf{u}, \mathbf{x})|_{v=u_i} = \mathbf{H}_i(\mathbf{u}, \mathbf{x}) \qquad (7.11)$$

it follows that the $G$-function is the **critical value** of the **G**-matrix

$$G(v, \mathbf{u}, \mathbf{x})|_{v=u_i} = \text{crit } \mathbf{G}(v, \mathbf{u}, \mathbf{x})|_{v=u_i}. \qquad (7.12)$$

**Theorem 7.6.1 (ESS – multistage)** *Let* $\mathbf{G}(v, \mathbf{u}, \mathbf{x})$ *be the* **G**-*matrix for a community satisfying (7.11). For a given* $\mathbf{u}$ *assume that there exists an ESE designated by* $\mathbf{x}^*$ *and let* $\mathbf{u} \in \mathcal{U}$ *be partitioned in such a way that the first* $n_{s^*} \leq n_s$ *components of* $\mathbf{u}$ *make up the coalition vector* $\mathbf{u}_c$

$$\mathbf{u} = [\, u_1 \mid \cdots \mid u_{n_s} \,] = [\, \mathbf{u}_c \mid \mathbf{u}_m \,].$$

*If the coalition vector* $\mathbf{u}_c$ *is an ESS for* $\mathbf{x}^*$ *then in terms of the $G$-function defined by (7.12)*

$$\max_{v \in \mathcal{U}} G(v, \mathbf{u}, \mathbf{x}^*) = G(v, \mathbf{u}, \mathbf{x}^*)\big|_{v=u_i} = 0 \qquad (7.13)$$

*for* $i = 1, \ldots, n_{s^*}$.

## 7.6 Maximum principle for multistage G-functions

*Proof.* Working in terms of the scalar fitness function, it follows from Lemma 6.7.2 that a necessary condition for an ecological equilibrium is given by

$$H_i(\mathbf{u}, \mathbf{x}^*) = 0 \quad \text{for} \quad i = 1, \ldots, n_{s^*}$$
$$H_i(\mathbf{u}, \mathbf{x}^*) \leq 0 \quad \text{for} \quad i = n_{s^*} + 1, \ldots, n_s.$$

It follows that for any $i = 1, \ldots, n_{s^*}$ and for any $j = n_{s^*} + 1, \ldots, n_s$

$$H_i(\mathbf{u}, \mathbf{x}^*) \leq H_i(\mathbf{u}, \mathbf{x}^*) = 0$$

or in terms of the scalar $G$-function

$$G(\mathbf{v}, \mathbf{u}, \mathbf{x}^*)\big|_{\mathbf{v}=\mathbf{u}_j} \leq G(\mathbf{v}, \mathbf{u}, \mathbf{x}^*)\big|_{\mathbf{v}=\mathbf{u}_i} = 0.$$

The ESS definition requires that $\mathbf{x}^*$ remains an ESE for any $\mathbf{u}_m \in \mathcal{U}$. Condition (7.13) satisfies this requirement. ∎

Solving for a multistage ESS requires working with both the **G**-matrix and the $G$-function.

**Example 7.6.1 (life cycle example)** *Example 4.8.1 introduced the following **G**-matrix*

$$\mathbf{G}(v, \mathbf{u}, \mathbf{x}) = \begin{bmatrix} f(v) - \sum_{j=1}^{n_x} u_j x_{j1} & v \\ v & -\left(\sum_{j=1}^{n_x} x_{j1}\right) \end{bmatrix}.$$

*As in Example 5.9.1 assume*

$$f(v) = -1 + 4v - v^2.$$

*In Example 6.7.1 we verified, using mutant strategies, that the coalition of one solution*

$$x_{11}^* = 1.781, \quad x_{12}^* = 4.562, \quad u_1^* = 4.562$$

*obtained using Darwinian dynamics satisfied the ESS definition. In this case we can also use the ESS maximum principle to analytically arrive at the above solution. At equilibrium the **G**-matrix is given by*

$$\mathbf{G}(v, \mathbf{u}, \mathbf{x}^*) = \begin{bmatrix} -1 + 4v - v^2 - u_1 x_{11}^* & v \\ v & -x_{11}^* \end{bmatrix}.$$

*Solving for the critical value of **G** (symbolic software is useful here)*

$$L = \text{crit}\left[\mathbf{G}(v, \mathbf{u}, \mathbf{x}^*)\right] = -\frac{1}{2} + 2v - \frac{1}{2}v^2 - \frac{1}{2}u_1 x_{11}^* - \frac{1}{2}x_{11}^* + \frac{1}{2}\sqrt{A + B}$$

where

$$A = 1 - 8vu_1x_{11}^* - 8v + 22v^2 + 2v^2u_1x_{11}^* + (x_{11}^*)^2 - 8v^3 - 2x_{11}^*v^2$$
$$B = v^4 - 2x_{11}^* + 8vx_{11}^* + 2u_1x_{11}^* - 2u_1(x_{11}^*)^2 + u_1^2(x_{11}^*)^2.$$

Setting $\frac{\partial L}{\partial v} = 0$ yields the following solution

$$v = \frac{1}{4}u_1x_{11}^* - \frac{1}{4}x_{11}^* + \frac{3}{2}$$
$$+ \frac{1}{4}\sqrt{u_1^2(x_{11}^*)^2 - 2u_1(x_{11}^*)^2 - 4u_1x_{11}^* + (x_{11}^*)^2 + 4x_{11}^* + 20}.$$

The equilibrium conditions yield

$$(-1 + 4v - v^2 - u_1x_{11}^*)x_{11}^* + vx_{12}^* = 0$$
$$vx_{11}^* - x_{11}^*x_{12}^* = 0.$$

Setting $v = u_1$ in the above three equations and solving these equations simultaneously

$$u_1 = \frac{1}{4}u_1x_{11}^* - \frac{1}{4}x_{11}^* + \frac{3}{2}$$
$$+ \frac{1}{4}\sqrt{u_1^2(x_{11}^*)^2 - 2u_1(x_{11}^*)^2 - 4u_1x_{11}^* + (x_{11}^*)^2 + 4x_{11}^* + 20}$$
$$0 = (-1 + 4u_1 - u_1^2 - u_1x_{11}^*)x_{11}^* + u_1x_{12}^*$$
$$0 = u_1x_{11}^* - x_{11}^*x_{12}^*$$

yields the result

$$x_{11}^* = \tfrac{3}{4} + \tfrac{1}{4}\sqrt{17} = 1.781$$
$$x_{12}^* = \tfrac{5}{2} + \tfrac{1}{2}\sqrt{17} = 4.562$$
$$u_1 = \tfrac{5}{2} + \tfrac{1}{2}\sqrt{17} = 4.562.$$

As shown in Figure 7.15, the adaptive landscape has a global maximum of zero at $u_1 = 4.562$.

While the above elementary model may not fit or address any specific question in life history evolution, it does show how a variety of life history scenarios can generate a similar projection matrix. In Example 4.8.1 we chose to interpret the stages as reproductive and non-reproductive individuals where density dependence and competition only occur among reproductive individuals. Alternatively, these stages can be interpreted as population densities within a source (reproductive state) and sink (non-reproductive state) habitat.

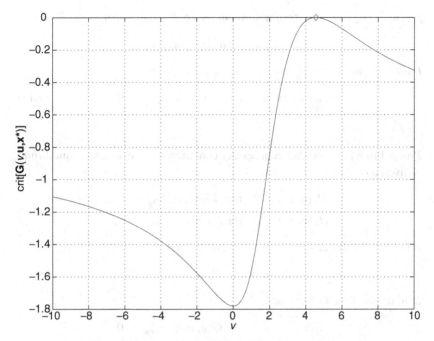

**Figure 7.15** A multistage ESS coalition of one strategy.

## 7.7 Maximum principle for non-equilibrium dynamics

Definition 6.8.1 describes an **ecological cycle** denoted by $\mathbf{x}°$, for the bounded motion of a population corresponding to a periodic orbit, limit cycle, $n$-cycle, or strange attractor. In order to state a maximum principle for these situations, we first define a **compound fitness generating function**.

**Definition 7.7.1 (compound fitness generating function)** *A function $G(\mathbf{v}, \mathbf{u}, \mathbf{x}°)$ is said to be a compound fitness generating function for the population dynamics if and only if*

$$G(\mathbf{v}, \mathbf{u}, \mathbf{x}°)|_{\mathbf{v}=\mathbf{u}_i} = H_i(\mathbf{u}, \mathbf{x}°)$$

*where $H_i(\mathbf{u}, \mathbf{x}°)$ is the compound fitness of Definition 6.8.3.*

Definition 6.8.2 describes an **ecologically stable cycle** (ESC) needed in the following theorem.

**Theorem 7.7.1 (ESS – non-equilibrium)** *Let $G(\mathbf{v}, \mathbf{u}, \mathbf{x}°)$ be the compound fitness generating function for a community defined by a G-function with vector strategies that have non-equilibrium dynamics. For a given $\mathbf{u} \in \mathcal{U}$, assume there exists an ESC designated by $\mathbf{x}°$ and let $\mathbf{u}$ be partitioned in such a way that the* first $n_{s^\circ} \leq n_s$ components of **u** make up the coalition vector $\mathbf{u}_c$

$$\mathbf{u} = \begin{bmatrix} \mathbf{u}_1 \mid \cdots \mid \mathbf{u}_s \end{bmatrix} = \begin{bmatrix} \mathbf{u}_c \mid \mathbf{u}_m \end{bmatrix}.$$

If the coalition vector $\mathbf{u}_c$ is an ESS for $\mathbf{x}^\circ$ then

$$\max_{\mathbf{v} \in \mathcal{U}} G(\mathbf{v}, \mathbf{u}, \mathbf{x}^\circ) = G(\mathbf{v}, \mathbf{u}, \mathbf{x}^\circ)|_{\mathbf{v}=\mathbf{u}_i} = 0 \qquad (7.14)$$

for $i = 1, \ldots, n_{s^\circ}$.

*Proof.* From Lemma 6.8.1, a necessary condition for an ecological equilibrium is given by

$$H_i(\mathbf{u}, \mathbf{x}^\circ) = 0 \quad \text{for} \quad i = 1, \ldots, n_s^\circ$$
$$H_j(\mathbf{u}, \mathbf{x}^\circ) \leq 0 \quad \text{for} \quad j = n_s^\circ + 1, \ldots, n_s.$$

It follows that for any $i = 1, \ldots, n_{s^\circ}$ and for any $j = n_{s^\circ} + 1, \ldots, n_s$

$$H_j(\mathbf{u}, \mathbf{x}^\circ) \leq H_i(\mathbf{u}, \mathbf{x}^\circ) = 0$$

or in terms of the $G$-function

$$G(v, \mathbf{u}, \mathbf{x}^\circ)|_{v=u_j} \leq G(v, \mathbf{u}, \mathbf{x}^\circ)|_{v=u_i} = 0.$$

The ESS definition requires that $\mathbf{x}^\circ$ remain an ESC for any $\mathbf{u}_m \in \mathcal{U}$. Condition (7.14) satisfies this requirement. ∎

A **compound adaptive landscape** based on a compound fitness generating function $G(\mathbf{v}, \mathbf{u}, \mathbf{x}^\circ)$ will have all of the features of plots obtained using the $G$-function with stable equilibrium dynamics. However, in order to plot a compound adaptive landscape the $G$-function must be evaluated at every point of the evolutionary cycle. Hence it is more difficult to generate than our usual adaptive landscape. Fortunately there are some short cuts that can be used.

Bounded non-equilibrium Darwinian dynamics is going either to provide an equilibrium value for **u** or not. If an equilibrium strategy $\mathbf{u}^*$ is obtained then this implies that the maximum value of $G(\mathbf{v}, \mathbf{u}^*, \mathbf{x}^\circ)$ with respect to **v** is independent of **x**. It follows that if $\mathbf{u}^*$ is an ESS then at every point of the cycle, say $\mathbf{x}^*$, the function $G(\mathbf{v}, \mathbf{u}^*, \mathbf{x}^*)$ will have a maximum with respect to **v** at every strategy contained in $\mathbf{u}^*$. However, $G(\mathbf{v}, \mathbf{u}^*, \mathbf{x}^*)$ need not be zero at such points. The advantage here is that the non-equilibrium ESS maximum principle can be checked by using an adaptive landscape based on $G(\mathbf{v}, \mathbf{u}^*, \mathbf{x}^*)$ with the assurance that this adaptive landscape has the same geometry as the compound adaptive landscape. The only difference is a shift in the landscape up or down. If a different $\mathbf{x}^*$ is used it will provide the same landscape shifted from the previous one.

## 7.7 Maximum principle for non-equilibrium dynamics

If an equilibrium strategy for **u** is not obtained from the Darwinian dynamics, then one may estimate a constant **u*** that may have ESS properties provided the fluctuations in **u** are not large. In this case one simply chooses a strategy that represents an average of the fluctuations.

**Example 7.7.1 (non-equilibrium L–V game)** *This is an example of a system in which the maximum value of $G(v, \mathbf{u}, \mathbf{x})$ with respect to $v$ is independent of $\mathbf{x}$. As in Example 6.8.2 the G-function is defined by*

$$G(v, \mathbf{u}, \mathbf{x}) = \frac{r}{K(v)} \left( K(v) - \sum_{j=1}^{n_s} a(v, u_j) x_j \right)$$

where

$$K(v) = K_m \exp\left(-\frac{v^2}{2\sigma_k^2}\right)$$

$$a(v, u_i) = 1 + \exp\left\{-\frac{(v - u_i + \beta)^2}{2\sigma_a^2}\right\} - \exp\left\{-\frac{\beta^2}{2\sigma_a^2}\right\}.$$

*and $r > 2$. Here we use the exponential difference equations, with the parameters*

$$r = 2.6$$
$$K_m = 100$$
$$\sigma_a^2 = \beta^2 = 4, \quad \sigma_k^2 = 12.5.$$

*Darwinian dynamics yields the same equilibrium solution as we obtained using the difference equation model in Example 6.8.2*

$$\mathbf{u}^* = \begin{bmatrix} 3.1294 & -0.2397 \end{bmatrix}.$$

*It is apparent from Figure 7.16 that the ecological cycle is a four-cycle. Figure 7.17 illustrates the adaptive landscape at each point of the four-cycle. A different $\mathbf{x}^*$ (at a different point on the n-cycle) results in the same figure with the plot shifted up or down. In other words, at the n-cycle solution, the strategies remain unchanged in value as they ride up and down on the peaks of the adaptive landscape in synchronization with the fluctuations in population dynamics. It is noted that within each frame the peaks corresponding to the equilibrium strategies have the same value for the G-function (noted at the top of each frame) and that the sum of these values is zero (to within the accuracy of the rounded data). If we replace the exponential difference equations with the difference equations we can again produce a 4-cycle by using the same parameters as above except setting $r = 2.5$ (using $r = 2.6$ results in an 18-cycle). The dynamics is*

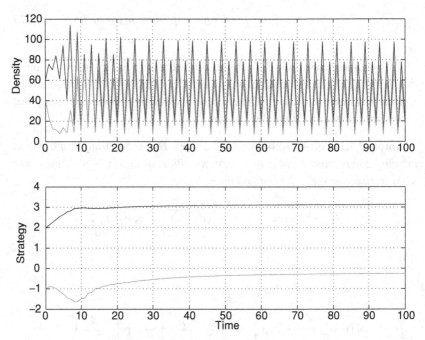

**Figure 7.16** The ecological cycle in this case is a 4-cycle.

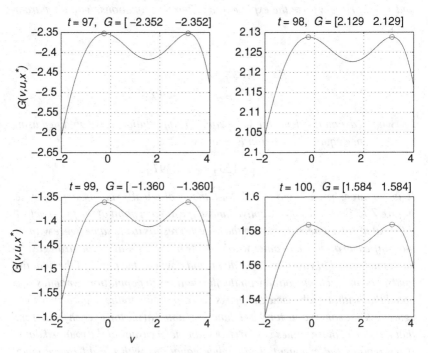

**Figure 7.17** The adaptive landscape at each of the four points of an ecological cycle. The time step and the value of the $G$-function at each peak are noted at the top of each graph.

## 7.7 Maximum principle for non-equilibrium dynamics

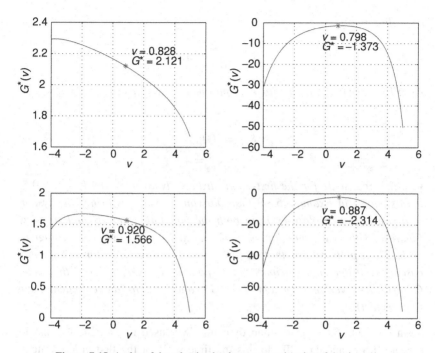

**Figure 7.18** A plot of the adaptive landscape at each point of the 4-cycle.

similar to Figure 7.16 except on the ecological cycle the fluctuations in density lie in a narrower range of approximately 20–60. The strategy dynamics curves are almost identical with the same $\mathbf{u}^*$ obtained as above. Likewise the adaptive landscapes are the same as Figure 7.17 except for vertical adjustments up or down. The following values are obtained for each point of the ecological cycle

$$1 + G\left(u_1, \mathbf{u}, \mathbf{x}^1\right) = 1 + G\left(u_2, \mathbf{u}, \mathbf{x}^1\right) = 2.160$$
$$1 + G\left(u_1, \mathbf{u}, \mathbf{x}^2\right) = 1 + G\left(u_2, \mathbf{u}, \mathbf{x}^2\right) = 0.606$$
$$1 + G\left(u_1, \mathbf{u}, \mathbf{x}^3\right) = 1 + G\left(u_2, \mathbf{u}, \mathbf{x}^3\right) = 1.747$$
$$1 + G\left(u_1, \mathbf{u}, \mathbf{x}^4\right) = 1 + G\left(u_2, \mathbf{u}, \mathbf{x}^4\right) = 0.437.$$

The product of these factors equals 1 (to within the accuracy of the rounded data), in agreement with the non-equilibrium maximum principle (i.e., $G(\mathbf{v}, \mathbf{u}, \mathbf{x}^\circ)|_{\mathbf{v}=\mathbf{u}_i} = 0$ so that $1 + G(\mathbf{v}, \mathbf{u}, \mathbf{x}^\circ)|_{\mathbf{v}=\mathbf{u}_i} = \prod [1 + H_i(\mathbf{u}, \mathbf{x})] = 1$).

**Example 7.7.2 (non-equilibrium L–V game with $\beta$ a function of x)** *This is an example of a system in which maximum values of $G(\mathbf{v}, \mathbf{u}, \mathbf{x})$ with respect to $\mathbf{v}$ are not independent of $\mathbf{x}$. In this case, we have the same model as in the previous example, except that, instead of $\beta$ constant, we introduce $\mathbf{x}$ dependence* *according to*

$$\beta = \frac{1}{50} \sum_{i=1}^{n_s} x_i.$$

*As in Example 6.8.3, we let the other parameters be given by*

$$r = 2.6$$
$$K_m = 100$$
$$\sigma_\alpha^2 = \sigma_k^2 = 4$$

*with the time scale for the first-order strategy dynamics given by $\sigma^2 = 0.5$. It was shown in Example 6.8.3 that Darwinian dynamics brings the system to a convergent stable n-cycle in both the population density and strategy. Figure 7.18 illustrates the adaptive landscape at the four points of the ecological cycle. These plots confirm that the fast-tracking strategy is not an ESS for this case since it does not maximize the G-function at every point of the cycle. Observe, however, that the sum of the $G^*$ values is zero (an ecological cycle requirement).*

In Example 6.8.3, Darwinian dynamics was used to estimate a constant strategy ESS candidate. To do this the time scale of the first-order strategy dynamics was decreased in order to avoid a fast-tracking strategy. This reduces the size of the fluctuations in strategy values allowing for the estimate. For those situations where it is not appropriate to change time scales arbitrarily, we have left some unanswered questions that the reader may be interested in pursuing on his/her own. In those situations where the strategies are fast-tracking:

- Can one always find a constant strategy that will have ESS properties with respect to fast-tracking mutant strategies?
- If so, how does one find it?
- What is the proper definition for an ESS in a world of fast-tracking strategies?

# 8
# Speciation and extinction

The identity of plants and animals rendered by pre-historic artists in cave paintings, rock art, and sculpture is usually recognizable. With the development of language, plants and animals were given names, but a systematic categorization arose more recently when Carl Linnaeus introduced the idea of species as a binomial nomenclature of grouping organisms by genus and species. These early forms of pictorial, vernacular, and formal means of identifying groups of animals and plants were unaltered by later knowledge of evolution and phylogeny. Categorization simply recognizes the following three obvious properties of nature.

- First, individuals (like matter in the universe) tend to be clumped rather than randomly or uniformly spread across the space of all imaginable morphologies, physiologies, and behaviors. This **clumping** of individuals around discrete types can to us be conspicuous – it's hard to misidentify an elephant. However, for some species, it can be really tricky. For example, both humans and male hummingbirds find it nigh impossible to distinguish the species identity of certain female hummingbirds. But, whether tightly clumped (as a planet or asteroid) or only vaguely clumped (more nebula-like), individuals can and seem to be naturally ordered as discrete kinds.
- Second, long before Darwin, heritability was recognized by the fact that kinds tended to breed among themselves (assortative mating). **Assortative mating** can be socially or geographically imposed, or it can be due to physical constraints (elephants and hummingbirds cannot breed in any circumstance). Of course, hybrids occur; yet the very concept of a "hybrid" implies that organisms can be grouped by heritable characteristics and that crossings between groups produce novel, yet predictable, mixes of heritable traits.

- Third, within a species, there can be distinct stages distinguishable by gender (male vs. female, or other mating types such as those in *Paramecium* or some plants), caste (distinct classes of workers and soldiers within ants), and morphology, and niche characteristics (interactions between the individual and the environment can establish particular trajectories of growth and development). It was recognized that gender, castes, or what we now refer to as reaction norms (trait by environment interactions) are in fact part of the **characteristics** that define a species. (Although, as casual bird watchers, many of us have had the embarrassment of identifying two species of birds only to find out later that female and male whydahs look quite different!)

The advent of evolutionary thinking and the idea that different species evolved from common ancestors threatened the notion that individuals group nicely as species. Heritability, and the immutability of species that characterized much pre-Darwinian thought, gave a comfortable permanence to species. The appearance of mutants or "sports" was explained away as imperfections, and in some cases, as abominations against the species's **archetype** or kind. With Darwinian evolution came the plausibility that organisms could exist as a continuum of characteristics. Missing links among species were not only plausible but seemed necessary as verification for Darwin's theory. But where was the evidence for a continuum of missing links? Why do we see individuals clumped around "archetypes" rather than a blurry continuum? In an evolutionary or Darwinian world view, species cannot be assumed as an empirical fact. They must be explained as a consequence of evolutionary processes with proposed verifiable mechanisms for either how species grade from one form to another over time, or how a hitherto single species gives rise to descendant species.

In this chapter we take up the subject of species, speciation, and extinction from the perspective of evolutionary game theory and Darwinian dynamics. We propose that natural selection provides the "gravity" that holds individuals together as discrete clumps in strategy space. This leads to our strategy-species concept. Natural selection becomes central and necessary to the definition and existence of species. We contrast this strategy-species concept with traditional species concepts and definitions. From our perspective, individuals having the same strategy define a species; and distinct forms such as gender, caste, and reaction norms become contingent strategies within the species. With respect to speciation, we will revisit the traditional mechanisms such as allopatric and sympatric speciation. Here, though, frequency-dependent selection is central to promoting diversity whether it be a diversity of species or a diversity of contingent strategies within a species. Natural selection becomes the driver of diversification. Speciation becomes the process by which the different peaks of an adaptive landscape become occupied. While subscribing to most of the traditional perspectives on speciation, evolutionary game theory necessitates making a distinction between speciation from a convergent stable minimum or a convergent stable maximum of the adaptive landscape (Cohen et al., 1999). Finally, we use evolutionary game theory and the $G$-function concept to distinguish between microevolution and macroevolution. Here we stray furthest from orthodoxy. Through the lens of the $G$-function, **microevolution** is thought of as repeatable and reversible evolutionary dynamics within a $G$-function. **Macroevolution** is the non-repeatable and irreversible heritable changes that produce new $G$-functions. From this perspective, the family level of taxonomy may fortuitously represent the rough cutoff between micro- and macroevolution. It would indeed be a happy irony if current taxonomic hierarchies reproduced somewhat faithfully the concept of the $G$-function, even as current phylogenetic approaches view as arbitrary all taxonomic levels above the species.

Extinction is the flip side of speciation from an evolutionary game perspective. The same cave paintings as illustrate the typology of species to ancient naturalists also reveal species that are now extinct. Fossils, the wrecks of past creations to some pre-evolutionary thinkers, provide ample evidence for the existence and disappearance of species. Thomas Jefferson interpreted the mammoth fossil teeth as those of giants who might just be discovered as Lewis and Clark ventured westwards. Sadly (for those of us that love large mammals), mammoths and mastodons are extinct, although only quite recently (some island populations seem to have persisted well into the Neolithic). Ecologically, extinction simply refers to the population size of a species reaching zero. The ecological causes for extinctions include habitat destruction, overexploitation, competitive exclusion, and stochastic extinction in small populations due to environmental, demographic, or genetic stochasticity. From the perspective of game theory we can examine extinctions within an evolutionary context. We will discuss three **contexts for extinctions** to occur: (1) the triaging of species within an invasion-structured community (many more species invade a community than can ecologically persist together), (2) an environmental shift that reduces the number of species at the ESS, and (3) an environmental shift in which a species is driven to extinction before it has time to evolutionarily respond with changes in strategy values that would have prevented extinction.

## 8.1 Species concepts

There is a rich and excellent literature on species concepts and we will only touch on some pertinent highlights here. Using evolutionary thinking, we tend to imbue species with three properties (Freeman and Herron, 2004): (1) species as groups of interbreeding individuals, (2) species as a fundamental unit of evolution, and (3) species as groups of individuals having the same evolutionary trajectory independent of other species.

The **biological species concept** captures these properties best. It simply states that species are individuals that can interbreed (actually or potentially) and that are reproductively isolated from other such populations (Mayr, 1970). A number of features make this definition less useful to evolutionary game theory. It applies only to organisms that exhibit sex or exchange genetic material. Yet asexuality, used by many species, can be recognized as a perfectly good survival strategy. In other words, sex itself is an adaptive strategy. Asexual species are just as likely as sexual species to show a tight aggregation of individuals around some mean phenotype. Also, both sexual and asexual species exhibit a continuum of heritable variation, making both types of species accessible to the methods we have presented in the previous chapters.

Sex may have arisen for its property that allows repair and proofreading of genetic material (Michod, 1998). With recombination, an additional benefit accrues from the exchange of genetic material coding for different traits. In this way novel traits can be swapped between individuals. This speeds strategy dynamics by increasing the rate of coadaptation of trait values within vector strategies. It amounts to the exchange of "ideas." With gametes come adaptations for **anisogamy** (asymmetrically sized gametes such as egg and sperm) where there is adaptive specialization between resource level (the large gamete) and dispersal (the small gamete). With anisogamy comes the inevitable evolution of gender specialization in the production of gametes. The evolution of genders is highly frequency dependent (the game of sex ratios is perhaps the oldest defined evolutionary game (Fisher, 1930), see also Subsection 9.3.1) and in many if not most cases results in an ESS that does not serve the good of the species (e.g., the "cost of males"). With genders come divergent objectives of quality versus quantity of matings between genders. This leads to whole suites of adaptive traits associated with sexual selection. In summary, a species concept built around interbreeding places the conceptual cart before the horse. Interbreeding and associated traits are themselves part of a species's adaptive strategy, not the sole definer of the species. Of course, interbreeding becomes a sure-fire indicator that the individuals do belong to the same species, and so the biological species concept has tremendous operational value, while providing perhaps little insight into why there are species.

The **phylogenetic species concept** (Brooks and McLennan, 1991) defines a species as the terminal nodes of phylogenetic trees as identified by shared derived characters. A species contains all of the descendants of a common ancestor that still share the same values for some set of derived traits. Differences among populations of individuals with respect to derived traits distinguish species. In this way, species are collections of individuals clumped in character space, but the ordering of species is hierarchical, based on monophyletic groupings of individuals into species and species into clades. It is a statistical definition of a species based on specific forms of cluster analysis that describe the distribution of characters among individuals, but does not ascribe evolutionary forces to explain the characters' states and their dynamics. The selection of characters is somewhat arbitrary and much lively debate occurs regarding good vs. bad characters. Characters may be traits under strong selection such as wing morphology in insects, or characters may be silent non-coding gene substitutions that cannot be subject to natural selection. The level of subdivision of the population into species and hence the number of species can depend on the selection of characters. With enough characters, literally every individual of an asexual species becomes its own terminal branch; and with mitochondrial DNA species can be defined to the level of any distinct female lineage. The phylogenetic species concept remains useful for the current statistical methodologies of systematics. But, to the extent that natural selection contributes to the branching of phylogenetic trees and to the collection of individuals around shared characters, it ignores what may be the defining process.

The **morphological species concept** (Mayr, 1970) groups individuals into species based upon recognizable morphologies. It relies on the clumping of individuals in morphological space with the assumption that there is little morphological overlap among different species. It uses the idea of shared characteristics, but, unlike phylogenetic analysis, it represents a less formal statistical definition of a species. Characters are not ordered as a hierarchy of branching and often morphological similarity can be based on gestalt. This concept provides for relatively easy assignment of individuals to species without the need for knowledge of interbreeding (biological species concept) or prescribed measurements of complete character matrices (phylogenetic species concept). Problems occur when distinct morphologies arise from conditional strategies such as gender or caste (remember the male and female whydahs!). It is essential that the morphologies be heritable and inclusive of all the forms that arise conditionally. Although this definition can be criticized as being the least formal or rigorously applicable, it does have some valuable operational properties. While not necessarily explicit, it is often presumed or known that the morphologies are adaptive strategies. Because these morphologies relate directly to the species' ecological niche, this species concept can be valuable to ecologists, linking species diversity to specific morphological traits relating to niche partitioning. Interestingly, while dated, the morphological species concept comes closest to a strategy species concept that we now present as our working definition of a species.

## 8.2 Strategy species concept

We draw on Darwin's postulates (Subsection 1.1.2) to derive a new useful species concept. A population's strategy (heritable phenotypes as provided by the first postulate) becomes an attractive starting point and we have already allowed such strategies to identify species. The second postulate recognizes heritable variation, variability in strategy values within the population. How are we to distinguish between heritable variation within species and variation among species? At this point, we have the equivalence of the morphological species concept. To go beyond, we must consider how variation influences fitness. In previous chapters, we have used the third postulate to justify the $G$-function, strategy dynamics, adaptive landscapes, and the various outcomes of strategy dynamics.

### 8.2.1 Species archetypes

The various equilibrium solutions obtained using Darwinian dynamics are essential. If a population's mean strategy converges to an equilibrium solution, then four types of interesting convergent stable points on the adaptive landscape are obtained: evolutionarily stable global maximum (a global ESS), evolutionarily stable local maximum (a local ESS), convergent stable local minimum, convergent stable saddle point. Any of these convergent stable points may be part of a coalition of one or more strategies. Under a fixed set of parameters, a given $G$-function or set of $G$-functions may have several convergent stable points. These points can vary in type and in the number of strategies in the coalition. Which convergent stable point is achieved by the Darwinian dynamics depends upon the starting conditions in terms of coalition size, strategies, and population sizes.

Suppose that a system has an ESS coalition of three. In accord with the first context for species extinctions, if the system begins with more than three separately evolving strategies, then some of the strategies will become extinct or coevolve to identical strategies. However, under appropriate initial conditions, the same three peaks will always be occupied. Extinctions can occur when two or more strategies climb the adaptive landscape towards the same convergent stable strategy value. For example, when two strategies find themselves on the same slope, and if they evolve at the same speed, the strategy lower down from the peak experiences lower fitness and hence its population declines at the expense of the species higher up the slope. On the other hand, if these two strategies were located on opposite sides of a peak, they could coexist for a long time, unless the landscape shifts to place both species on the same slope. It is also possible for two strategies to coevolve to the same peak. In this case, they become indistinguishable.

We find it useful to define the strategy corresponding to a convergent stable point on the adaptive landscape (see Section 6.2.4) as a **species archetype**. Species archetypes are properties of the $G$-function and the strategy constraint set. Species archetypes exist whether or not a convergent stable point is presently occupied by a strategy in the population; and they exist whether or not any current population is on an evolutionary trajectory towards a given convergent stable point. As convergent stable points, species archetypes serve as attractors that group and clump the strategies of a population around specific values. Natural selection becomes the "gravity" that causes individuals to form distinct clumps in strategy space. The presence and properties of species archetypes are illustrated in the following example.

**Example 8.2.1 (species archetypes)** *Consider the Lotka–Volterra competition model in Example 4.3.1. We can choose a set of parameters such that the model has three different combinations of species archetypes with very different configurations of the adaptive landscape. Using the following parameter values*

$$r = 0.25, \quad K_m = 100$$
$$\sigma_\alpha^2 = 4, \quad \sigma_k^2 = 20, \quad \beta = 2$$

*we begin with and permit just one strategy. Darwinian dynamics results in a convergent stable minimum located at $u = 6.065$ (at a population density of $x^* = 39.864$) as illustrated in Figure 8.1. While this solution is not an ESS, the corresponding strategy represents a species archetype because it is a convergent stable point on the adaptive landscape. The clumping of strategies as a result of this archetype is illustrated in Figures 8.2 and 8.3 where population density is plotted vs. strategy. These figures were obtained by integrating the population dynamics for a time span of 5000 units with 41 strategies uniformly distributed over an interval. The strategies are not allowed to evolve and a population density of 5 is assigned as an initial condition for each strategy.*

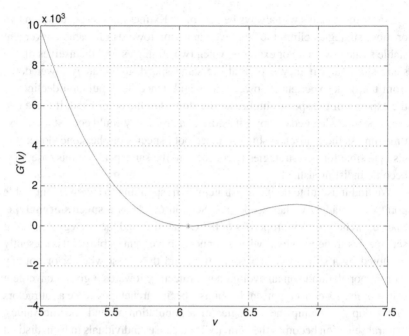

**Figure 8.1** Adaptive dynamics can result in a stable minimum that is not an ESS.

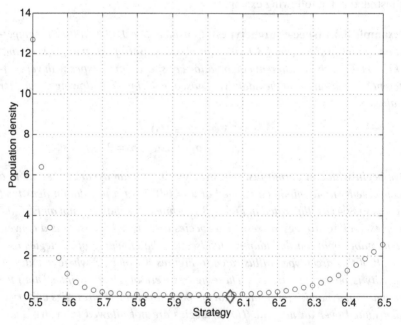

**Figure 8.2** Using a narrow distribution of strategies about the archetype results in the clumping of strategies at the ends of the distribution.

## 8.2 Strategy species concept

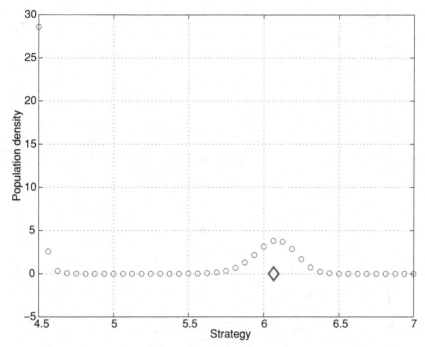

**Figure 8.3** A wider distribution of strategies results in a clumping in the vicinity of the archetype as well as at the left end of the distribution.

*The species archetype is identified by the diamond. The interval* [5.5, 6.5] *used in Figure 8.2 is sufficiently narrow that the clumping of strategies occurs at the endpoints of the interval. This occurs because the landscape "seen" by the strategies is basically the same shape as the valley landscape of Figure 8.1 in the vicinity of the species archetype. The resulting higher fitness at the endpoints of the interval maintains the separation shown. However, if we increase the interval to* [4.5, 7] *as illustrated in Figure 8.3 it is now sufficiently wide that the landscape "seen" by the strategies is warped in such a fashion that higher fitness is now available not only in the vicinity of the species archetype, but at the left end of the interval as well. Note that in each case the archetype provides two clumps of strategies and that the clumps need not be located at the archetype. Figure 8.3 also provides another illustration as to why a single evolving strategy is able to achieve a convergent stable minimum. As we have shown before, as the strategy approaches the minimum, it is still climbing a hill on the continuously changing adaptive landscape until it actually reaches the minimum. If we begin with two separately evolving strategies, the Darwinian dynamics results in strategies that occupy two distinct convergent*

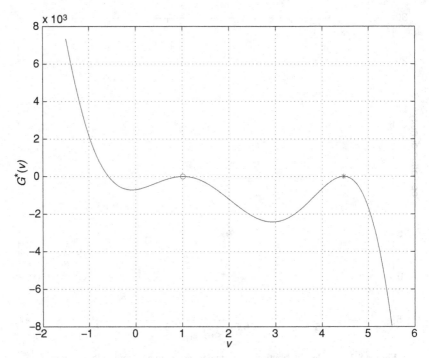

**Figure 8.4** The two species are at evolutionarily stable maxima but they do not compose an ESS.

stable maxima (with $\mathbf{u} = \begin{bmatrix} 1.0139 & 4.4628 \end{bmatrix}$ and $\mathbf{x}^* = \begin{bmatrix} 52.1209 & 38.9977 \end{bmatrix}$). These archetypes are located at local convergent stable maxima as illustrated in Figure 8.4. They are not an ESS, as evidenced by the higher fitness values to the left of the local maximum. The clumping of strategies about these archetypes is evident in Figure 8.5. This figure was generated by starting with 61 strategies uniformly distributed over the interval [0, 6] each with a corresponding starting density of 5. The population dynamic equations (with no strategy dynamics) were integrated for 5000 time units. Comparing this result with Figures 8.2 and 8.3, note how the final configuration for the clumping is totally dependent on the interval over which the strategies are distributed. Many of the rarer strategies would have or will become extinct if we require a strategy to maintain a threshold population density to persist in the community. If we begin with three or more separately evolving strategies, Darwinian dynamics can result in a three strategy archetype (with $\mathbf{u} = \begin{bmatrix} -3.1015 & 0.7692 & 4.2147 \end{bmatrix}$ and $\mathbf{x}^* = \begin{bmatrix} 12.3837 & 47.7584 & 39.3033 \end{bmatrix}$) that together form an ESS as illustrated in Figure 8.6. Performing the same simulation as in the previous case except using

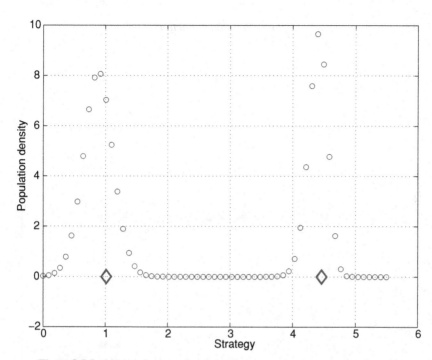

**Figure 8.5** In this case the strategies clump about a two species archetype denoted by the diamonds.

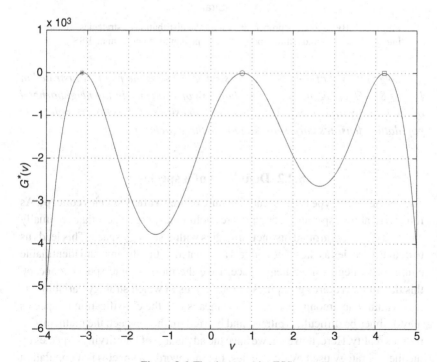

**Figure 8.6** The three-species ESS.

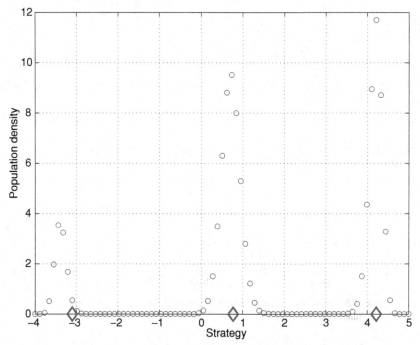

**Figure 8.7** By choosing a proper interval for the distribution of strategies, clumping is obtained around a three species archetype that together form an ESS.

a strategy interval of $[-4, 5]$ with 81 strategies yields the results illustrated in Figure 8.7. Once again we have a clumping of strategies in the neighborhood of the strategy archetypes. The population densities of the clusters reflect the population densities corresponding to the three archetypes.

### 8.2.2 Definition of a species

The species archetype is an essential feature of the **strategy species concept**. As illustrated above, species archetypes are evolutionary organizers that eventually clump individuals around distinct strategies within strategy space. This leads us to define a species as those individuals associated with a distinct and identifiable clump of strategies in strategy space. Like the morphologic species concept, this definition requires groupings of individuals in which strategy variability is less within than among species. It also means that the classification of species may at times be difficult, arbitrary, and blurred. In this respect, the mathematics mimics reality. Here, however, we have the advantage of identifying a species by the mean strategy used by the species. In other words, a **species** is a population of individuals identified by their mean strategy, sharing an identifiable range of strategies, a $G$-function, and a strategy set. This definition is compatible with and builds upon the looser strategy equals species notion we have used in previous chapters. It is also consistent with the strategy distribution requirement used to develop strategy dynamics in Chapter 5.

The mean strategy used by a species may or may not be identical to a species archetype. When they are identical, it is important to keep in mind that the difference between an archetype and a species is that an archetype is a point in strategy space, whereas an extant species exhibits some variation in strategy values [this distinction fits well with pre-Darwinian notions of archetypal species and their manifestations in the real world; (Ruse, 1979)].

Even when a species strategy is far from any species archetype, it is still identifiable as a clump of evolving strategies. As we have shown above, if one begins with a population of individuals with a continuum of strategies strung out along the slope of an adaptive landscape, those individuals higher up the slope will increase in number while those lower down will decrease in number and likely become extinct. So, as the species evolves up the landscape, mean strategy values change directionally and, as we show below, the strategies in the distribution about the mean tighten up around the mean as it approaches a peak. Once a peak is occupied, the mean species strategy will then approximate the archetype associated with the peak.

We have seen how species can occupy convergent stable points both by using an invasion of strategies (a process of triaging species or strategies through extinction) and by the strategy dynamics of pre-defined species. The two procedures can lead to the same result because they are both manifestations of the same process driving evolution. Loosely stated, this is survival of the fittest. More accurately, evolution results from the reproductive advantage of strategies with higher fitness over strategies with lower fitness.

## 8.3 Variance dynamics

We have identified a species by a mean strategy with an identifiable range of strategies **clumped** about the mean. This raises the issue of how this distribution of strategies about the mean changes with time. In particular, does the variance change with time and how might this variance dynamics affect speciation and extinction?

### 8.3.1 Strategies over a fixed interval

A simple simulation illustrates how a clumping of strategies representing a species tends to form and narrow as the species approaches an ESS.

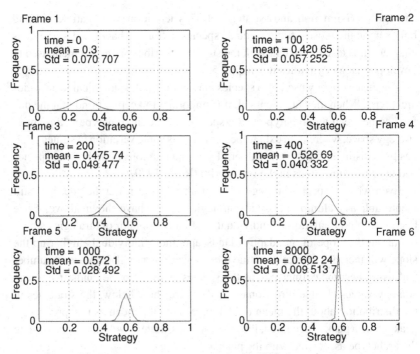

**Figure 8.8** As the mean strategy approaches the ESS, the variance narrows.

**Example 8.3.1 (L–V competition game – ESS coalition of one)** *Consider again the L–V competition game of Example 8.2.1 using the parameters $r = 2.5$, $K_m = 100$, and $\sigma_k^2 = \beta = 2$, $\sigma_\alpha^2 = 4$. With these parameters, the model has an ESS coalition of one with $u_1 = 0.6065$ at an equilibrium population of $x_1^* = 91.21$. The first frame of Figure 8.8 is obtained by distributing a population size of 90 over 41 evenly spaced strategies in the interval $[0, 1]$. The strategies are plotted on the horizontal axis and the frequency of the population using a particular strategy is plotted on the vertical axis. The last five frames show how this distribution changes with time. Note that the distribution changes shape and the standard deviation (Std) decreases as it approaches the ESS solution. After a time interval of $t = 8000$, the distribution is almost a spike with a mean strategy of $u_1 = 0.602\,24$ and a mean density $x_1 = 91.33$; very close in value to the ESS solution. The exact ESS solution cannot be obtained using this procedure since the ESS is not contained in the original distribution of strategies.*

By distributing the population density over the interval $[0, 1]$ in the above example any of these strategies is available to the population at any time. Strategy

## 8.3 Variance dynamics

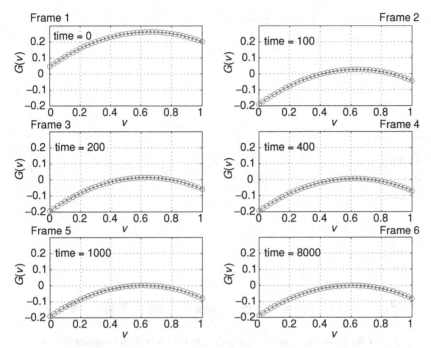

**Figure 8.9** Shortly after the simulation starts, only those strategies in the neighborhood of the ESS have a positive fitness.

dynamics (in the sense of Chapter 5) is not required to achieve the ESS approximately since a value very close to the ESS is already available to the population. All that is necessary for the initial distribution to evolve to an ESS is enough time for the differential changes in fitness to sort out the more fit strategies. This process is illustrated in Figure 8.9 where the adaptive landscape is plotted at various times and where small circles show the fitness of each strategy on these landscapes. The 41 strategies present in the population do not greatly affect the shape of the adaptive landscape. Initially, as seen in the first frame, all strategies have positive fitness and all individuals using these strategies will increase in number. However, very quickly, the adaptive landscape sinks so that the majority of strategies have negative fitness. From $t = 100$ to $t = 8000$, the shape of the adaptive landscape changes very little. This allows individuals with strategies in the neighborhood of the ESS to increase in number while individuals further from the ESS decrease in number. The net effect is to move the distribution as shown in Figure 8.8.

Figure 8.10 plots the mean strategy with time for the same simulation as used in the example. It is very similar to the results obtained using Darwinian

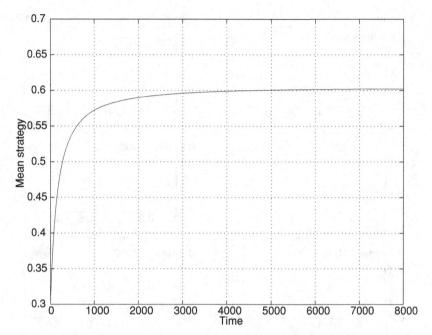

**Figure 8.10** The mean strategy changes with time in a fashion similar to that obtained using strategy dynamics with $\sigma^2 = 0.001$.

dynamics (Chapter 5) with the variance term $\sigma^2 = 0.001$. This is not unexpected, because the first-order strategy dynamics equations of Chapter 5 were derived using the idea of a finite distribution of strategies clustered about a mean strategy. Using first-order strategy dynamics of Chapter 5, the speed at which the mean strategy approaches an equilibrium solution depends on the product of the slope of the adaptive landscape and the variance term $\sigma^2$, whereas the speed at which the mean strategy approaches equilibrium, as in Figure 8.10, is dependent upon the width of the initial distribution as well as the shape and position of the adaptive landscape. These factors are related and the net result is basically the same.

It is of interest to learn how the variance of the distribution of strategies about the mean decreases as the mean strategy approaches the ESS and further decreases with time if a species is allowed to remain at the ESS. This was not a feature evident in the development of the first-order strategy dynamics of Chapter 5. In fact, under the assumptions used to derive the first-order strategy dynamics, it was shown in Section 5.12 that the variance remained constant. In all previous simulations, using strategy dynamics, we have assumed a constant $\sigma^2$. This does not invalidate any of our previous results, as they have been independent of the choice of $\sigma^2$ (provided that $\sigma^2$ is not so large as to produce non-equilibrium dynamics).

For some situations, it is useful to vary $\sigma^2$ when using first-order strategy dynamics. For example, suppose, in the above problem, we use Darwinian dynamics (with first-order strategy dynamics) to arrive at the $u_1 = 0.6065$, $x_1^* = 91.21$ solution. Now suppose that, after a long time, environmental conditions change so that $\sigma_k^2 = 4$. The old species archetype no longer exists and a new species archetype occurs at $u_1 = 1.213$, $x_1^* = 83.20$ (see Example 5.4.1). Will the species using $u_1 = 0.6065$ evolve to this ESS archetype or will it become extinct? The answer to this question depends on several factors. Once the ESS archetype has changed, the same issues as tended to narrow the variance will tend to broaden it again. Given enough time and with no competition from other species, the species will be able to adapt and achieve a new ESS. However, if a mutant species (with a broad variance) is introduced before the original species has time to broaden its own variance, the new species will most likely evolve to the new ESS archetype and the original species will become extinct.

### 8.3.2 Clump of strategies following a mean

Using strategies over a fixed interval illustrates evolution over a relatively small interval. Due to the large number of strategies required for computation, this is an impractical numerical procedure for finding an ESS over a large interval or when the ESS is a coalition of two or more strategies. The following examples illustrate an alternative procedure, where we use the definition of a species as a cluster of strategies about a mean to model evolution. It not only has the advantage of mimicking the evolutionary process but requires fewer computations and is applicable over large intervals as well.

**Example 8.3.2 (L–V competition game – ESS coalition of one)** *Using the same parameters as in the previous example and starting with the same initial conditions, we re-run the simulation except that now the only strategies available are those clustered about a mean strategy. In particular, an initial mean strategy of $u = 0.3$ is chosen with a discrete distribution of 15 strategies on a 0.4 interval with a frequency distribution as shown in the first frame of Figure 8.11. The remaining frames of Figure 8.11 illustrate how the mean strategy and its corresponding distribution evolve with time. The variance remains fairly constant until the mean strategy arrives at the species archetype. With time, the variance again becomes small. Figure 8.12 shows how the clump of strategies approaches the ESS on the adaptive landscape. Note that the clump overshoots the ESS and then returns.*

## Speciation and extinction

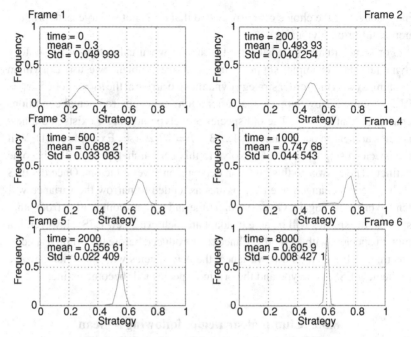

**Figure 8.11** A clump of strategies evolves to the ESS.

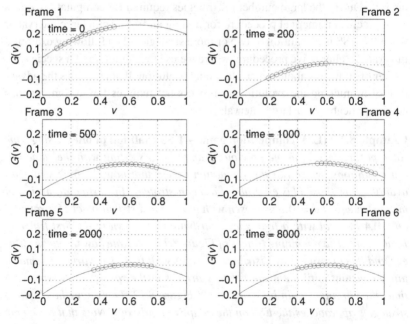

**Figure 8.12** As time goes on, the clump of strategies straddles the ESS as given by $u_1 = 0.6065$.

## 8.3 Variance dynamics

Unlike using a distribution of strategies over a fixed interval, having a clump of strategies follow the mean introduces the possibility that the mean strategy will oscillate about a species archetype before it settles down to an equilibrium solution. This occurs in the above example. The rate at which the periodic solution for the mean strategy decreases in amplitude as it approaches the equilibrium solution is a function of the time lag between the location of the mean strategy and the location of the center of the clump of strategies. The simulation has a built-in time delay before the clump "catches up" with the mean. In the above example, after every 150 time units the clump again centers on the current mean. Decreasing the time delay increases the speed at which the mean strategy moves and the magnitude of the oscillation. Making the time delay very small results in an unstable solution.

This situation with time delay is like the behavior obtained when using first-order strategy dynamics with a large variance term $\sigma^2$. Recall that $\sigma^2$ is the speed term that determines the time scale of the strategy dynamic equations. When the time scale of these equations approaches or exceeds the time scale of the population dynamics equations, periodic, chaotic, or unstable solutions can be obtained (Abrams, 2001b, 2003) as in the non-equilibrium dynamics examples in Chapters 5, 6, and 7. If we have the clump of strategies always centered about the mean, this implies that the individual phenotypes associated with each strategy in the clump could instantaneously change value (e.g., equivalent to setting $\sigma^2$ very large). By introducing the time delay, we are allowing the population dynamics to sort out winning strategies through differential changes in fitness.

**Example 8.3.3 (L–V competition game – coalition of two)** *Using the same parameters as in the previous example except for $\sigma_k^2 = 12.5$, we have shown in Example 5.4.1 that, with just a single species, Darwinian dynamics resulted in a non ESS local minimum at $u_1 = 3.639$, $x_1^* = 57.59$. This corresponds to a single species archetype. With two species, Darwinian dynamics resulted in an ESS coalition of two with $u_1 = 3.036$, $u_2 = -0.3320$, $x_1^* = 52.27$, $x_2^* = 37.58$. This corresponds to a two-species archetype. In this case, we choose an initial mean strategy of $u = 2$ with a discrete distribution of 15 strategies over a unit interval of 1 with a time delay of 100 time units. The first frame of Figure 8.13 shows the initial frequency distribution. The remaining frames of Figure 8.13 illustrate how the mean strategy and its corresponding distribution evolve with time. Note how the distribution becomes bimodal as it approaches the single-species archetype with a mean strategy very close to the archetype value. Because of this bimodal distribution, the variance in the last frame is considerably larger than the variance of the first frame. Figure 8.14 illustrates how the clump*

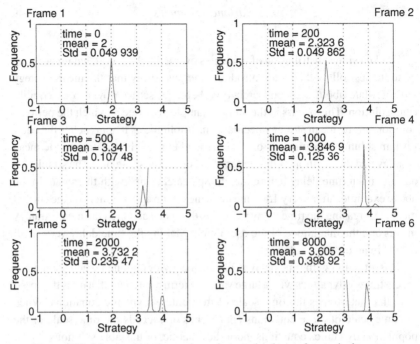

**Figure 8.13** After reaching the species archetype, the clump of strategies becomes a bimodal distribution.

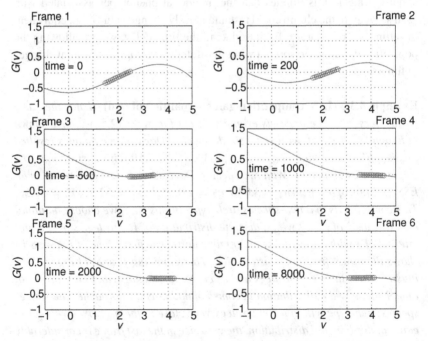

**Figure 8.14** How the clump of strategies approaches the species archetype.

*of strategies approaches the single-species archetype on the adaptive landscape. Of course the number of individuals using these strategies changes with time. By comparing each frame of Figure 8.13 with the corresponding frame of Figure 8.14 we see how the frequency of each strategy in the clump changes.*

When the clump of strategies arrives at the single-species archetype $u_1 =$ 3.639, the distribution becomes bimodal due to the fact that the mean strategy is located at a local minimum. Other strategies in the clump have a higher fitness, with those at the endpoints of the clump at the highest fitness. Further evolution is not possible due to the restricted width imposed on the clump of strategies. The ending situation suggests replacing the two spikes in the bimodal distribution by two mean distributions located at the spikes, each with its own cluster of strategies. Allowing this constitutes a speciation event.

## 8.4 Mechanisms of speciation

Speciation is the manner by which one species grades into another, or splits into two or more species. There are two major classic models of speciation: **allopatric speciation**[1] defined as the divergence of a population's strategy as a consequence of a geographical barrier and **sympatric speciation** defined as the divergence of a population's strategy as a consequence of disruptive selection acting on the population's current strategy. We examine both sympatric and allopatric speciation from the perspective of the adaptive landscape.

### 8.4.1 Sympatric speciation at an evolutionarily stable minimum

Various models for sympatric speciation have been proposed (Thoday and Gibson, 1962; Maynard Smith, 1966). Here we propose that, as a point of disruptive selection, an evolutionarily stable minimum provides a mechanism of sympatric speciation. As in the last example, adaptive dynamics may take a species to an evolutionarily stable minimum. Such a point is convergent stable but not resistant to invasion by alternative strategies. Escaping from an evolutionarily stable minimum can occur by one of two processes.

The first process involves the introduction of a novel species with a different strategy value from the resident species (this, of course, raises the question of where this second species originated from – more on this when we discuss

---

[1] We take advantage of the strategy-species concept to define both allopatric speciation and sympatric speciation.

allopatric speciation). We introduced a novel species in Example 5.4.1. As in the first two frames of Figure 5.5, the two species now have a valley between them, so that they will evolve up opposite slopes of the adaptive landscape. The two species can persist together, and both species will diverge in their strategy values away from the species archetype defined by the evolutionarily stable minimum.

The second process involves mutation or random mating in the resident species's population that maintains variation in the clump of strategy values among the individuals. If the individuals breed asexually, the process is analogous to introducing novel species, albeit with strategies very close to the population's mean. However, this initial closeness does not matter as long as there exists any variation in strategy values. Strategies contained in each end of a bimodal distribution (as in the last example) will pull away and evolve up the opposite slopes. As they evolve, there are now two clumps of strategy values, that eventually become identifiable as separate species. What if the species are interbreeding? With completely random mating, a population would remain at the evolutionarily stable minimum as a single species. However, under **assortative mating** where like individuals (e.g., individuals at one end of the bimodal distribution) are more likely to breed with each other than with the entire population, the result is the same as in the asexual case, resulting in speciation. Generally there will always be some level of assortative mating (Dieckmann and Doebeli, 1999; Doebeli and Dieckmann, 2000). In fact, there will be strong selection for assortative mating when an interbreeding population finds itself "trapped" at an evolutionarily stable minimum. This is because the two daughter species have a considerable fitness advantage over the ancestral species.

The combination of heritable variation in the clump of strategies associated with a species, along with some level of assortative mating at an evolutionary stable minimum, provides an opportunity for **sympatric speciation**. Natural selection first drives the species to a point of disruptive selection (i.e., a convergent stable minimum), splits the population by assortative mating, and then causes divergence by driving separate groups of strategies up opposite sides of the valley of the adaptive landscape. This mechanism of sympatric speciation conforms closely to that described by Rosenzweig (1978) as **competitive speciation**.[2]

---

[2] Sympatric speciation in the literature is often ascribed to any speciation event that occurs within a population in the absence of geographic barriers or physical barriers to population mixing and interbreeding. Under competitive speciation, Rosenzweig (1978) was explicit regarding how flexibility in the adaptive landscape combined with natural selection could produce sympatric speciation. Rosenzweig noted that a valley between two daughter species may not actually exist until the original species has actually speciated.

## 8.4 Mechanisms of speciation

Introducing assortative mating allows a continuation of Example 8.3.3. This is done by replacing the bimodal distribution by two clumps of strategies and continuing the simulation. The two clumps will diverge rapidly from each other, ultimately arriving at the ESS coalition of two given by $u_1 = 3.036$, $u_2 = -0.3320$, $x_1^* = 52.27$, $x_2^* = 37.58$ (see also Example 7.1.2).

The next example is used to illustrate sympatric speciation. The model follows the structure of Brown and Pavlovic (1992) and Meszena et al. (1997). It uses a **G**-matrix to addresses evolution within a structured meta-population (Metz and Gyllenberg, 2001). Two habitats are coupled by migration that is a fixed property of the organism's ecology and environment (the individuals cannot control their migration rates). This allows the model to provide an evolutionary game perspective on gene flow models from population and quantitative genetics (Kisdi and Geritz, 1999). The model's intellectual roots trace back to Levins's (1968, p. 10) question regarding the evolution of generalists and specialists within fine- (high-migration-rate) or coarse-grained (low-migration-rate) environments.

**Example 8.4.1 (sympatric speciation in a gene flow model)** *Consider two habitats where populations grow logistically, there is migration between patches, and carrying capacity is influenced by an individual's strategy. There is explicit dependence on population size and a focal individual's strategy, but not on the strategies of others. Changes in the population sizes of species i within habitats 1 and 2, respectively, are given by*

$$\dot{x}_{i1} = \left[ \frac{r}{K_1(u_i)} \left( K_1(u_i) - \sum_{j=1}^{n_x} x_{j1} \right) - m \right] x_{i1} + m x_{i2}$$

$$\dot{x}_{i2} = \left[ \frac{r}{K_2(u_i)} \left( K_2(u_i) - \sum_{j=1}^{n_x} x_{j2} \right) - m \right] x_{i2} + m x_{i1}.$$

*It is assumed that all inter- and intra-specific competition coefficients equal 1. Let a normal curve describe the relationship between carrying capacity and the individual's strategy. Habitats can vary in the maximum attainable carrying capacity, $B_j$, and in the strategy value that achieves this maximum, $\gamma_j$*

$$K_j(v) = B_j \exp\left( -\frac{(\gamma_j - v)^2}{2\sigma_k^2} \right).$$

*Under this functional relationship between carrying capacity and strategy, habitats vary in quality depending on the value of $B_j$. If the strategies that maximize K in the two habitats vary, $\gamma_1 \neq \gamma_2$, then an individual's strategy represents a trade-off between fitness in habitat 1 and fitness in habitat 2. Assuming that*

$\gamma_2 > \gamma_1$, then $u_i \in [\gamma_1, \gamma_2]$ represent the active edge of Levins's fitness set (those strategies that actually trade-off $K_1$ and $K_2$). All other values for u are on the interior of the fitness set in that they could never be favored by natural selection because such strategies can be replaced by those that simultaneously improve fitness in both habitats (Levins, 1962, 1968). The population projection **G**-matrix in this case is given by

$$\mathbf{G}(v, \mathbf{u}, \mathbf{x}) = \begin{bmatrix} \frac{r}{K_1(v)}\left(K_1(v) - \sum_{j=1}^{n_x} x_{j1}\right) - m & m \\ m & \frac{r}{K_2(v)}\left(K_2(v) - \sum_{j=1}^{n_x} x_{j2}\right) - m \end{bmatrix}.$$

Here the fitness of a focal individual appears directly influenced by its own strategy, the population densities of each species, and the distribution of each species's population among the two habitats. While the strategies of others do not directly influence fitness, frequency dependence enters the model via the effect of each species's strategy on population sizes and distributions (see Heino et al. (1998), for a discussion of this type of frequency-dependent selection). The critical value of this **G**-matrix provides a simple G-function. We use Darwinian dynamics to obtain convergent stable points and then inspect the adaptive landscape to reveal whether or not they conform to the ESS maximum principle. Consider the following parameters: $r = 0.2$, $B_1 = 100$, $B_2 = 75$, $m = 0.1$, $\gamma_1 = 0$, $\gamma_2 = 1$, $\sigma_k^2 = 4$. Using a single strategy with an initial value of $u(0) = 2$, with initial conditions for the two stages $\mathbf{x}(0) = \begin{bmatrix} 10 & 20 \end{bmatrix}$ and speed term $\sigma^2 = 0.5$, we obtain the equilibrium solution $u^* = 0.4629$ and $\mathbf{x}^* = \begin{bmatrix} 90.6401 & 78.1339 \end{bmatrix}$. Figure 8.15 illustrates that this equilibrium value for $u^*$ is indeed an ESS. In this model, $m$ is a bifurcation parameter determining whether the ESS has a single compromise strategy or whether the ESS is a coalition of two strategies. For example, keeping all the parameters the same as above but changing $m = 0.005$ and rerunning the simulation with the same initial conditions we obtain the equilibrium solution $u^* = 0.3540$ and $\mathbf{x}^* = \begin{bmatrix} 97.7922 & 71.8315 \end{bmatrix}$. However, this solution is not an ESS, as illustrated in Figure 8.16. In fact, if only a single species is allowed to evolve, it will evolve to a local minimum on the adaptive landscape. This suggests the possibility of an ESS coalition of two strategies. Once again using Darwinian dynamics with the same parameters as before, but with initial conditions that simulate assortative mating by choosing $u_1(0) = 0.3540$, $u_2(0) = 0.35$, $\mathbf{x}_1(0) = \begin{bmatrix} 97.7845 & 71.8254 \end{bmatrix}$, $\mathbf{x}_2(0) = \begin{bmatrix} 5 & 3 \end{bmatrix}$, and $\sigma_1^2 = \sigma_2^2 = 0.5$, the equilibrium solutions $u_1^* = 0.0440$, $\mathbf{x}_1^* = \begin{bmatrix} 86.6525 & 17.5526 \end{bmatrix}$ and $u_2^* = 0.9560$,

## 8.4 Mechanisms of speciation

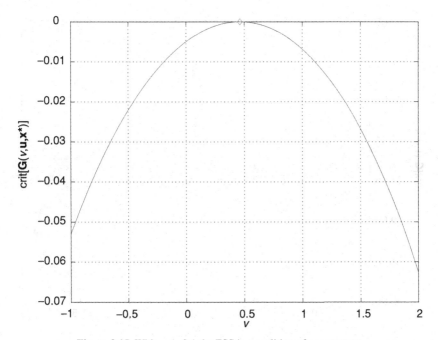

**Figure 8.15** With $m = 0.1$ the ESS is a coalition of one strategy.

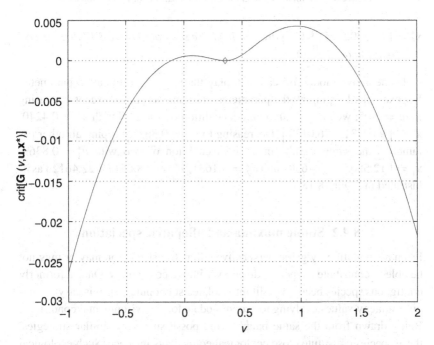

**Figure 8.16** With $m = 0.005$ a single strategy evolves to a local minimum on the adaptive landscape.

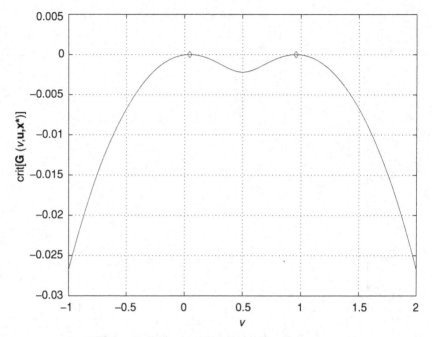

**Figure 8.17** With $m = 0.005$ the ESS is a coalition of two strategies.

$\mathbf{x}_2^* = \begin{bmatrix} 11.3302 & 55.9345 \end{bmatrix}$ *are obtained. We see from Figure 8.17 that the ESS is a coalition of two strategies.*

In the above model, $\sigma_k^2$ can also play the role of a bifurcation parameter. For example, keeping all the parameters the same as above, except increasing to $m = 0.05$, we once again obtain a coalition of one ESS with $u^* = 0.4240$, $\mathbf{x}^* = \begin{bmatrix} 93.2731 & 76.0123 \end{bmatrix}$. Decreasing to $\sigma_k^2 = 0.5625$, keeping all other parameters the same, results in an ESS coalition of two with $u_1^* = 0.8369$, $\mathbf{x}_1^* = \begin{bmatrix} 12.5865 & 38.4630 \end{bmatrix}$ and $u_2^* = 0.1631$, $\mathbf{x}^* = \begin{bmatrix} 68.6517 & 22.4612 \end{bmatrix}$ as illustrated in Figure 8.18.

### 8.4.2 Stable maxima and allopatric speciation

From an evolutionary game perspective, allopatric speciation may or may not be able to contribute to species diversity within a community. Once a formerly contiguous species becomes split into isolated subpopulations it is easy to imagine strategy values evolving to accommodate local environmental conditions. Being drawn from the same bauplan and possessing very similar strategies, these species constitute close ecological equivalents in space. Such ecological

## 8.4 Mechanisms of speciation

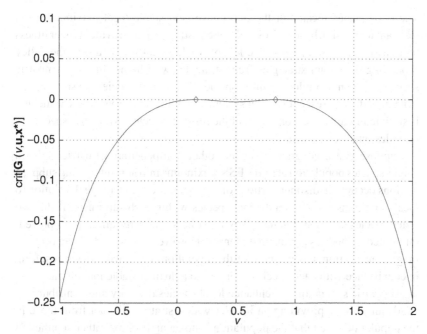

**Figure 8.18** Decreasing $\sigma_k^2$ can also result in an ESS coalition of two strategies.

equivalents are common along latitudinal gradients (Bengal tiger and Siberian tiger), elevation gradients (rock doves and Himalayan pigeons of Nepal), east–west sides of mountain ranges or continents (Atlantic versus Pacific salmon, or the eastern and western gray squirrel of North America), and among continents sharing recent exchanges of species (North American versus European beaver). Hence, there is no question but that allopatric speciation can and does produce large amounts of biodiversity in space. But how can allopatric speciation enhance local species diversity?

One could argue that, as ecological equivalents, these sister taxa should be occupying the same or similar peaks of the adaptive landscape. If the species are reunited through migration or changes in geographical barriers, then we expect one to outcompete the other as they both "struggle" for the same peak. Competitive exclusion may occur rapidly if the two species find themselves on the same side of a peak, particularly so if one species already occupies the peak. If the two species find themselves on opposite sides of the same peak, stable coexistence can occur for a while as the species evolve towards the single peak.

Consider again the situation depicted in Figure 8.4. This two-species system is not an ESS (for the particular example, the ESS is a coalition of three species).

The two species located at the two convergent stable local maxima are not resistant to invasion by a species with a mean strategy that provides higher fitness (e.g., a strategy somewhat to the left of the left peak). Because of the valley separating the extant strategies from strategies with higher fitness, sympatric speciation cannot produce a third species. This third species must be added from outside the present evolving system, and it must have a strategy value that is sufficiently different from those of the resident species. Allopatric speciation provides one such mechanism.

Suppose that a geographic (or any other non-penetrating) barrier splits a single-species population at a non-ESS maximum[3] into non-interacting subpopulations occupying distinct portions of the species' original range. Evolution to local conditions results in a distinct species within each sub-range. If the two sister species are then reunited, they must be sufficiently different to prevent interbreeding and loss of distinct identity. However, if reunited, in the original range, competition may promote further evolution of the two species, resulting in coexistence with each species evolving to a distinct stable maximum.

Allopatric speciation can enhance local species diversity and contribute towards an ESS by providing an invader with a strategy distinct from those in the population given that the population is not at an ESS but rather at either an evolutionarily stable minimum or a non-ESS evolutionarily stable maximum. If the population is at an evolutionarily stable minimum, the invading strategy may help in providing diversity through speciation, but an evading strategy is not necessary because sympatric speciation is sufficient for diversification. When the population is not at an evolutionarily stable maximum, sympatric speciation cannot be expected to contribute towards diversification. In order to occupy that portion of the landscape containing higher fitness, an invader from the outside is required with a strategy quite distinct from that of the resident species. Allopatric speciation can provide such an invader.

We use the Lotka–Volterra example for the species archetypes to illustrate how allopatric speciation might shift a community from a non-ESS evolutionarily stable maximum of two species to the ESS community of three species.

**Example 8.4.2 (allopatric speciation in the Lotka–Volterra model)** *Let the community begin as the one in Figure 8.4. Sympatric speciation cannot produce the three-species ESS. Let a geographical barrier split the range of the community into two. Each new range differs ecologically somewhat from the other and from the community's original range. The difference is reflected in*

---

[3] Similar to Figure 8.4, but with a single species located at a single local maximum that is not an ESS.

## 8.4 Mechanisms of speciation

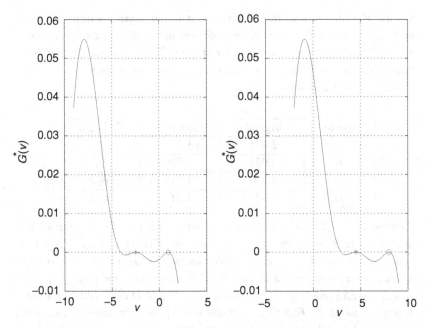

**Figure 8.19** The four species resulting from the two environmental conditions ($E_1$ to the left and $E_2$ to the right). Each figure shows the two co-existing species that have evolved to evolutionarily stable maxima.

the strategy value, $v$, that maximizes carrying capacity

$$K(v) = K_m \exp\left\{-\frac{(v+E)^2}{2\sigma_k^2}\right\}$$

where $E$ represents the state of the environment within each of the species ranges. Let $E_0 = 0$, $E_1 = -3.5$, $E_2 = 3.5$ be the environmental conditions of the original range, and the two new ranges, respectively. Figure 8.19 shows that, upon splitting, the original strategies evolve to new and distinct values that reflect the prevailing conditions in the new ranges. Each range has two species occupying evolutionarily stable maxima. In neither range are these strategies ESSs and the total system now has four species. Each of the original species has speciated allopatrically into a pair of daughter species. Now, let the geographical barrier disappear and let the community's range return to its original configuration of $E_0$. The four species come together, compete, and evolve. The end result is a community composed of an ESS coalition of three as shown in Figure 8.6. The two species of $E_1$ evolve to occupy the first two peaks (on the left). The two species of $E_2$ undergo convergent evolution towards the third peak. The two species evolve up the same slope of the adaptive landscape.

*Because the species with the smaller u has a head start towards that peak, the larger species suffers a precipitous decline in population size and becomes effectively outcompeted by the smaller species. The geographical barrier increased diversity from two species to four, the removal of the barrier reduced diversity to three species.*

Allopatric speciation

| Species | Original $u$ | Ending $u$ | Ending $x$ |
|---|---|---|---|
| A | −2.481 | −3.1 | 12.4 |
| B | 0.927 | 0.769 | 39.2 |
| C | 4.519 | 4.21 | 39.2 |
| D | 7.927 | 4.21 | 0.07 |

*The table shows the four species that resulted from the allopatric speciation (A and B are from $E_1$ and C and D are from $E_2$). The columns give their strategy values at their respective stable maxima (Original u), their strategy values after reuniting and achieving the ESS (Ending u), and their population sizes at the ESS (Ending x). Because species C and D converge on the same strategy value at the ESS they, by definition, become the same species, even though the descendants of D are virtually extinct.*

The above example also illustrates the second context for extinction through a change in the environment. With the geographical barrier, the whole community contains four species (diversity obtained from different environmental conditions within areas). In fact, with the barrier, there is a potential for coexistence of six species (an ESS coalition of three for each area). When the environment changes back to its original condition and the barrier is removed, the system ESS collapses back to three species.

### 8.4.3 Adaptive radiation

Traditionally, **adaptive radiation** has been seen as a species entering a new, unexploited environment and then speciating to fill a variety of empty **niches**. This view introduces the problem of defining and knowing the number of niches that exist in a community. As an alternative view, we propose that adaptive radiation is the process of filling out the available species archetypes. This creation process can continue until the number of species is equal to the number of strategies in an ESS for the system under consideration. A set of species archetypes corresponding to an ESS can be thought of as a set of "niches," in agreement with the traditional view. However, the archetype perspective yields both the number and the character of well-defined niches. In other words, adaptive

## 8.4 Mechanisms of speciation

radiation represents the sequences of sympatric and/or allopatric speciation events that fill the niches defined by the species archetypes. It can continue (through invasion and/or sympatric speciation) until all the ESS archetypes are filled.

Without invasion or sympatric speciation, the species, located at a minimum point on the adaptive landscape, is doomed to a kind of evolutionary purgatory, where it exists perpetually at a point of minimum individual fitness and maximum disruptive selection. It remains an empirical question whether species can persist for very long at an evolutionarily stable minimum (Brown and Pavlovic, 1992; Abrams *et al.*, 1993*b*; Metz *et al.*, 1996; Rees and Westoby, 1997; Mitchell, 2000). If not, a species at a convergent stable minimum is open to invasion and/or sympatric speciation, so that speciation is a likely outcome. Sympatric speciation at evolutionarily stable fitness minima along with allopatric speciation provides the mechanism for an adaptive radiation by which a community beginning with a single species can go through a succession of speciation events to achieve a multistrategy ESS. The theory of competitive speciation (Rosenzweig, 1978, 1987*b*) and associated demonstrations with *Drosophila* (del Solar, 1966; Rice and Salt, 1990), other invertebrates (Feder *et al.*, 1988; Schluter, 2000), and bacteria (Lenski and Travisano, 1994) provide conceptual and empirical support for the notion that natural selection can drive speciation. Furthermore, we will show how natural selection determines both the disposition and number of strategies within the ESS.

To see how an adaptive radiation can take place, consider a community of just one species that happens to have a strategy corresponding to the bottom of a valley in the adaptive landscape. Rees and Westoby's (1997, see their Figure 6) evolutionary model of plant competition within a single $G$-function illustrates what may happen following sympatric speciation. The two species evolve towards new maxima on the adaptive landscape or one of them once again evolves to a new minimum on the landscape that promotes the speciation of still more species. Any species that evolves to a new evolutionary minimum allows for sympatric speciation to increase the number of species. If all species achieve maximum points on the adaptive landscape, sympatric speciation can no longer produce speciation. If there are unoccupied peaks of higher fitness on the adaptive landscape speciation can continue by allopatric speciation until all the peaks are occupied.

The Lotka–Volterra competition model of Example 8.4.2 illustrates an adaptive radiation. In this example there is a three-strategy ESS. Radiation begins with a single species evolving to an evolutionarily stable minimum (Figure 8.1). Following competitive speciation the two species evolve to the non-ESS, evolutionarily stable maxima (Figure 8.4). At this point further radiation

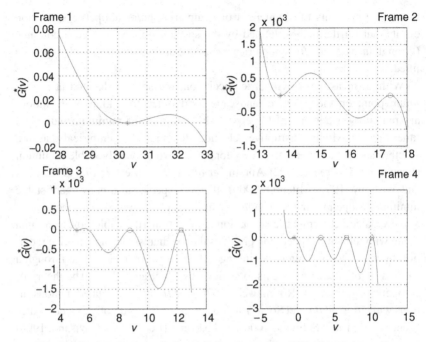

**Figure 8.20** An adaptive radiation towards a five-species ESS. Sympatric speciation carries the system from a single species up to four species that converge on non-ESS evolutionarily stable maxima.

towards the three species of the ESS requires allopatric speciation. By expanding the breadth of the niche axis of this model the adaptive radiation can continue.

**Example 8.4.3 (adaptive radiation in the Lotka–Volterra model)** *Figure 8.20 shows the sequence using the same parameters as before except using $\sigma_k^2 = 100$ allows for an ESS coalition of five. In the first frame, the starting species has evolved to an evolutionarily stable minimum. It then speciates sympatrically, with one species evolving to a stable maximum and the other species evolving to a minimum, as shown in frame 2. The species at the minimum speciates sympatrically and, with three species, the ones with the larger strategies evolve to stable maxima while the species with the smallest strategy evolves to a minimum as shown in frame 3. Again the species at the minimum speciates sympatrically and the resulting four species all evolve to convergent stable maxima as shown in frame 4. These strategies do not form an ESS and sympatric speciation can no longer advance diversity. This is because simply adding a fifth species somewhere in the vicinity of the resident*

## 8.4 Mechanisms of speciation

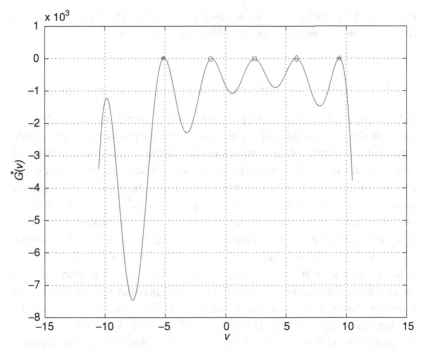

**Figure 8.21** An ESS coalition of five strategies for the Lotka–Volterra competition model.

strategies will not bring about the five-strategy ESS that exists for this model. To achieve the ESS, allopatric speciation must introduce a species with a strategy far smaller than the smallest $u$ among the resident species. With such an introduction, all of the species evolve new strategies that occupy the peaks of the ESS coalition of five as illustrated in Figure 8.21.

In the above example, species with strategy values much less than zero are competition avoiders. They sacrifice the resource-rich portion of the niche axis, in order to avoid competition from the species with very large $u$s, and they exert weak competitive effects on other species. Species with strategy values near zero are resource competitors. They occupy the richest portion of the niche axis, experience strong competitive effects from larger species, and exert strong competitive effects on smaller species. Species with much larger $u$s are interferers. They sacrifice the rich portion of the niche axis, exert strong competitive effects on each other, and experience little competition from other species with lower $u$s. As the adaptive radiation progresses, species are added as competition avoiders, and the resident species evolve to become greater interferers. In Figure 8.21 there is a local maximum (below $G = 0$) to the left of the smallest species. If $\sigma_k^2$ is increased in value (broader niche axis) it will rise to become the next species of an ESS coalition of six.

## 8.5 Predator–prey coevolution and community evolution

In Example 7.4.1 the number of prey and predator species at an ESS depends on the predator's niche breadth (aptitude of individual predators at successfully capturing a wide range of prey strategies). When the predator has a very broad niche breadth the ESS contains just a single prey and predator species. For ever narrower predator niche breadths, the number of prey and predator species in the ESS is added in the order: two prey species and one predator species, two prey and two predator species, three prey and two predators, etc. The predator's niche breadth in this model is a bifurcation parameter determining the extent of speciation and the diversity of species at the ESS. Typically, if the number of predator and prey species is originally less than required by an ESS coalition, then some of the prey and predator species first evolve to evolutionarily stable minima. Sympatric speciation can then produce the number of species at the ESS. If the number of species is originally greater than the number in the ESS coalition, then convergent evolution merges or displaces the surplus species.

The predator–prey example illustrates the distinction between ecologically keystone species and evolutionarily keystone species (Brown and Vincent, 1992). A species is **ecologically keystone** if its presence is necessary to promote the persistence, in ecological time, of other species in the community (Paine, 1966). Removal of an ecologically keystone species will result in the extirpation of other species from the community prior to evolutionary change. A species is **evolutionarily keystone** if its presence increases the number of species at an ESS. In the above predator–prey scenario both the prey and predators are evolutionarily keystone. The absence of the prey would result in the absence of predator species at the ESS, and in the absence of the predators only a single-prey strategy composes the ESS. Hence, once the community contains at least two prey species, the predator species are evolutionarily keystone. However, at an ESS the various prey and predator species are not necessarily ecologically keystone.

For illustration, consider an ESS with two prey and one predator species. Over an ecological time interval, removing the predator will allow natural selection to produce a convergence of the two prey species towards a single peak on the adaptive landscape. Generally this results in the extinction of whichever prey species lags in the evolutionary race towards the predator-free ESS. Hence

## 8.5 Predator–prey coevolution and community evolution

the predator species is evolutionarily keystone. However, over an ecological time interval, if the predator is removed, the two prey species appear to co-exist as competitors using their current strategies, implying that the predator is not ecologically keystone. An ecologist might conclude, in this case, that the predator is unnecessary for controlling or maintaining the diversity of prey species. However, only in ecological time is this a valid conclusion.

The character of the ESS brought about by a two-$G$-function system illustrates the third context for species extinction. The introduction of a new $G$-function into a system, such as a predator, may suddenly impose a new source of predation on a prey species. Evolutionarily, this may change the number and/or disposition of strategies within the prey's ESS. Ecologically, the prey may find itself with a strategy that leaves it so compromised that it has a negative population growth rate. In fact, its only hope for positive population growth rate may require large changes in strategy towards the new ESS. However, the disparity between the ecological time scale for changes in population size and the evolutionary time scale for changes in strategy may render the prey species extinct prior to its reaching an ESS (Holt and Gaines, 1992; Gomulkiewicz and Holt, 1995). A flightless species of wren on a small island in New Zealand experienced rapid extinction from the introduction of a single cat. Perhaps with time the wrens could have evolved a successful anti-predatory strategy, but the cat's appetite and aptitude removed all of the wrens within weeks. A similar and more dramatic event may have played out with the Late Pleistocene extinction of large mammals in North America. Whereas African mammals had a million or more years to adapt and respond to the evolution and technological improvements of humans, North America received its inoculum of humans prior to or during the last ice age. These humans came fully equipped with Paleolithic technologies for subduing large mammals. The giant sloths, mastodons, mammoths, and other large mammals had no such head start on responding to human predation, and humans likely drove them to extinction through over-exploitation (Martin, 1984; Lyons *et al.*, 2004). Some species obviously evolved capacities to avoid humans (present-day bison, white-tailed deer, elk, etc.) while others did not have sufficient time to evolve to humans' activity. Given enough time to adapt, might mammoths have simply become elephants?

Adaptive radiation involving insects and plants may explain much tropical diversity. Plants may be able to avoid herbivory by adopting novel suites of chemical and structural defenses, that in turn select for new insect species specializing in overcoming these defenses. The selection in plants to adopt novel defenses and the selection in the insects to specialize in overcoming particular plant defenses are a powerful engine of diversity.

**Example 8.5.1 (adaptive radiation in the predator–prey model)** *Let us reconsider the $\sigma_b^2 = 0.75$ case of the two G-function predator–prey coevolution, Example 7.4.1, that has an ESS with two predator strategies and three prey strategies. Figure 8.22 shows the adaptive radiation sequence that can produce this result starting with a single prey species and then moving through successive increases in diversity of prey and predators. The ESS has just a single prey at $u_1 = 0$ when there is no predator. Upon introducing the predator, the prey remains at $u_1 = 0$ and the predator evolves a matching strategy of $u_2 = 0$. The predator is at a maximum and the prey is now at an evolutionarily stable minimum. At this point, the predator exerts disruptive selection on the prey, resulting in sympatric speciation. Following sympatric speciation of the prey, the two prey diverge to two stable maxima, while the predator evolves to a stable minimum. This leads to sympatric speciation of the predator, the two predator species evolve to stable maxima while the two prey evolve to convergent stable inflection points. There is selection against the prey evolving more extreme strategy values. However, the position of the predator strategy blocks selection for less extreme strategy values that would actually provide higher fitness. Sympatric speciation cannot produce any further diversity. Invasion of a prey species from the outside with a strategy in between those of the predators is needed for the community to advance to an ESS coalition of three prey and two predator species (the ESS is illustrated in Figure 7.12).*

The evolutionarily stable inflection point illustrates another configuration of species archetypes that is convergent stable yet not an ESS. Like the evolutionarily stable maximum, allopatric speciation may be necessary to produce an invader species that has a strategy sufficiently distinct from the resident population to escape from the current community and achieve the ESS community.

## 8.6 Wright's shifting balance theory and frequency-dependent selection

Wright's (1931) shifting balance theory of evolution (Wright, 1969) combines **genetic drift** (random sampling error in the transmission of genes) and natural selection to explain how multiple peaks on an adaptive landscape could be occupied or how a population can achieve a superior peak from an inferior peak of the landscape. This theory, based on the assumption of both density-independent and density-dependent selection as the only two principal forms of selection, results in an adaptive landscape that is relatively rigid. A **rigid landscape** is one that does not change in response to even large changes in

## 8.6 Wright's shifting balance theory

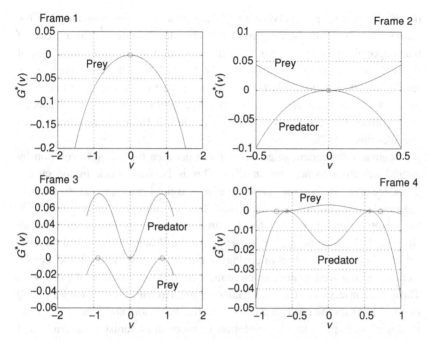

**Figure 8.22** The adaptive radiation of the predator–prey model from a single prey and a single predator species to a non-ESS community of two prey and two predator species.

the position and composition of strategies along the landscape. However, with frequency-dependent selection, the shape of the adaptive landscape can change drastically in response to even small changes in **u** and **x** (Nowak and Sigmund, 2004). This is particularly true when the number and frequency of strategies are far from the ESS. While genetic drift is sufficient for a species or population to move from one peak on the adaptive landscape to another, it is not necessary. However, as the number and frequency of strategies approach the ESS the landscape tends to become more rigid and the landscape changes much less in response to comparable changes in strategy position and frequency (Brown and Pavlovic, 1992; Vincent et al., 1993).

The flexibility of frequency-dependent adaptive landscapes provides several ways for natural selection to cross valleys and occupy multiple peaks without having to make large jumps in strategy value. First, as previously noted, when the number of strategies in a community is below the number corresponding to an ESS, a species may evolve to a minimum on the adaptive landscape. Once there, sympatric speciation introduces another strategy and the pair of strategies can now diverge as they evolve on a modified adaptive landscape. Second, even if different species are evolving up the same side of a peak (one ahead of the other) a minimum in the adaptive landscape may be moving toward one of the subpopulations and overtake it, leaving a valley between the strategies of the two species (see for example Vincent et al., 1993). If this happens, each species will then begin evolving towards a separate maximum. Third, a valley may not even exist in the adaptive landscape until one or more other species have evolved to the strategy values that produce it (Rosenzweig, 1978; Rees and Westoby, 1997).

Frequency-dependent selection is not a panacea that allows evolution by natural selection to achieve an ESS. This is because weak frequency dependence can exhibit the same relatively rigid landscapes as frequency-independent selection. Also, adaptive behaviors can create discontinuities in the adaptive landscape of a fixed morphological or physiological strategy (Brown, 1990; Gomulkiewicz and Holt, 1995). When an individual behaves selectively towards a subset of ecological circumstances, there is no longer selection for improved aptitude on circumstances that fall outside of the behavior. The niche conservatism (Holt and Gaines, 1992; Holt and Gomulkiewicz, 1997) that results from the coadaptation of selective behaviors and specialized morphologies may preclude the evolution of more opportunistic behaviors and generalized morphologies, even when such a combination of strategies would form all or part of the ESS. Nevertheless, frequency dependence greatly increases the chances that evolution by natural selection will arrive at a multi-strategy ESS.

## 8.7 Microevolution and macroevolution

Microevolution is repeatable and reversible evolutionary changes that takes place within different evolutionarily identical populations. Recall that a population of individuals that have the same set of evolutionarily feasible strategies and have the same ecological consequences of possessing particular strategies are evolutionarily identical and can be represented by a single $G$-function. **Microevolution** may be thought of as the change in strategy frequencies within existing $G$-functions.

Evolutionary change may result in the formation of new and different $G$-functions and their strategy sets. These kinds of evolutionary changes have variously been described in the literature as macro-mutations, new adaptive zones, or constraint-breaking adaptations. They all describe evolutionary changes that are not necessarily predictable, reversible, or repeatable. Changes

## 8.7 Microevolution and macroevolution

that result in new sets of evolutionarily identical individuals are known as **macroevolution** and may be thought of as the introduction of new $G$-functions and their associated strategy sets.

Evolution, speciation, and extinction predicted by the single $G$-function Lotka–Volterra model represent microevolution. Furthermore, the coevolution, adaptive radiation, and speciation within the two-$G$-function predator–prey model also represent microevolution. Even though the sweep of species, speciation events, and extinctions can be quite extensive within the predator–prey model, it is still microevolution because, given the extant $G$-functions, all of the events are repeatable and reversible.

When and how does a species within one $G$-function give rise to a new $G$-function? It may simply be a matter of scale, in which case the assignment of a new $G$-function may be somewhat arbitrary. Some traits have high variability (e.g., body size) and would be considered as strategies in a single $G$-function while another trait (tooth number in mammals) appears fixed and hence part of the defining character of a bauplan. Different $G$-functions would be required for species with different fixed traits. The taxonomic hierarchies, as revealed by phylogenies, may reveal hierarchies of evolutionary flexibility among traits. These hierarchies may be the useful starting point for defining a different $G$-function, and for crossing the threshold from microevolution to macroevolution.

Just what constitutes feasible, repeatable, and reversible evolutionary changes within a population remains an open question. To some, the answer is as narrow as extant genetic variability. Hence, the current interest in preserving genetic variability as a means of maintaining the evolutionary flexibility of a species. This view sees mutations as rare and limited in scope, and sees natural selection as a narrow finishing school for an extremely small strategy set. Many practitioners of quantitative genetics allow for an expanded set of evolutionarily feasible strategies. The strategies may not be present in the population and may crop up sufficiently frequently as recurrent mutation. But, a rather static genetic-covariance matrix (with numerous positive off-diagonal elements) representing linkage disequilibria, epistasis, and pleiotropy among genes presents a formidable barrier to extensive evolutionary change. As a quantitative trait experiences directional selection, deleterious and correlated changes in other traits retard and eventually stop further adaptive changes in the character under selection. Developmental constraints and canalizations, linkages, and epistatic and pleiotropic effects allow for an expanded yet still smallish attainable strategy set around each individual or population.

Several lines of evidence support an expansive view of what is evolutionarily feasible. This is one in which the strategy set of a population is huge and encompasses whole taxa up through the levels of genera, families, and maybe some orders. This is supported by the fact that truncation experiments or artificial selection on most quantitative strategies (behavioral, morphological, and physiological) result in directional evolutionary change, and this change appears to continue well beyond the existing variability within the starting population (Endler, 1986). Moreover, many of the non-zero, off-diagonal elements of quantitative genetic variance-covariance matrices may represent co-adaptations or temporary linkages that can change relatively quickly under selection. In the short run, these coadaptations and linkages create correlated evolutionary changes. Like the adaptive landscape under frequency dependence, the variance-covariance matrix of quantitative genetics may evolve faster than the frequencies and values of the strategies themselves. Perhaps our preoccupation with the recipe of inheritance as the lingua franca of evolutionary discussions has handicapped us with a restrictive view of feasible heritable variation.

How big is an evolutionarily feasible strategy set? Breeds of domestic dogs provide an instructive forum for crystallizing a range of views. One may argue that the great range of variation in domestic dogs is itself de facto proof of the unusualness of wolves or jackals and the power of artificial selection over natural selection; witness the lack of such phenomenal variability in other domestic animals (but do not think too hard about plants such as *Brassica*, or birds such as domestic pigeons). In contrast, even domestic dogs, which encompass more variability than exists naturally in the family Canidae, may be a smallish subset of the evolutionarily feasible strategy set found in each individual of the family Canidae. Canids may not be unusually evolutionarily flexible. Rather, the large number of dog breeds results from their varied usefulness to humans. And, because we have selected from the strategy set of wolves or jackals the strategies of interest to us, we have probably rejected or left unexplored an even larger set of evolutionarily feasible strategies. The strategy set around each individual or population of species may be vast compared with extant genetic variability.

The extent of $G$-functions and the size of strategy sets determine the scope of microevolution. If $G$-functions and strategy sets encompass whole taxa at the levels of genera and families, then microevolution produces the species within a multi-strategy ESS. Suppose the natural species of canids (e.g., wolves, coyotes, red foxes, and gray foxes in the Midwestern part of the USA) represent the peaks of a frequency-dependent adaptive landscape and the multi-strategy ESS of a single $G$-function and strategy set (that includes all breeds of domestic dog).

## 8.7 Microevolution and macroevolution

Then, because of natural selection, what we see in nature is a rather dull subset of what is evolutionarily feasible.

But individuals are not infinitely evolutionarily plastic. There are qualitative characters and quantitative characters that cannot or will not change repeatedly and reversibly in response to truncation or directional selection. Tooth number in terrestrial mammals appears to be such a character. A relatively constant tooth number of 42 connects all members of Canidae, and indubitably dog breeds such as bulldogs would have been favored if they occurred with fewer teeth. (Interestingly, tooth number is probably much more evolutionarily plastic in other taxa such as sharks.) In fact, characters that are fixed within a $G$-function and strategy set define the bauplan and distinguish it from others. Such constraints (physical, developmental, genetic, etc.) define the strategy set and determine the fitness consequences of pitting a strategy against particular abiotic and biotic environments. Qualitative and quantitative changes in these constraints that result in changes to the $G$-function and its strategy set constitute macroevolution. Characters that define the deeper nodes of phylogenies (distinguishing among orders and families) may define hierarchies of distinct bauplans and associated $G$-functions, whereas characters associated with distinguishing among species and genera within families may represent species diversity within bauplans and $G$-functions.

Many of the great moments in evolution probably represent macroevolutionary changes: cell organelles, the nucleus, multi-cellularity, organ systems, vascular plants, exoskeletons, endoskeletons, etc. Rosenzweig and McCord (1991) discuss several in fascinating detail (neck morphology in turtles, and heat sensory pits in vipers). Following the establishment of a macroevolutionary change, there should be rapid evolutionary changes to the ESS for the existing bauplan and the establishment of an ESS for the new bauplan. The new $G$-function with its strategy set may completely replace an old one from which it arose, in which case the new ESS may have no species or strategies from the old $G$-function. Or the new ESS may contain strategies from the old $G$-function as well as from the new. In the former case the macroevolutionary change replaces older evolutionary technology with a new one. In the latter case the new and old $G$-functions are complementary and the macroevolutionary change actually increases the pool of evolutionary technology available to natural selection via microevolution. The process that replaces the strategies and species of an old $G$-function with the strategies and species of a new one represents incumbent replacement (Rosenzweig and McCord, 1991). And the new ESS resulting from the establishment of a new $G$-function represents evolutionary progress and the procession of life.

## 8.8 Incumbent replacement

**Incumbent replacement** is the process of shifting from current species archetypes to new species archetypes as a result of a macroevolutionary change. Because evolution by natural selection produces an ESS composed of species archetypes, the direction of evolutionary change is dictated by the species archetypes created with the introduction of a new $G$-function. The prior bauplan contains a smaller number of $G$-functions and strategy sets while the new bauplan contains an expanded number of $G$-functions and/or strategy sets. Species from the prior bauplan will likely impede evolution and slow the replacement of old species by new species available under the new bauplan.

Rosenzweig and McCord (1991) elaborate on several convincing examples of incumbent replacement. Their example with pit (New World) vipers and non-pit (Old World) vipers is wonderful. These two groups of vipers can be thought of in terms of two separate $G$-functions, the crucial difference being that the new $G$-function (pit vipers) possesses within its strategy set a separate set of infrared sensors, in addition to its eyes. This strategy appears to be absent in the old $G$-function (non-pit vipers). Since their appearance, pit viper have replaced non-pit vipers throughout North America (e.g., rattlesnakes) and South America (e.g., bushmaster). They have crossed the Bering Strait and are now face to face with non-pit vipers in parts of Southeast Asia. Non-pit vipers (e.g., vipers and adders) exist in the absence of pit vipers in the rest of Asia, and throughout Europe and Africa. The direction of replacement is foreordained. The ability to sense the heat of their prey and the lack of this strategy within the prior $G$-function's strategy set means that the new ESS will include pit vipers and may exclude non-pit vipers entirely. However, the incumbency of non-pit vipers ensures that the replacement process will be slow. For example, the presence of a forest-dwelling non-pit viper (appropriately colored, sized, and arboreal) may stall the evolution of a forest-dwelling pit viper from an extant grassland pit viper (inappropriately colored, sized, and shaped for the forest). An environmental change or chance local extinction may be necessary to knock the incumbent off its peak, at which time the pit viper's adaptive landscape may slope favorably towards its destiny (a new ESS).

Along the zone of contact, environmental changes or extinctions can result in pit vipers replacing pit vipers, pit vipers replacing non-pit vipers, or even non-pit vipers replacing non-pit vipers, but not non-pit vipers replacing pit vipers. Incumbent replacement provides an explanation for how macroevolutionary change results in the establishment of new $G$-functions and the often-stalled replacement of old ones. It is possible that the extinction of the dinosaurs and the radiation of mammals provides the most famous example of incumbent replacement. Incumbent replacement also probably explains why many evolutionary changes occur just once or only a few times and are tightly associated with monophyletic clades. Once a novel and adaptive suite of strategies establishes within one $G$-function, the extant members of the $G$-function probably inhibit or preclude the evolution of similar types of strategies from another $G$-function.

## 8.9 Procession of life

The macroevolutionary introduction of a new $G$-function can have a direct effect on the adaptive landscape and an existing ESS by replacing the current $G$-function or by rearranging the numbers and dispositions of strategies within an extant $G$-function. One way this can be done is by changing the physical environment. For instance, the advent of photosynthesis precipitated profound global climate change, and our present climate, hydrology, and geology have been established as a by-product of past and present biotas. Macroevolutionary changes to $G$-functions and their consequences for an ESS across the globe may define and direct the procession of life. The existing pool of $G$-functions and their associated extant strategies may determine the pool of evolutionarily feasible macroevolutionary changes (a kind of macrostrategy set). For instance, macroevolutionary changes to the $G$-functions of multicellular organisms cannot occur until multicellularity itself establishes as a macroevolutionary change from unicellular organisms (this leaves open speculative but interesting discussion as to whether unicellularity was a necessary precursor to multicellularity). Once a new $G$-function is established, it may reduce or increase the pool of extant $G$-functions either by replacing existing ones or by coexisting with the extant $G$-functions. The strategies which evolve and establish from a new $G$-function may either reduce or increase biodiversity by increasing or decreasing the numbers of strategies possible in an ESS for each $G$-function. The establishment of a new $G$-function may result in changes to the physical environment that may further influence the character and diversity of the ESS.

In the context of evolutionary game theory, the **procession of life** is the addition and replacement of $G$-functions over evolutionary time. Natural selection operates at both a macroevolutionary and a microevolutionary level. At the macroevolutionary scale, natural selection determines when a new $G$-function is established and when old $G$-functions are replaced. At the microevolutionary scale, natural selection determines what ESS will emerge from within and among the extant pool of $G$-functions and strategy sets.

These two levels may explain evolutionary patterns of **punctuated equilibria** (Eldridge and Gould, 1972) over geological time. Periods of stasis may be the product of natural selection holding life at an enduring and robust ESS. The punctuated periods of rapid evolutionary change may be thought of as the result of natural selection equilibrating to a new ESS following a significant macroevolutionary change.

# 9
# Matrix games

Matrix games, introduced in Subsection 3.1.2, formed the core of the early work on evolutionary games. Most game theoretic models, notions of strategy dynamics, solution concepts and applications of ESS definitions occurred explicitly in the context of matrix games. For continuous strategies, modelers relied on either Nash solutions (Auslander *et al.*, 1978), or model-specific interpretations of the ESS concept (Lawlor and Maynard Smith, 1976; Eshel, 1983). The bulk of developments in evolutionary game theory associated with matrix games pre-date the $G$-function, strategy dynamics, and the ESS maximum principle. For a review of these developments see Hines (1987), Hofbauer and Sigmund (1988), and Cressman (2003). In this chapter, we place matrix games within the context of $G$-functions and the more general theory of continuous evolutionary games. We reformulate the ESS frequency maximum principle developed in Section 7.5 for application to matrix games. This reformulation requires some additional terminology and new definitions.

Fitness for a matrix game is expressed in terms of strategy frequency and a matrix of payoffs. As with continuous games, the $G$-function in the matrix game setting must take on a maximum value at all of the strategies which make up the ESS coalition vector. The reformulated maximum principle is applicable to both the traditional **bi-linear matrix game** and a more general **non-linear matrix game**. In a bi-linear matrix game, a strategy is chosen from a strategy set composed of probabilities, an individual interacts at random with other individuals within the population and receives a payoff determined by constant elements of a payoff matrix. After many such interactions, the expected payoff to an individual is the sum of the products of strategy frequencies multiplied by the appropriate entries from the payoff matrix. In a non-linear matrix game, the strategy set need not be probabilities and/or the elements of the matrix may be functions of **u**, instead of constants.

The strategy set for a bi-linear game consists of a finite number of choices of which row or column to pick from a given matrix. In this situation, the frequency of strategies within the population can be interpreted in two ways. In one view, the frequency of a given strategy gives the proportion of individuals within the population that possess the strategy (Zeeman, 1980). In this case, strategy frequencies describe a **polymorphic population** of individuals, each playing a particular **pure strategy** from the finite strategy set. Alternatively, strategy frequencies may describe a **mixed strategy** vector that represents the probabilities that an individual will play each of the separate pure strategies (Maynard Smith, 1982). In this case, there exists a **monomorphic population** of individuals playing strategies from a continuous set of probabilities between zero and 1. In other words, when more than one strategy exists within the population, the population can be a mixture of individuals using pure strategies, or it can be a population of individuals using mixed strategies. The distinction between a mix of pure strategies on the one hand and mixed strategies on the other may appear subtle, but the two represent quite different games with different strategy dynamics and different consequences for the ESS.

In fact, discrepancies in the behaviors of mixed strategies and populations with mixtures of pure strategies put into question Maynard Smith's original ESS definition and illustrate some limitations. The mixed strategies identified by Maynard Smith's original definition neither conform with the ESS maximum principle nor satisfy conditions of convergent stability. From this, one might conclude that the ESS maximum principle is not relevant for matrix games, suggesting that a theory for continuous games must remain separate from that of matrix games. This is not so. Here we develop a matrix-ESS maximum principle that is directly applicable to matrix games and demonstrate its usefulness with a number of examples for both bi-linear and non-linear matrix games. We apply the $G$-function approach to some familiar matrix games: prisoner's dilemma, game of chicken, and rock–scissors–paper. In addition we include the sex ratio game, kin selection, and reciprocal altruism as matrix games formulated in terms of $G$-functions. These examples show that the $G$-function approach, developed in the previous chapters, is equally relevant for matrix games.

When employing the matrix-ESS maximum principle to solve a bi-linear matrix game, one $G$-function is used to find both pure and mixed strategy solutions. The distinction becomes one of specifying the appropriate strategy space. Furthermore, the $G$-function and matrix-ESS maximum principle can be used to solve matrix games that are not in the usual bi-linear form (Vincent and Cressman, 2000). We use the $G$-function approach to both illustrate and resolve some of the shortcomings of Maynard Smith's original ESS definition. This approach preserves the spirit of Maynard Smith's pioneering ESS concept.

## 9.1 A maximum principle for the matrix game

We start with the usual population dynamics in terms of fitness functions as given by the following

Population dynamics in terms of fitness functions

$$\begin{aligned} \text{Difference:} \quad & x_i(t+1) = x_i\left[1 + H_i[\mathbf{u}, \mathbf{x}]\right] \\ \text{Exp. Difference:} \quad & x_i(t+1) = x_i \exp H_i[\mathbf{u}, \mathbf{x}] \\ \text{Differential:} \quad & \dot{x}_i = x_i H_i[\mathbf{u}, \mathbf{x}]. \end{aligned}$$

(9.1)

But, in most matrix games, there is no explicit consideration of population size. Hence, the requirements of the previous chapters for an ecological equilibrium or even a bounded solution for $\mathbf{x}^*$ may not be relevant. The standard way to avoid this difficulty is to express the ecological dynamics in terms of the frequency of individuals in the population using a particular strategy rather than the number of individuals in a population using them.

### 9.1.1 Frequency formulation

Strategy-frequency dynamics were introduced by Taylor and Jonker (1978) for the bi-linear matrix game. What follows includes Taylor and Jonker's approach as a special case (Vincent and Cressman, 2000). Let

$$p_i = \frac{x_i}{N}$$

and

$$N = \sum_{k=1}^{n_s} x_k$$

be the total population number. Under this transformation, the **frequency vector** $\mathbf{p} = \begin{bmatrix} p_1 & \cdots & p_{n_s} \end{bmatrix}$ lies in the **frequency space** defined by

$$\Delta^{n_s} = \{(p_1, \ldots, p_{n_s}) \mid \sum_{i=1}^{n_s} p_i = 1, \ p_i \geq 0\}.$$

As in Section 4.7, Equations (9.1) may be written in terms of changes in strategy frequencies (recall that we drop the exponential difference case when expressing the population dynamics in terms of frequency)

$$\begin{aligned} \text{Difference:} \quad & p_i(t+1) = p_i \frac{1 + H_i(\mathbf{u}, \mathbf{p}, N)}{1 + \bar{H}} \\ \text{Differential:} \quad & \dot{p}_i = p_i \left[ H_i(\mathbf{u}, \mathbf{p}, N) - \bar{H} \right] \end{aligned}$$

(9.2)

with the total population number given by

$$\boxed{\begin{array}{ll}\text{Difference:} & N(t+1) = N\left(1+\bar{H}\right) \\ \text{Differential:} & \dot{N} = N\bar{H}\end{array}} \tag{9.3}$$

where

$$\bar{H} = \sum_{j=1}^{n_s} p_i H_i(\mathbf{u}, \mathbf{p}, N)$$

is the average fitness.

### 9.1.2 Strategies

When a solution to the bi-linear matrix game is sought in terms of mixed strategies, they are probabilities associated with an individual choosing a row of a matrix $\mathbf{M}$. The bi-linear matrix game is a **symmetric matrix game** (see Subsection 3.1.2) so that only one matrix is needed to define payoffs. For example, if $\mathbf{M}$ has two rows and two columns and if player one has the strategy $\mathbf{u}_1 = \begin{bmatrix} 0.25 & 0.75 \end{bmatrix}$ then, over many contests, this player is choosing row one 25% of the time and row two 75% of the time. On the other hand, player two with the strategy $\mathbf{u}_2 = \begin{bmatrix} 0 & 1 \end{bmatrix}$ plays row two in every contest. The first case is an example of a mixed strategy, where player one is choosing different rows during many plays of the game. The second case is an example of a mixed strategy that is also a pure strategy. Player two chooses the same row for each play of the game. There can be many other players using various other strategies, but players always play in pairs to determine their payoffs. For a given play of the game, we know that player two chooses row two and suppose it turns out that player one chooses row one. For this play, the payoff to player one is the element $m_{12}$ of $\mathbf{M}$ corresponding to the intersection of row one with column two. The payoff to player two is the element $m_{21}$ corresponding to the intersection of row two with column one. Mixed strategies are chosen from the continuous strategy space defined by

$$\mathcal{U}_c = \{\mathbf{u}_i \in \mathcal{R}^{n_u} \mid 0 \leq u_{ij} \leq 1 \; \forall \; j = 1, \ldots, n_u \text{ and } \sum_{j=1}^{n_u} u_{ij} = 1\}.$$

Note that $\mathcal{U}_c$ includes pure strategies (probabilities 0 and 1) as a possible mixed strategy choice.

When a solution to the bi-linear matrix game is sought in only pure strategies, they are chosen from a subset of $\mathcal{U}_c$ where the components of the strategy vector can only have the values 0 and 1. The discrete strategy space for the pure strategy

## 9.1 A maximum principle for the matrix game

version of this matrix game is the two-element space defined by

$$\mathcal{U}_p = \left\{ \mathbf{u}_i \in \mathcal{R}^{n_u} \mid u_{ij} \in \{0, 1\} \forall \, j = 1, \ldots, n_u \text{ and } \sum_{j=1}^{n_u} u_{ij} = 1 \right\}.$$

The non-linear matrix game may have other restrictions. The strategy set for the general situation is still designated by $\mathcal{U}$.

### 9.1.3 Payoff function

Let $E(\mathbf{u}_i, \mathbf{u}_j)$ be the **expected payoff** to an individual using strategy $\mathbf{u}_i$ when interacting with an individual using strategy $\mathbf{u}_j$. The standard evolutionary model (Cressman, 1992) associated with an $n_u \times n_u$ matrix game assumes that

$$E(\mathbf{u}_i, \mathbf{u}_j) = \mathbf{u}_i \mathbf{M} \mathbf{u}_j^T \tag{9.4}$$

and $\mathbf{M}$ is an $n_u \times n_u$ payoff matrix

$$\mathbf{M} = \begin{bmatrix} m_{11} & \cdots & m_{1n_u} \\ \vdots & \ddots & \vdots \\ m_{n_u 1} & \cdots & m_{n_u n_u} \end{bmatrix}.$$

The payoff function is **bi-linear** in the strategies when the elements of $\mathbf{M}$ are constants, since $E(\mathbf{u}_i, \mathbf{u}_j)$ is linear in the components of the strategy vectors $\mathbf{u}_i$ and $\mathbf{u}_j$.

The fitness of an individual playing strategy $\mathbf{u}_i$ is the expected payoff in a random contest where $p_j$ is the frequency of players using strategy $\mathbf{u}_j$. Thus by definition

$$H_i[\mathbf{u}, \mathbf{p}] = \sum_{j=1}^{n_s} E(\mathbf{u}_i, \mathbf{u}_j) \, p_j. \tag{9.5}$$

Note that fitness does not depend on $N$. Because of this, the frequency dynamics as given by (9.2) are decoupled from the population dynamics as given by (9.3) and allows us to seek non-trivial equilibrium solutions for $p^*$ regardless of $N$. An equilibrium solution for $N^*$ need not even exist. Because of the decoupling, the $N$ equations are usually ignored when dealing with matrix games. In what follows, we assume that $N$ is not of interest and we drop these equations.

### 9.1.4 Frequency dynamics

Substituting (9.5) into the expressions for average fitness, we obtain

$$\bar{H} = \sum_{i=1}^{n_s} p_i \sum_{j=1}^{n_s} E\left(\mathbf{u}_i, \mathbf{u}_j\right) p_j = \mathbf{p}\mathbf{E}(\mathbf{u})\mathbf{p}^T$$

where $\mathbf{E}(\mathbf{u})$ is the expected payoff matrix whose elements are given by $E\left(\mathbf{u}_i, \mathbf{u}_j\right)$, and $\mathbf{p}$ is the frequency vector (defined above) whose components are given by $p_i$. Using this definition for $\bar{H}$ and substituting the fitness function into (9.2) we obtain the following frequency dynamics

$$\text{Difference:} \quad p_i(t+1) = p_i \frac{1+\sum_{j=1}^{n_s} E(\mathbf{u}_i,\mathbf{u}_j)p_j}{1+\mathbf{p}\mathbf{E}(\mathbf{u})\mathbf{p}^T}$$

$$\text{Differential:} \quad \dot{p}_i = p_i \left[ \sum_{j=1}^{n_s} E\left(\mathbf{u}_i, \mathbf{u}_j\right) p_j - \mathbf{p}\mathbf{E}(\mathbf{u})\mathbf{p}^T \right].$$

(9.6)

In a fashion similar to what we have done before we use the following definitions to arrive at a matrix-ESS maximum principle.

**Definition 9.1.1 (matrix ecological equilibrium)** *Given a strategy vector* $\mathbf{u} \in \mathcal{U}$, *a point* $\mathbf{p}^* \in \Delta^{n_s}$ *is said to be a matrix ecological equilibrium point for (9.6) provided that there exists an index* $n_{s^*}$ *with* $1 \le n_{s^*} \le n_s$ *such that*

$$\sum_{j=1}^{n_s} E\left(\mathbf{u}_i, \mathbf{u}_j\right) p_j^* = \mathbf{p}^* \mathbf{E}(\mathbf{u}) \mathbf{p}^{*T}, \quad p_i^* > 0 \quad \text{for} \quad i = 1, \cdots, n_{s^*}$$
$$p_i^* = 0 \qquad\qquad\qquad\qquad\qquad\qquad \text{for} \quad i = n_{s^*}+1, \cdots, n_s.$$

Assume that, for every strategy vector $\mathbf{u} \in \mathcal{U}$, a matrix ecological equilibrium solution $\mathbf{p}^*$ exists. For this equilibrium point to be stable, we require that every trajectory which begins in frequency space near the equilibrium point remain in frequency space for all $t$, and converge to the equilibrium point as $t \to \infty$.

**Definition 9.1.2 (matrix-ESE)** *Given a strategy vector* $\mathbf{u} \in \mathcal{U}$, *a matrix ecological equilibrium point* $\mathbf{p}^* \in \Delta^{n_s}$ *is said to be a matrix ecologically stable equilibrium (matrix-ESE) if there exists a ball* $\mathcal{B}$ *centered at* $\mathbf{p}^*$ *such that for any* $\mathbf{p}(0) \in \Delta^{n_s} \cap \mathcal{B}$ *the solution generated by (9.6) satisfies* $\mathbf{p}(t) \in \Delta^{n_s}$ *for all* $t > 0$ *and asymptotically approaches* $\mathbf{p}^*$ *as* $t \to \infty$. *If the radius of the ball can be made arbitrarily large, the ecological equilibrium point is said to be a* **global matrix-ESE**, *otherwise it is said to be a* **local matrix-ESE**.

**Theorem 9.1.1 (matrix-ESE)** *Given* $\mathbf{u} \in \mathcal{U}$, *if a matrix ecological equilibrium point* $\mathbf{p}^*$ *is a local matrix-ESE then*

$$\sum_{j=1}^{n_s} E\left(\mathbf{u}_i, \mathbf{u}_j\right) p_j^* = \mathbf{p}^* \mathbf{E}(\mathbf{u}) \mathbf{p}^{*T} \quad \text{for} \quad i = 1, \cdots, n_{s^*}$$

$$\sum_{j=1}^{n_s} E\left(\mathbf{u}_k, \mathbf{u}_j\right) p_j^* \leq \mathbf{p}^* \mathbf{E}(\mathbf{u}) \mathbf{p}^{*T} \quad \text{for} \quad k = n_{s^*} + 1, \cdots, n_s.$$

*Proof.* The first condition follows from the definition of a matrix ecological equilibrium. Suppose that $\sum_{j=1}^{n_s} E\left(\mathbf{u}_k, \mathbf{u}_j\right) p_j^* > \mathbf{p}^* \mathbf{E}(\mathbf{u}) \mathbf{p}^{*T}$ for some $k \in \{n_{s^*} + 1, \cdots, n_s\}$. By continuity, we know that there exists a ball of non-zero radius centered at $\mathbf{p}^*$ such that $\sum_{j=1}^{n_s} E\left(\mathbf{u}_k, \mathbf{u}_j\right) p_j > \mathbf{p}^* \mathbf{E}(\mathbf{u}) \mathbf{p}^{*T}$ in the neighborhood $\Delta^{n_s} \cap \mathcal{B}$. This implies that any $p_k$ located in this neighborhood is increasing with time. This contradicts the assumption that $\mathbf{p}^*$ is a matrix-ESE, hence the second condition follows. ∎

### 9.1.5 Matrix-ESS

Our definition of a matrix-ESS is similar to Definition 6.2.4 only it is now stated in terms of vector strategies, frequency $\mathbf{p} \in \Delta^{n_s}$, and the matrix-ESE definition.

**Definition 9.1.3 (matrix-ESS)** *A coalition vector* $\mathbf{u}_c \in \mathcal{U}$ *is said to be a matrix-ESS for the equilibrium point* $\mathbf{p}^*$ *if, for all* $n_s > n_{s^*}$ *and for all strategies* $\mathbf{u}_m \in \mathcal{U}$, *the equilibrium point* $\mathbf{p}^*$ *is a matrix-ESE.*

It follows from (9.5) that the $G$-function for the bi-linear game is given by

$$G(\mathbf{v}, \mathbf{u}, \mathbf{p}) = \sum_{j=1}^{n_s} E(\mathbf{v}, \mathbf{u_j}) p_\mathbf{j}. \tag{9.7}$$

In terms of the $G$-function the frequency dynamics becomes

$$\boxed{\begin{aligned} \text{Difference:} \quad & p_i(t+1) = p_i \frac{1 + G(\mathbf{v},\mathbf{u},\mathbf{p})|_{\mathbf{v}=\mathbf{u}_i}}{1+\mathbf{p}^T\mathbf{E}(\mathbf{u})\mathbf{p}} \\ \text{Differential:} \quad & \dot{p}_i = p_i \left[ G(\mathbf{v}, \mathbf{u}, \mathbf{p})|_{\mathbf{v}=\mathbf{u}_i} - \mathbf{p}^T \mathbf{E}(\mathbf{u})\mathbf{p} \right] \end{aligned}} \tag{9.8}$$

Using the above definitions, we obtain the following maximum principle.

**Theorem 9.1.2 (matrix-ESS maximum principle)** *Let* $G(\mathbf{v}, \mathbf{u}, \mathbf{p})$ *be the fitness generating function for a community defined by (9.8). For a given* $\mathbf{u}$ *assume* that there exists a matrix-ESE designated by $\mathbf{p}^*$. Let $\mathbf{u}$ be partitioned in such a way that the first $n_{s^*} \leq n_s$ components of $\mathbf{u}$ make up the coalition vector $\mathbf{u}_c$

$$\mathbf{u} = [\, \mathbf{u}_1 \mid \cdots \mid \mathbf{u}_s \,] = [\, \mathbf{u}_c \mid \mathbf{u}_m \,].$$

If the coalition vector $\mathbf{u}_c \in \mathcal{U}$ is a matrix-ESS for $\mathbf{p}^*$ then

$$\max_{\mathbf{v} \in \mathcal{U}} G(\mathbf{v}, \mathbf{u}, \mathbf{p}^*) = G(\mathbf{v}, \mathbf{u}, \mathbf{p}^*)\big|_{\mathbf{v}=\mathbf{u}_i} = \mathbf{p}^* \mathbf{E}(\mathbf{u}) \mathbf{p}^{*T} \qquad (9.9)$$

for $i = 1, \ldots, n_{s^*}$.

*Proof.* Let $n_{s^*} \leq n_s$, it follows from the matrix-ESE theorem that for any $i = 1, \ldots, n_{s^*}$ and for any $k = n_{s^*} + 1, \ldots, n_s$ we have

$$\sum_{j=1}^{n_s} E\left(\mathbf{u}_k, \mathbf{u}_j\right) p_j \leq \sum_{j=1}^{n_s} E\left(\mathbf{u}_i, \mathbf{u}_j\right) p_j^* = \mathbf{p}^* \mathbf{E}(\mathbf{u}) \mathbf{p}^{*T}$$

which, from the definition of the $G$-function, implies

$$G(\mathbf{v}, \mathbf{u}, \mathbf{p}^*)\big|_{\mathbf{v}=\mathbf{u}_k} \leq G(\mathbf{v}, \mathbf{u}, \mathbf{p}^*)\big|_{\mathbf{v}=\mathbf{u}_i} = \mathbf{p}^* \mathbf{E}(\mathbf{u}) \mathbf{p}^{*T}$$

which in turn implies (9.9). ∎

The matrix-ESS maximum principle states that $G(\mathbf{v}, \mathbf{u}, \mathbf{p}^*)$ must take on a maximum with respect to $\mathbf{v} \in \mathcal{U}$ and that the maximum value is equal to the average fitness. If $G(\mathbf{v}, \mathbf{u}, \mathbf{p}^*)$ should take on a proper maximum when the matrix-ESS is a coalition of one (i.e., $n_s^* = 1$), it is possible to state a stronger result.

**Corollary 9.1.1 (sufficient condition for a matrix-ESS)** *Given a matrix game defined by (9.8), if $n_s^* = 1$ (i.e., $\mathbf{p}^* = \begin{bmatrix} 1 & 0 & \cdots & 0 \end{bmatrix}$) and $\mathbf{u}_1$ satisfies the matrix-ESS maximum principle such that $G(\mathbf{v}, \mathbf{u}, \mathbf{p}^*)$ takes on a proper maximum with respect to $\mathbf{v} \in \mathcal{U}$, then $\mathbf{u}_1$ is a matrix-ESS.*

*Proof.* From continuity, there exists a ball $\mathcal{B}$ centered at $\mathbf{p}^*$ such that the fitnesses of all strategies $u_2, \cdots, u_{n_s}$ at any point $\mathbf{p} \in \Delta^{n_s} \cap \mathcal{B}$ are less than average fitness. This implies that $\sum_{j=1}^{n_s} E\left(\mathbf{u}_1, \mathbf{u}_j\right) p_j > \mathbf{p}^* \mathbf{E}(\mathbf{u}) \mathbf{p}^{*T}$. Consequently $p_1(t)$ is convergent stable in the neighborhood of $\mathbf{p}^*$. ∎

### 9.1.6 Maynard Smith's original ESS definition

We use the term single-mutant-ESS to mean the ESS of a bi-linear matrix game that satisfies the original ESS definition as given by Maynard Smith (1982), see also Hofbauer and Sigmund (1988), and Cressman (1992). In this formulation

## 9.1 A maximum principle for the matrix game

the expected fitness of individuals in a two-person matrix game is given by

$$H_1 = p_1 E(u_1, u_1) + p_2 E(u_1, u_2)$$
$$H_2 = p_1 E(u_2, u_1) + p_2 E(u_2, u_2)$$

where $E(u_i, u_j)$ represents the expected payoff as given by (9.4) to an individual playing strategy $u_i$ against an opponent playing $u_j$. The expected fitness for the fraction of players $p_i$ using strategy $u_i$ is given by $H_i$. Maynard Smith described the ESS as a strategy which when common cannot be invaded by any rare alternative strategy. Maynard Smith's original ESS formulation assumes random pairwise interactions and provides a definition based on static inequalities of frequency-dependent payoffs (Maynard Smith, 1976).

**Definition 9.1.4 (single-mutant-ESS)** *Let $u_1$ be a common strategy in a monomorphic population; then the strategy $u_1$ is a single-mutant-ESS for a two-player game if for any other strategy $u_j$ either*

$$E(u_1, u_1) > E(u_2, u_1)$$

*or*

$$E(u_1, u_1) = E(u_2, u_1)$$

*and*

$$E(u_1, u_2) > E(u_2, u_2).$$

A strategy satisfying Maynard Smith's original definition may not necessarily satisfy the definition of a matrix-ESS given above. As defined in this book, the ESS must be resistant to invasion by any number of simultaneous, rare alternative strategies and must be convergent stable. A strategy satisfying the single-mutant-ESS definition need not be resistant to invasion by multiple mutants, and it may not satisfy the matrix-ESS maximum principle (examples will be given below). However, Maynard Smith's original ESS definition follows from, and is a special case of, the matrix-ESS definition when applied to a monomorphic population (every player is using the same strategy) that experiences invasions from only a single mutant strategy. For this reason we refer to this original definition as the single-mutant-ESS.

In contrast, the more general Definition 9.1.2 allows for a distribution of strategy frequencies (the population may be either monomorphic or polymorphic) and it is stable against simultaneous invasion by small subpopulations with any distribution of allowable mutant strategies. A matrix-ESS is stable for the specific population dynamics under consideration, which may include nonlinear matrix games as well as the traditional bi-linear game. For this reason, the matrix-ESS has more stringent conditions than does the single-mutant-ESS. A strategy satisfying the matrix-ESS must also be a single-mutant-ESS. However, there can be strategies that are a single-mutant-ESS but that are not a matrix-ESS.

Vincent *et al.* (1996) points out that, under the standard approach to bi-linear matrix games and evolutionary stability, a mixed-strategy single-mutant-ESS can never be evolutionarily stable according to the matrix-ESS definition. This happens because the $G$-function is flat for a mixed-strategy single-mutant-ESS. That is, it does not take on a proper maximum at the single-mutant-ESS. Furthermore, it has been shown by Brown and Vincent (1987c) that in a bi-linear matrix game the mixed-strategy single-mutant-ESS can always be invaded by more than one mutant using mixed strategies, and hence does not have the necessary property of a matrix-ESS.

We will now use the above results to examine the bi-linear matrix game and demonstrate the problems associated with mixed-strategy solutions. We first sort out the potentially confusing distinction between a matrix game played using mixed strategies and the same game played with pure strategies. We do not need two theories to do this. The essential difference between the two games lies in the definition of the strategy set $\mathcal{U}$ so that the matrix-ESS solution for both games may be sought using the matrix-ESS maximum principle and its corollary. The $G$-function approach allows for this unification.

## 9.2 The 2 × 2 bi-linear game

A $2 \times 2$ bi-linear game is a symmetric matrix game in which the payoff matrix $\mathbf{M}$ has two rows and two columns. The four elements of this matrix are constants independent of population size, existing strategies or the frequencies of existing strategies. Let $u$ be the proportion of time an individual chooses the first row. It follows that strategies are vectors $\mathbf{u}_i$ with two components of the form $[\, u_i \quad 1 - u_i \,]$, but only one scalar value $u_i$ is needed to specify both components. Each strategy is at a frequency $p_i$. There can be any number of strategies $n_s$ with a corresponding $G$-function given by (9.7).

$$G(\mathbf{v}, \mathbf{u}, \mathbf{p}) = \sum_{j=1}^{n_s} E(\mathbf{v}, \mathbf{u_j}) p_j$$
$$= E(\mathbf{v}, \mathbf{u}_1) p_1 + E(\mathbf{v}, \mathbf{u}_2) p_2 + \cdots + E(\mathbf{v}, \mathbf{u}_{n_s}) p_{n_s}.$$

The expected payoff to an individual playing $\mathbf{v}$ against an opponent playing $\mathbf{u}_i$

is given by

$$E(\mathbf{v}, \mathbf{u}_i) = \mathbf{v}\mathbf{M}\mathbf{u}_i^T$$

$$= \begin{bmatrix} v & 1-v \end{bmatrix} \mathbf{M} \begin{bmatrix} u_i \\ 1-u_i \end{bmatrix}$$

$$= m_{11} u_i v + m_{12}(1-u_i)v + m_{21} u_i(1-v) + m_{22}(1-u_i)(1-v).$$

The basic approach for solving any evolutionary stability problem using the matrix-ESS maximum principle is to first examine whether there exists a coalition of one with $\mathbf{p}^* = \begin{bmatrix} 1 & 0 & \cdots & 0 \end{bmatrix}^T$. In this case,

$$G(\mathbf{v}, \mathbf{u}, \mathbf{p}^*) = E(\mathbf{v}, \mathbf{u}_1) \tag{9.10}$$

and

$$\mathbf{p}^* \mathbf{E}(\mathbf{u}) \mathbf{p}^{*T} = m_{11}.$$

If $G(\mathbf{v}, \mathbf{u}, \mathbf{p}^*)$ takes on a proper maximum with respect to $\mathbf{v}$ at $\mathbf{u}_1$, then $\mathbf{u}_1$ is a matrix-ESS by Corollary 9.1.1. Furthermore, if $\mathbf{p}^*$ is a global ecologically stable equilibrium point (for any $n_s > 1$), then one need look no further since, by definition, there cannot exist a matrix-ESS coalition of two or more strategy vectors. If conditions are such that $G(\mathbf{v}, \mathbf{u}, \mathbf{p}^*)$ does not take on a proper maximum with respect to $\mathbf{v}$ at $\mathbf{u}_1$, then one must look for an ESS coalition of two. In seeking an ESS coalition of two, $\mathbf{p}^* = \begin{bmatrix} p_1^* & p_2^* & 0 & \cdots & 0 \end{bmatrix}$ so that

$$G(\mathbf{v}, \mathbf{u}, \mathbf{p}^*) = E(\mathbf{v}, \mathbf{u}_1) p_1^* + E(\mathbf{v}, \mathbf{u}_2) p_2^*. \tag{9.11}$$

Similar arguments can be used to see whether an ESS coalition of two exists. If not, one can then seek an ESS coalition of three, etc. We will now demonstrate the procedure for finding the matrix-ESS solution in the $2 \times 2$ bi-linear game using the two different strategy sets: $\mathcal{U}_p$ corresponding to pure strategies, and $\mathcal{U}_c$ corresponding to continuous mixed strategies.

### 9.2.1 Pure strategies

A common, restricted version of the bi-linear matrix game is based on the model of Taylor and Jonker (1978). In this case, the strategy choices are restricted to the pure strategy set

$$\mathcal{U}_p = \{\mathbf{u}_i = \begin{bmatrix} u_i & 1-u_i \end{bmatrix} \in \mathcal{R}^2 \mid u_i \in \{0, 1\}\}.$$

#### 9.2.1.1 Coalition of one

If a matrix-ESS coalition of one exists for the pure strategy case, it is given by either $u_1 = 0 \implies \mathbf{u}_1 = \begin{bmatrix} 0 & 1 \end{bmatrix}$ or $u_1 = 1 \implies \mathbf{u}_1 = \begin{bmatrix} 1 & 0 \end{bmatrix}$. We will examine whether either of these solutions can satisfy the matrix-ESS maximum principle. From (9.10) we have

$$G(\mathbf{v}, \mathbf{u}, \mathbf{p}^*) = vm_{11} + (1 - v)m_{21}.$$

There are only two points to check. Clearly, when $v = u_1 = 0$, $G(\mathbf{v}, \mathbf{u}, \mathbf{p}^*) = m_{11}$, and, when $v = u_1 = 1$, $G(\mathbf{v}, \mathbf{u}, \mathbf{p}^*) = m_{21}$. Thus, $\mathbf{u}_1$ satisfies the matrix-ESS maximum principle if and only if

$$m_{11} \geq m_{21}.$$

Furthermore by Corollary 9.1.1, if the condition $m_{11} > m_{21}$ is satisfied, then $\mathbf{u}_1 = \begin{bmatrix} 1 & 0 \end{bmatrix}$ is, in fact, a matrix-ESS. The equality case requires further investigation to see whether $\mathbf{p}^*$ is convergent stable. We have from (9.8) that $p_1$ will be convergent stable provided that

$$\sum_{j=1}^{2} E(\mathbf{u}_1, \mathbf{u}_j) p_j - \mathbf{p}^T \mathbf{E}(\mathbf{u}) \mathbf{p}$$
$$= m_{11} p_1 + m_{12} p_2 - p_1 m_{11} p_1 - p_1 m_{21} p_2 - p_2 m_{12} p_1 - p_2 m_{22} p_2$$
$$= p_1 (1 - p_1) m_{11} + p_2 (1 - p_1) m_{12} - p_2 p_1 m_{12} - p_2 p_2 m_{22}$$

is positive. Replacing $(1 - p_1)$ with $p_2$ and factoring out $p_2$ results in

$$\sum_{j=1}^{2} E(\mathbf{u}_1, \mathbf{u}_j) p_j - \mathbf{p}^T \mathbf{E}(\mathbf{u}) \mathbf{p} = p_1 (m_{11} - m_{12}) + p_2 (m_{12} - m_{22}).$$

Thus, if $m_{11} = m_{21}$, we still have convergent stability provided that $m_{12} > m_{22}$. In summary then, $\mathbf{u}_1 = \begin{bmatrix} 1 & 0 \end{bmatrix}$ is a matrix-ESS if and only if

$$m_{11} > m_{21}$$

or

$$m_{11} = m_{21} \text{ and } m_{12} > m_{22}.$$

Note that if $m_{11} = m_{21}$, the fitness generating function $G(\mathbf{v}, \mathbf{u}, \mathbf{p}^*)$ is independent of $\mathbf{v}$. That is, $G(\mathbf{v}, \mathbf{u}, \mathbf{p}^*)$ is "flat." Nevertheless we have just shown that convergent stability is still possible without $G(\mathbf{v}, \mathbf{u}, \mathbf{p}^*)$ taking on a proper maximum. Because these are exactly the same results as obtained using the single-mutant-ESS definition, we arrive at the conclusion that $\mathbf{u}_1$ is a matrix-ESS if and only if it is a single-mutant-ESS. However, there can be payoff matrices $\mathbf{M}$ for which $\mathbf{u}_1$ satisfies the matrix-ESS maximum principle but $\mathbf{u}_1$ is not a matrix-ESS and that this can only occur when $G(\mathbf{v}, \mathbf{u}, \mathbf{p}^*)$ is independent of $\mathbf{v}$.

A similar analysis shows the second pure strategy, $\mathbf{u}_1 = \begin{bmatrix} 0 & 1 \end{bmatrix}$, is an ESS coalition of one if and only if either

$$m_{22} > m_{12}$$

or

$$m_{22} = m_{12} \text{ and } m_{21} > m_{11}.$$

In particular, $\begin{bmatrix} 0 & 1 \end{bmatrix}$ is a matrix-ESS if and only if it is a single-mutant-ESS. If both $m_{11} > m_{21}$ and $m_{22} > m_{12}$ are satisfied, then there exists two ESS coalitions of one. Each of these must be a local solution. That is, solving (9.8) for some initial conditions in $\Delta^2$ will result in players using $\mathbf{u}_1 = \begin{bmatrix} 1 & 0 \end{bmatrix}$ having an equilibrium frequency of 1 and for other initial conditions in $\Delta^2$ players using $\mathbf{u}_1 = \begin{bmatrix} 0 & 1 \end{bmatrix}$ will have an equilibrium frequency of 1.

#### 9.2.1.2 Coalition of two

In this case, the only possible solution is the coalition $\mathbf{u}_1 = \begin{bmatrix} 1 & 0 \end{bmatrix}$, $\mathbf{u}_2 = \begin{bmatrix} 0 & 1 \end{bmatrix}$. From (9.11), $G(\mathbf{v}, \mathbf{u}, \mathbf{p}^*)$ reduces to

$$G(\mathbf{v}, \mathbf{u}, \mathbf{p}^*) = \begin{bmatrix} v & 1-v \end{bmatrix} \begin{bmatrix} m_{11} \\ m_{21} \end{bmatrix} p_1^* + \begin{bmatrix} v & 1-v \end{bmatrix} \begin{bmatrix} m_{12} \\ m_{22} \end{bmatrix} p_2^*.$$

From the equilibrium requirements that $G^*(\mathbf{v}, \mathbf{u}, \mathbf{p}^*) = \mathbf{p}^* \mathbf{M} \mathbf{p}^{*T}$ at $\mathbf{v} = \mathbf{u}_1$ and $\mathbf{v} = \mathbf{u}_2$,

$$m_{11} p_1^* + m_{12} p_2^* = \mathbf{p}^* \mathbf{M} \mathbf{p}^{*T}$$

and

$$m_{21} p_1^* + m_{22} p_2^* = \mathbf{p}^* \mathbf{M} \mathbf{p}^{*T}.$$

Solving these equations for $p_1^*$ and $p_2^*$ (using the fact that $p_1^* + p_2^* = 1$) yields

$$p_1^* = \frac{m_{22} - m_{12}}{m_{11} + m_{22} - m_{12} - m_{21}} = \frac{a}{a+b}$$

$$p_2^* = \frac{m_{11} - m_{21}}{m_{11} + m_{22} - m_{12} - m_{21}} = \frac{b}{a+b}$$

where

$$a = m_{22} - m_{12}$$
$$b = m_{11} - m_{21}.$$

Thus an ecological equilibrium will exist if and only if $a$ and $b$ are both positive or both negative (i.e., if and only if $ab > 0$). These requirements also guarantee that the matrix-ESS maximum principle is satisfied trivially since, in this case,

$G(\mathbf{v}, \mathbf{u}, \mathbf{p}^*)$ is a flat maximum with respect to $\mathbf{v}$ at both $\mathbf{u}_1$ and $\mathbf{u}_2$ (i.e., $G(\mathbf{v}, \mathbf{u}, \mathbf{p}^*)$ is identically equal to the average fitness for all $\mathbf{v}$). However, an analysis of the dynamic shows $\mathbf{p}^*$ is convergent stable if and only if $a$ and $b$ are both negative – the exact conditions for $\mathbf{p}^*$ to be a single-mutant-ESS (Maynard Smith, 1982) for a polymorphic population. That is, $\mathbf{u}_c = [\mathbf{u}_1, \mathbf{u}_2]$ is a matrix-ESS coalition of two if and only if $\mathbf{p}^*$ is a single-mutant-ESS. It is again clear that there are payoff matrices $\mathbf{M}$ for which $\mathbf{u}_c$ satisfies the matrix-ESS maximum principle, but $\mathbf{u}_c$ is not a matrix-ESS and in this case $G^*(\mathbf{v}, \mathbf{u}, \mathbf{p}^*)$ is identically equal to the average fitness as a function of $\mathbf{v}$ (Bishop and Cannings, 1976).

**Example 9.2.1 (Prisoner's Dilemma)** *Reconsider Example 3.1.3. If both prisoners cooperate they get two years in prison, a cooperative prisoner playing against a defecting prisoner gets five years in prison, a defecting prisoner playing against a cooperating prisoner goes free and a defecting prisoner playing against a defecting prisoner gets four years in prison.*

|   | C  | D  |
|---|----|----|
| C | −2 | −5 |
| D | 0  | −4 |

*Thus*

$$\mathbf{M} = \begin{bmatrix} -2 & -5 \\ 0 & -4 \end{bmatrix}$$

*and since $m_{22} > m_{12}$ a coalition of one matrix-ESS solution exists, given by $\mathbf{u}_1 = \begin{bmatrix} 0 & 1 \end{bmatrix}$. This solution is to defect with four years in prison. What a pity! If the prisoners cooperated, they would get only two years in prison. This result illustrates again that an ESS solution is not a group optimal solution requiring cooperation subject to cheating. Figure 9.1 illustrates that the solution to (9.8) results in an equilibrium frequency of $\mathbf{p}^* = \begin{bmatrix} 1 & 0 \end{bmatrix}$ when $\mathbf{u}_1$ is played against $\mathbf{u}_2 = \begin{bmatrix} 1 & 0 \end{bmatrix}$ even when $p_1$ is initially small. The adaptive landscape generated by $G(\mathbf{v}, \mathbf{u}, \mathbf{p}^*)$, as illustrated in Figure 9.2, is a straight line segment, clearly demonstrating that $\mathbf{u}_1$ satisfies the matrix-ESS maximum principle. This solution also satisfies the single-mutant-ESS definition because $E(u_1, u_1) = -4$ and $E(u_2, u_1) = -5$. That is $E(u_1, u_1) > E(u_2, u_1)$.*

As required by the matrix-ESS maximum principle, the maximum value of $G(\mathbf{v}, \mathbf{u}, \mathbf{p}^*)$ is equal to the average fitness. In general, the average fitness at equilibrium will not be zero, as in the above example where $G(\mathbf{v}, \mathbf{u}, \mathbf{p}^*)|_{\mathbf{v}=\mathbf{u}_1} = -4$. This result is of no consequence for the matrix game as long as the total population number as determined from (9.3) is not a factor in the game. In the above example, $N$ continually decreases with time (but can never reach zero). In other examples, when the average fitness is positive, $N$ continually increases

## 9.2 The 2 × 2 bi-linear game

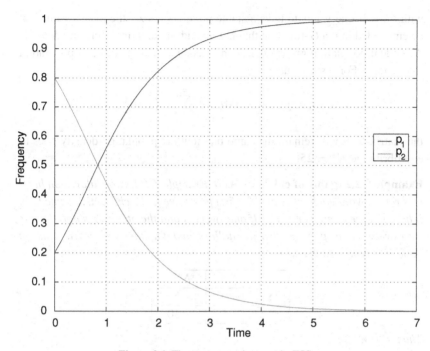

**Figure 9.1** The strategy $u_1$ is a matrix-ESS.

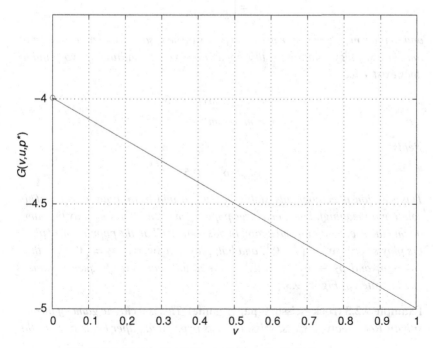

**Figure 9.2** The adaptive landscape is linear for the bi-linear game.

with time. If one wants to have a stable population size, the matrix game can be embedded into a $G$-function that has an additional term describing density-dependent population growth that is independent of strategies and population frequencies. For instance using

$$G(v, u, p, N) = 100/N + \sum_{j=1}^{n_s} E\left(\mathbf{v}, \mathbf{u}_j\right) p_j$$

the matrix-ESS will remain the same but now the population density goes to equilibrium at $N^* = 25$.

**Example 9.2.2 (game of chicken)** *As in Example 3.1.4, two children on bicycles race toward each other and the first to swerve is the chicken. Assume that, if both swerve, there is no cost. If one swerves and the other does not, the one who swerves must pay the other five dollars, and if neither swerves, it costs ten dollars to repair each bike*

|     | S | NS  |
| --- | - | --- |
| S   | 0 | −5  |
| NS  | 5 | −10 |

*Thus*

$$\mathbf{M} = \begin{bmatrix} 0 & -5 \\ 5 & -10 \end{bmatrix}$$

*and neither $m_{11} \geq m_{21}$ nor $m_{22} \geq m_{12}$ is satisfied, so there is no coalition of one pure strategy solution for this game. However, a coalition of two solution does exist since*

$$a = m_{22} - m_{12} = -5$$
$$b = m_{11} - m_{21} = -5$$

*yields*

$$p_1^* = p_2^* = 0.5.$$

*This solution is ecologically stable because $a$ and $b$ are both negative. This solution is meaningful only in a large population of children who play the game among themselves again and again. In this context if half the population of players play swerve, $\mathbf{u}_1 = \begin{bmatrix} 1 & 0 \end{bmatrix}$, and half play non-swerve, $\mathbf{u}_2 = \begin{bmatrix} 0 & 1 \end{bmatrix}$, then $p_1 \to p_1^*$ and $p_2 \to p_2^*$ regardless of the initial (non-zero) frequency values, as illustrated in Figure 9.3.*

**Example 9.2.3 (rock–scissors–paper game)** *This is a classic game of intransitivity. Rock beats scissors, scissors beats paper, and paper beats rock. In the*

## 9.2 The 2 × 2 bi-linear game

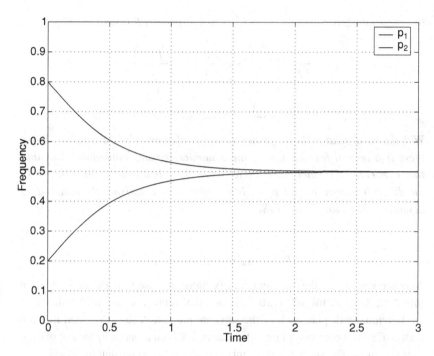

**Figure 9.3** A coalition of two pure strategies exists for the game of chicken.

*standard formulation, the game is zero sum. Winning yields 1, losing yields −1 and a tie yields the player 0.*

|   | R  | S  | P  |
|---|----|----|----|
| R | 0  | 1  | −1 |
| S | −1 | 0  | 1  |
| P | 1  | −1 | 0  |

*Obviously an ESS with a coalition of one pure strategy cannot exist. The corresponding winning strategy could always invade. Similarly, any coalition of two pure strategies cannot be ESS, as one strategy would drive the other to extinction. The only candidate solution for an ESS has a coalition of three pure strategies where each occurs at a frequency of a third. However, this solution does not satisfy the conditions for a bi-linear ecological equilibrium. The strategy dynamics does not coverage on the equilibrium frequencies of $p^* = \begin{bmatrix} \frac{1}{3} & \frac{1}{3} & \frac{1}{3} \end{bmatrix}$. The frequency dynamics corresponds to a neutrally stable limit cycle. The resulting limit cycle depends upon the initial starting frequencies. However, the solution has the property of maintaining the persistence of all three strategies within the population at a long-term average frequency of $p^*$. One can make* *the game non-zero-sum by adding a reward or penalty for ties*

|   | R | S | P |
|---|---|---|---|
| R | $\varepsilon$ | 1 | $-1$ |
| S | $-1$ | $\varepsilon$ | 1 |
| P | 1 | $-1$ | $\varepsilon$ |

*With this modification the frequency dynamics changes with the sign of $\varepsilon$. When there is a reward for ties, $\varepsilon > 0$, the dynamics remains non-equilibrium and they converge on a stable limit cycle. When there is a penalty for ties, $\varepsilon < 0$, the dynamics converges on $p^*$ and this solution now satisfies the matrix-ESS definition (Maynard Smith, 1982).*

### 9.2.2 Mixed strategies

Another version of the bi-linear matrix game is based on the concept of a mixed strategy. A **mixed strategy** is one that represents the probability that an individual will play each of the separate pure strategies. Provided that the matrix-ESS is a coalition of one, the matrix-ESS is a monomorphic population of individuals. The strategy set for mixed strategies is a continuous set given by

$$\mathcal{U}_c = \left\{ \mathbf{u}_i = \begin{bmatrix} u_i & 1 - u_i \end{bmatrix} \in \mathcal{R}^2 \mid 0 \leq u_i \leq 1 \right\}.$$

#### 9.2.2.1 Coalition of one

Suppose $\mathbf{u}_1 = \begin{bmatrix} u_1 & 1 - u_1 \end{bmatrix}$ is an ESS coalition of one in mixed strategies. Then $\mathbf{p}^* = \begin{bmatrix} 1 & 0 \end{bmatrix}^T$ and

$$G(\mathbf{v}, \mathbf{u}, \mathbf{p}^*) = \begin{bmatrix} v & 1 - v \end{bmatrix} \begin{bmatrix} m_{11} & m_{12} \\ m_{21} & m_{22} \end{bmatrix} \begin{bmatrix} u_1 \\ 1 - u_1 \end{bmatrix}$$
$$= v\left[(a + b)u_1 - a\right] + (m_{21} - m_{22})u_1 + m_{22}$$

where

$$a = m_{22} - m_{12}$$
$$b = m_{11} - m_{21}.$$

Since the gradient is given by

$$\left. \frac{\partial G(\mathbf{v}, \mathbf{u}, \mathbf{p}^*)}{\partial v} \right|_{v = u_1} = (a + b)u_1 - a$$

## 9.2 The 2 × 2 bi-linear game

and $0 \leq u_1 \leq 1$, it follows that the necessary condition for $G(\mathbf{v}, \mathbf{u}, \mathbf{p}^*)$ to take on a maximum with respect to $v$ at $u_1$ results in

$$u_1 = \begin{cases} 1 & \text{if } b > 0 \\ 0 & \text{if } a > 0 \\ \dfrac{a}{a+b} & \text{if } ab > 0 \end{cases}. \qquad (9.12)$$

While the necessary conditions of the matrix-ESS maximum principle are satisfied by any of the three cases in (9.12), when we plot $G(\mathbf{v}, \mathbf{u}, \mathbf{p}^*)$ as a function of $v$, it is discovered that only for the cases of $u_1 = 1$ or $u_1 = 0$ does $G(\mathbf{v}, \mathbf{u}, \mathbf{p}^*)$ take on a proper maximum with respect to $v$. With $u_1 = a/(a+b)$, the plot of $G(\mathbf{v}, \mathbf{u}, \mathbf{p}^*)$ is flat for all values of $v$. This suggests that there may not be an ESS coalition of one solution for this case. In fact, it has been shown (Brown and Vincent, 1987c) that not only is there no matrix-ESS coalition of one when $u_1 = a/(a+b)$, there is never a matrix-ESS coalition of two or more.

**Example 9.2.4 (game of chicken)** *Let us re-examine the game of chicken in mixed strategies. Since*

$$a = b = -5$$

*we have the solution*

$$u_1 = 0.5.$$

*However, due to the fact that $G(\mathbf{v}, \mathbf{u}, \mathbf{p}^*)$ does not take on a proper maximum, this solution is not a matrix-ESS, but it is a single-mutant-ESS. If we play $u_1$ against any number of strategies less than $u_1$ (or greater than $u_1$), it appears to have the ESS properties. However, if $u_1$ is played against strategies that are both smaller and larger than $u_1$ then more than one strategy will coexist along with $u_1$. This clearly demonstrates that $u_1$ is not a matrix-ESS.*

For 2 × 2 bi-linear matrix games, the only matrix-ESS for the mixed-strategy model are pure strategy coalitions of one that also correspond to a single-mutant-ESS.

### 9.2.3 Evolution of cooperation

Kin selection and reciprocal altruism represent two models for the evolution of cooperation. Kin selection is a game among relatives where the assumption of random interactions is relaxed. Reciprocal altruism is an iterated game that allows for learning and the introduction of new strategies based on what individuals learn about each other. For instance, in reciprocal altruism the strategy tit-for-tat can be used. In the following examples, both models of cooperation are formulated as a game based on the prisoner's dilemma.

**Example 9.2.5 (kin selection)** *We use a special case of the prisoner's dilemma to model the evolution of cooperation. Let $b > c > 0$. A cooperator incurs a cost $-c$ of bestowing a benefit $b$ on the other player. A defector incurs no costs and bestows no benefits. With these assumptions the matrix for the prisoner's dilemma becomes*

|   | C     | D  |
|---|-------|----|
| C | $b-c$ | $-c$ |
| D | $b$   | 0  |

*When the elements of the matrix satisfy $m_{22} > m_{12}$, the pure strategy D is a matrix-ESS. We model* **kin selection** *by relaxing the assumption of random interactions. Let $r$ be the probability of like interacting with like, and let $(1-r)$ represent the remaining random interactions. The payoff matrix becomes*

|   | C       | D                        |
|---|---------|--------------------------|
| C | $b-c$   | $r(b-c)+(1-r)(-c)$       |
| D | $(1-r)b$| 0                        |

*Cooperation is now a pure strategy matrix-ESS when $m_{11} > m_{21} \Longrightarrow r > c/b$ (**Hamilton's rule** (Hamilton, 1963)) and D is a pure strategy matrix-ESS when $m_{22} > m_{12} \Longrightarrow r < c/b$.*

**Example 9.2.6 (reciprocal altruism)** *We can model* **reciprocal altruism** *in a similar fashion by modifying the strategy C to include an iterated version of prisoner's dilemma. The iterated game is played many times with the assumption that an individual knows the strategy identity of its opponent with probability, $a$, either through prior experience or from watching other plays of the game. Cooperation is modified into a form of tit-for-tat (TFT) strategy, where TFT plays C with strangers, plays C with players known to use TFT and plays D with known defectors. This modified iterated prisoner's dilemma produces the following payoff matrix*

|     | TFT      | D            |
|-----|----------|--------------|
| TFT | $b-c$    | $(1-a)(-c)$  |
| D   | $(1-a)b$ | 0            |

*When familiarity, as determined by $a$, is high enough, TFT can be an ESS. In this game, TFT is a pure strategy matrix-ESS when $m_{11} > m_{21} \Longrightarrow a > c/b$ (note the similarity to Hamilton's rule). However, regardless of the value of $a$ ($0 \leq a \leq 1$) it follows that $m_{22} > m_{12}$ so that the pure strategy of D is always* a matrix-ESS. Thus, depending on the value of a there is either one or two matrix-ESS solutions. When each pure strategy is an ESS, then there are two solutions, and each is local with respect to initial conditions.

## 9.3 Non-linear matrix games

The matrix-ESS definition is in no way restricted to bi-linear matrix games. Specifically, the linear relationships that produce flat adaptive landscapes in the previous section can be eliminated by relaxing the assumption that the elements of the matrix $\mathbf{M}$ are constants or that the strategies need to be interpreted as probabilities. Rather the components of $\mathbf{M}$ may be functions of strategies and/or $\mathcal{U}$ need not be restricted to either $\mathcal{U}_c$ or $\mathcal{U}_p$. For non-linear matrix games, application of the matrix-ESS maximum principle will more often yield interior candidate solutions that are proper maxima of the $G$-function. These results are more akin to those obtained in previous chapters. The following example, chosen for mathematical clarity, illustrates this point. By changing the strategy set for the game of chicken, the matrix-ESS maximum principle results in an interior solution with a proper maximum which is now a matrix-ESS coalition of one. In the bi-linear game $\mathcal{U}_c$ is a straight line segment. In the modified game of chicken, $\mathcal{U}$ is a convex curve.

**Example 9.3.1 (modified game of chicken)** *This $2 \times 2$ game has a constant matrix $\mathbf{M}$ equivalent to the previous game of chicken but with a different constraint set defined by*

$$\mathcal{U} = \{\mathbf{u}_i \in \mathcal{R}^2 \mid 0 \leq u_{ij} \leq 1 \ \forall \ j = 1, 2 \text{ and } u_{i1}^2 + u_{i2} = 1\} \ .$$

*We have from (9.10)*

$$G(\mathbf{v}, \mathbf{u}, \mathbf{p}^*) = E(\mathbf{v}, \mathbf{u}_1)$$

*with $\mathbf{v} = \begin{bmatrix} v_1 & v_2 \end{bmatrix}$ and $\mathbf{u}_1 = \begin{bmatrix} u_{11} & u_{12} \end{bmatrix}$. However, from the constraints*

$$v_1^2 + v_2 = 1$$
$$u_{11}^2 + u_{12} = 1$$

*the components of the vector strategies may be written in terms of the scalars $v = v_1$ and $u_1 = u_{11}$*

$$\mathbf{v} = \begin{bmatrix} v & 1 - v^2 \end{bmatrix}$$
$$\mathbf{u}_1 = \begin{bmatrix} u_1 & 1 - u_1^2 \end{bmatrix}$$

so that

$$G(\mathbf{v}, \mathbf{u}, \mathbf{p}^*) = vm_{11}u_1 - vm_{12}u_1^2 + vm_{12} - m_{21}u_1v^2 + m_{21}u_1 \\ + m_{22}v^2u_1^2 - m_{22}v^2 - m_{22}u_1^2 + m_{22}. \quad (9.13)$$

Thus

$$\frac{\partial G(\mathbf{v}, \mathbf{u}, \mathbf{p}^*)}{\partial v} = m_{11}u_1 - m_{12}u_1^2 + m_{12} - 2m_{21}u_1v + 2m_{22}vu_{11}^2 - 2m_{22}v$$

and replacing $v$ with $u_1$ yields

$$\left.\frac{\partial G(\mathbf{v}, \mathbf{u}, \mathbf{p}^*)}{\partial v}\right|_{v=u_1} = m_{11}u_1 - m_{12}u_1^2 + m_{12} - 2m_{21}u_1^2 + 2m_{22}u_1^3 - 2m_{22}u_1.$$

If an ESS coalition of one exists in the interior of $\mathcal{U}$, we may find it by setting this derivative equal to zero. Using the game of chicken matrix with the value of 10 added to each term[1]

$$\mathbf{M} = \begin{bmatrix} 10 & 5 \\ 15 & 0 \end{bmatrix}$$

and setting $\left.\frac{\partial G(\mathbf{v}, \mathbf{u}, \mathbf{p}^*)}{\partial v}\right|_{v=u_1} = 0$ results in two solutions, $u_1 = 0.5469$ and $u_1 = -0.2612$. Thus $\mathbf{u}_1$ with components

$$u_{11} = 0.5469, \quad u_{12} = 0.7009$$

is the only candidate matrix-ESS coalition of one in the interior of $\mathcal{U}$. We can check to see if this solution represents a proper maximum for $G(\mathbf{v}, \mathbf{u}, \mathbf{p}^*)$ by substituting $u_1 = 0.5469$ into (9.13) to obtain

$$G(\mathbf{v}, \mathbf{u}, \mathbf{p}^*)\big|_{u_1 = 0.5469} = -8.2035v^2 + 8.9735v + 8.2035.$$

As shown in Figure 9.4, a plot of this quadratic function yields a proper maximum at $v = 0.5469$ with the maximum value equal to the average fitness

$$E(\mathbf{u}_1, \mathbf{u}_1) = \mathbf{u}_1 \mathbf{M} \mathbf{u}_1^T = 10.657.$$

In the above example, $\mathbf{p}^* = \begin{bmatrix} 1 & 0 & \cdots & 0 \end{bmatrix}^T$ is a convergent stable equilibrium for all possible mutant strategies and hence $\mathbf{u}_1^*$ is a matrix-ESS. Compare this result with the same bi-linear matrix game played on the strategy set $\mathcal{U}_c$. In that case, we obtain the single-mutant-ESS $u_{11} = 0.5$, $u_{12} = 0.5$ which is not a proper maximum point of $G(\mathbf{v}, \mathbf{u}, \mathbf{p}^*)$ and not a matrix-ESS with respect to all possible mutant strategies.

---

[1] This is done in order to avoid complex roots. The bi-linear game defined with this $\mathbf{M}$ matrix is equivalent to the game of chicken and has the same solution.

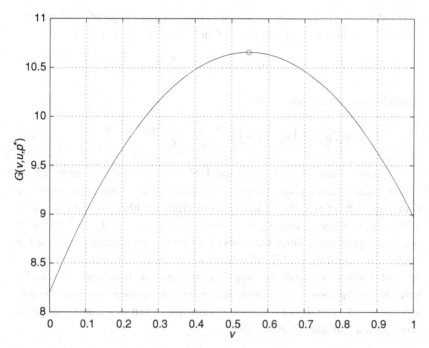

**Figure 9.4** The function $G(v, u, p^*)$ takes on a proper maximum at $v = 0.5469$.

### 9.3.1 Sex ratio game

We will use an elementary (non-genetic) model to explain why the numbers of males and females are approximately equal in most animal populations. It is assumed that the sex ratio of offspring is determined by their mother and that a female's fitness is measured by the expected number of grandchildren. The strategy of females of type $i$ is specified by $\mathbf{u}_i = \begin{bmatrix} u_{i1} & u_{i2} \end{bmatrix}$ where $u_{i1}$ is the number of male children and $u_{i2}$ is the number of female children. Thus $k_i = u_{i1} + u_{i2}$ is the total number of her children. Under random mating, each male is assumed to mate with $\bar{r}_{sex}$ females where $\bar{r}_{sex}$ is the current sex ratio (total number of females/total number of males) of the population as a whole. In this case the expected number of grandchildren from a female of type $i$ is

$$k_i u_{i2} + \bar{k}\bar{r}_{sex} u_{i1}$$

where $\bar{k}$ is the average number of children per female in the population as a whole. In the special case of only two strategy types, $i$ and $j$, with $i$ in small numbers and $j$ in large numbers, the sex ratio may be approximated by $\bar{r}_{sex} = \dfrac{u_{j2}}{u_{j1}}$ and $\bar{k}$ may be approximated by $k_j$. In this case the expected number of grandchildren from a female of type $i$ competing in a population of females of type $j$ may be written as (Maynard Smith, 1982; Cressman, 1992)

$$E(\mathbf{u}_i, \mathbf{u}_j) = k_i u_{i2} + k_j \frac{u_{j2}}{u_{j1}} u_{i1} \qquad (9.14)$$

which, in non-linear matrix form, is

$$E(\mathbf{u}_i, \mathbf{u}_j) = \begin{bmatrix} u_{i1} & u_{i2} \end{bmatrix} \begin{bmatrix} 0 & k_j/u_{j1} \\ k_i/u_{j1} & 0 \end{bmatrix} \begin{bmatrix} u_{j1} \\ u_{j2} \end{bmatrix}.$$

The above assumption for $\bar{r}_{sex}$ and $\bar{k}$ allows us to put this game into a matrix game format. While these assumptions are not valid when there are many different types in the population away from equilibrium conditions, they will be valid when examining the conditions for a matrix-ESS coalition of one near equilibrium. Since the matrix-ESS maximum principle is applied at equilibrium and since the ESS of interest will turn out to be a coalition of one, we will continue our analysis using this expected payoff matrix. It should be noted that, if this problem is formulated without the above assumptions, it may still be solved using the ESS maximum principle, albeit as a continuous game rather than a non-linear matrix game.

### 9.3.1.1 The politically correct solution

If the "costs" of producing a male or female child is the same, then the number of children produced will simply depend on the total resources available to each female. If we assume these resources to be the same for all types it follows that $k_i = k_j = k$. The strategy set in this case may be written as

$$\mathcal{U} = \left\{ \mathbf{u}_i \in \mathcal{R}^2 \mid 0 \leq u_{ij} \leq k \; \forall \; j = 1, 2 \text{ and } u_{i1} + u_{i2} = k \right\}.$$

From (9.14), the expected payoff in this case becomes

$$E(\mathbf{u}_i, \mathbf{u}_j) = k \left[ u_{i2} + \frac{u_{j2}}{u_{j1}} u_{i1} \right].$$

Assuming an ESS coalition of one we have from (9.10)

$$G(\mathbf{v}, \mathbf{u}, \mathbf{p}^*) = E(\mathbf{v}, \mathbf{u}_1)$$
$$G(\mathbf{v}, \mathbf{u}, \mathbf{p}^*) = k[v_2 + \frac{u_{12}}{u_{11}} v_1].$$

Using the constraints to eliminate $v_2$ and $u_{12}$ we have

$$G(\mathbf{v}, \mathbf{u}, \mathbf{p}^*) = k[k - v_1 + \frac{k - u_{11}}{u_{11}} v_1].$$

## 9.3 Non-linear matrix games

A necessary condition for $G(\mathbf{v}, \mathbf{u}, \mathbf{p}^*)$ to take on a maximum with respect to $v_1$ in the interior of $\mathcal{U}$ is given by

$$\left.\frac{\partial G(\mathbf{v}, \mathbf{u}, \mathbf{p}^*)}{\partial v_1}\right|_{v_1=u_{11}} = -1 + \frac{k - u_{11}}{u_{11}} = 0$$

which implies $u_{11} = k/2$. That is, half of the offspring should be males.

Although the result is the expected one that $\bar{r}_{sex} = 1$, it follows that

$$\left.G(v_1, u_{11}, \mathbf{p}^*)\right|_{u_{11}=k/2} = k^2$$

does not depend on $v_1$ and so $G(\mathbf{v}, \mathbf{u}, \mathbf{p}^*)$ does not have a proper maximum when $\mathbf{u}_1 = \begin{bmatrix} \frac{k}{2} & \frac{k}{2} \end{bmatrix}^T$. Thus, it is unclear, without further analysis, whether this is a matrix-ESS where each female will have the same number of sons as daughters or whether this may be the average behavior of the female population. This problem may be avoided if the constraint set $\mathcal{U}$ is modified so that the components of the strategy vector are not restricted to lie on a straight line. In fact, this possibility is considered in the literature on sex ratio games as a special case of resource allocation models where the allocation decision is how many sons and how many daughters to produce (see p. 44 Maynard Smith, 1982).

### 9.3.1.2 Other possible solutions

If the cost of producing females is different from the cost of producing males, then a relationship must exist between the two such as $u_{i2} = f(u_{i1})$. If we again assume the resources available for producing offspring are the same for all types, there will be an upper limit, $k$, to the total number of children produced when they are all males. The strategy set for this case is

$$\mathcal{U} = \left\{\mathbf{u}_i \in \mathcal{R}^2 \mid 0 \leq u_{i1} \leq k \text{ and } u_{i2} = f(u_{i1})\right\}$$

where the total number of children produced is given by

$$k_i = u_{i1} + u_{i2} = u_{i1} + f(u_{i1}).$$

Clearly $f(u_{i1})$ must satisfy the requirement that $f(k) = 0$. Note that we return to the politically correct strategy set if we take $f(u_{i1}) = k - u_{i1}$.

In order to simulate a convex curve, we will use an arc of a circle

$$u_{i2} = f(u_{i1}) = \sqrt{(k^2 - u_{i1}^2)}. \tag{9.15}$$

In this case $k$ children will be produced if they are either all females or all males. A female who uses a mixed strategy will produce a greater number of children with the maximum number corresponding to $u_{i1} = \sqrt{2}\frac{k}{2}$.

From (9.14) the expected payoff in this case is

$$E(\mathbf{u}_i, \mathbf{u}_j) = [u_{i1} + f(u_{i1})] f(u_{i1}) + [u_{j1} + f(u_{j1})] \frac{f(u_{j1})}{u_{j1}} u_{i1}.$$

Assuming an ESS coalition of one we have from (9.10)

$$G(\mathbf{v}, \mathbf{u}, \mathbf{p}^*) = [v_1 + f(v_1)] f(v_1) + [u_{11} + f(u_{11})] \frac{f(u_{11})}{u_{11}} v_1.$$

Thus

$$\frac{\partial G(\mathbf{v}, \mathbf{u}, \mathbf{p}^*)}{\partial v_1} = \left[1 - \frac{v_1}{f(v_1)}\right] f(v_1) - [v_1 + f(v_1)] \frac{v_1}{f(v_1)}$$
$$+ [u_{11} + f(u_{11})] \frac{f(u_{11})}{u_{11}}$$

which reduces to

$$\frac{\partial G(\mathbf{v}, \mathbf{u}, \mathbf{p}^*)}{\partial v_1} = f(v_1) - 2v_1 - \frac{v_1^2}{f(v_1)} + f(u_{11}) + \frac{f^2(u_{11})}{u_{11}}.$$

We again seek an interior solution for $v_1$ from

$$\left.\frac{\partial G(\mathbf{v}, \mathbf{u}, \mathbf{p}^*)}{\partial v_1}\right|_{v_1 = u_{11}} = 2f(u_{11}) - 2u_{11} - \frac{u_{11}^2}{f(u_{11})} + \frac{[f(u_{11})]^2}{u_{11}} = 0$$

that has the solution

$$u_{11} = \sqrt{2}\frac{k}{2}.$$

This is also the value for $u_{11}$ which produces the maximum number of children. To see whether this is a proper maximum we plot $G^*$ as a function of $v_1$. Choosing $k = 10$ we obtain the results shown in Figure 9.5. Thus, by Corollary 9.1.1, $u_{11} = 7.071$ corresponds to a matrix-ESS coalition of one. Furthermore, simulation may be used to demonstrate that this solution is in fact a globally convergent stable equilibrium and so it is the unique matrix-ESS.

It follows from (9.15) that $u_{12} = u_{11}$. Thus, in this case, the solution is again the politically correct solution of producing the same numbers of males and females. However, this result differs from the previous case in that this solution is now a matrix-ESS. We get the same result owing to the symmetry in the trade-off curve between males and females. Seeking a matrix-ESS with a non-symmetric trade-off can yield other solutions.

Figure 9.5 also reminds us that neglecting total population dynamics as given by (9.3) is not always a good idea. It would make the *Zero Population Growth* organization very unhappy. A mean fitness of 200 is clearly worrisome (a result of choosing $k = 10$).

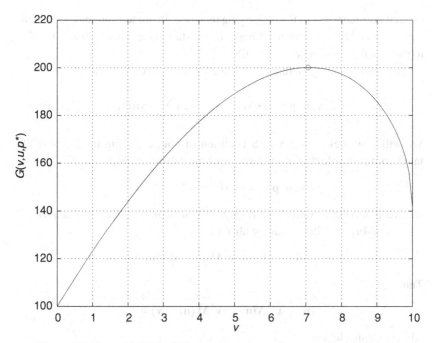

**Figure 9.5** The matrix-ESS solution produces the maximum number of children.

### 9.3.2 Kin selection

In what follows we provide a more formal and expansive analysis of kin selection than presented in Example 9.2.5. The static version of the model considered here (Grafen, 1979; Hines and Maynard Smith, 1979) originated as a means to incorporate the relatedness concept of Hamilton (1964) into an "inclusive fitness" function as opposed to a fitness function based explicitly on the underlying genetics. It is assumed that the rate of interaction between relatives reflects the degree of relatedness. The analysis follows that presented by Vincent and Cressman (2000).

In place of (9.5), suppose the individual fitness of someone using strategy $\mathbf{u}_i$ is

$$H_i[\mathbf{u}, \mathbf{p}] = \alpha \mathbf{u}_i \mathbf{M} \mathbf{u}_i^T + (1 - r) \sum_{j=1}^{n_s} \mathbf{u}_i \mathbf{M} \mathbf{u}_j^T p_j$$

where $\mathbf{M}$ is a payoff matrix of constants and $0 < r < 1$ is the strategy-independent rate at which games occur between relatives. This model has been investigated by Hines and Maynard Smith (1979) from the frequency perspective (see also Grafen (1979) and Taylor (1989) for the case $n = 2$) without explicit consideration of the underlying dynamic. In our terminology, these references found ESS coalitions of one through static fitness comparisons similar to those defining a single-mutant-ESS.

The corresponding $G$-function for the frequency dynamic model is

$$G(\mathbf{v}, \mathbf{u}, \mathbf{p}) = r\mathbf{v}\mathbf{M}\mathbf{v}^T + (1-r)\sum_{j=1}^{n_s} \mathbf{v}\mathbf{M}\mathbf{u}_j^T p_j. \tag{9.16}$$

We will now seek a matrix-ESS coalition of one, $\mathbf{u}_1$, using the matrix-ESS maximum principle. In this case it is necessary that

$$G(\mathbf{v}, \mathbf{u}, \mathbf{p}^*) = r\mathbf{v}\mathbf{M}\mathbf{v}^T + (1-r)\mathbf{v}\mathbf{M}\mathbf{u}_1^T$$

take on a maximum value with respect to $\mathbf{v}$ at $\mathbf{u}_1$ that is equal to the average fitness $\mathbf{u}_1\mathbf{M}\mathbf{u}_1^T$. It follows that for all $\mathbf{v} \in \mathcal{U}_c$

$$r\mathbf{v}^T\mathbf{M}\mathbf{v} + (1-r)\mathbf{v}^T\mathbf{M}\mathbf{u}_1 - \mathbf{u}_1^T\mathbf{M}\mathbf{u}_1 \leq 0. \tag{9.17}$$

Thus

$$(\mathbf{u}_1 - \mathbf{v})^T\mathbf{M}\mathbf{u}_1 + \alpha\mathbf{v}^T\mathbf{M}(\mathbf{u}_1 - \mathbf{v}) \geq 0$$

which is equivalent to

$$(\mathbf{u}_1 - \mathbf{v})^T\mathbf{M}\mathbf{u}_1 + r\mathbf{v}^T\mathbf{M}(\mathbf{u}_1 - \mathbf{v}) + \left[r\mathbf{u}_1^T\mathbf{M}(\mathbf{u}_1 - \mathbf{v}) - r\mathbf{u}_1^T\mathbf{M}(\mathbf{u}_1 - \mathbf{v})\right] \geq 0$$
$$(\mathbf{u}_1 - \mathbf{v})^T(\mathbf{M} + r\mathbf{M}^T)\mathbf{u}_1 - r(\mathbf{u}_1 - \mathbf{v})^T\mathbf{M}(\mathbf{u}_1 - \mathbf{v}) \geq 0.$$

It follows that (9.17) can be rewritten as

$$(\mathbf{u}_1 - \mathbf{v})^T(\mathbf{M} + r\mathbf{M}^T)\mathbf{u}_1 \geq r(\mathbf{u}_1 - \mathbf{v})^T\mathbf{M}(\mathbf{u}_1 - \mathbf{v}). \tag{9.18}$$

Suppose that $\mathbf{u}_1$ lies in the interior of $\mathcal{U}_c$. Unless

$$(\mathbf{u}_1 - \mathbf{v})^T(\mathbf{M} + r\mathbf{M}^T)\mathbf{u}_1 = 0$$

for all $\mathbf{v} \in \mathcal{U}_c$ it would always be possible to find a $\mathbf{v} \in \mathcal{U}_c$ such that $\mathbf{u}_1 - \mathbf{v}$ would result in a negative value for the left-hand side of (9.18). This would violate the inequality since by making the length of $\mathbf{u}_1 - \mathbf{v}$ arbitrarily small, the quadratic nature of the right-hand side of (9.18) would result in its value being greater than the left-hand side. Thus

$$(1+r)(\mathbf{u}_1 - \mathbf{v})^T\mathbf{M}(\mathbf{u}_1 - \mathbf{v}) = (\mathbf{u}_1 - \mathbf{v})^T(\mathbf{M} + r\mathbf{M}^T)(\mathbf{u}_1 - \mathbf{v}) \leq 0.$$

In fact, this last inequality must be strict or else there is a continuum of equilibrium points through $\mathbf{u}_1$ in the "direction" $\mathbf{v}$ which contradicts the fact that $\mathbf{u}_1$ is an ESS. This shows that $\mathbf{u}_1$ is a single-mutant-ESS for the adjusted payoff matrix $\mathbf{M} + r\mathbf{M}^T$.

## 9.3 Non-linear matrix games

**Theorem 9.3.1 (game against relatives)** *Suppose $0 < r < 1$ is fixed and $\mathbf{u}_1$ is in the interior of $\mathcal{U}_c$. Then $\mathbf{u}_1$ is a matrix-ESS for the game against relatives with payoff matrix $\mathbf{M}$ if and only if $\mathbf{u}_1$ is a single-mutant-ESS for the matrix $\mathbf{M} + \alpha \mathbf{M}^T$.*

*Proof.* The "only if" direction is proved above. For the other direction, assume $\mathbf{u}_1$ is a matrix-ESS of $\mathbf{M} + \alpha \mathbf{M}^T$. For a fixed set of mutant strategies $\mathbf{u}_2, \ldots, \mathbf{u}_{n_s}$ in $\Delta^{n_u}$, we have $G(\mathbf{u}_i, \mathbf{u}_1, \mathbf{p}^*) < G(\mathbf{u}_1, \mathbf{u}_1, \mathbf{p}^*)$ for $2 \le i \le n_s$ by the above analysis. Thus, by Corollary 9.1.1, $\mathbf{u}_1$ is a matrix-ESS. ∎

Let us apply the above result to the $2 \times 2$ payoff matrix $M = \begin{bmatrix} a & b \\ c & d \end{bmatrix}$. An interior ESS must be a single-mutant-ESS of $\mathbf{M} + \alpha \mathbf{M}^T = \begin{bmatrix} (1+r)a & b+rc \\ c+rb & (1+r)d \end{bmatrix}$ and so the first component of $\mathbf{u}_1$ is given by

$$\frac{b - d + r(c - d)}{(1+r)(b+c-a-d)}$$

which must be between 0 and 1 and the denominator must be positive (Taylor, 1989). In this case, the ecological equilibrium point $\mathbf{p}^* = \begin{bmatrix} 1 & 0 & \cdots & 0 \end{bmatrix}^T$ is a global matrix-ESE. Therefore, we conclude that there is a unique global matrix-ESS coalition of one in the interior of $\mathcal{U}_c$ if and only if $c + rb > (1+r)a$ and $b + rc > (1+r)d$.

# 10
# Evolutionary ecology

The **ecological theater** and **evolutionary play** were Hutchinson's words for evoking the interplay between ecological and evolutionary dynamics. Yet evolutionary ecology, that harmonious blend between what is evolutionarily feasible and ecologically acceptable, has often been difficult to achieve conceptually. The genes of traditional genetic models of evolution are not easily integrated with the individuals of population dynamic models. The $G$-function provides a conceptual bridge between population dynamics and evolutionary changes in strategy frequency. The structure of the $G$-function and the underlying population dynamics provide the ecological theater while the strategy dynamics, species archetypes, and macroevolutionary production of novel $G$-functions provide the evolutionary theater. Evolutionary game theory is broadly applicable to modeling questions in evolutionary ecology. Here we explore a subset of topics. The topics are chosen because of familiarity, broad interest in ecology and evolution, and for illustrating the formulation and application of the $G$-function approach.

## 10.1 Habitat selection

Behaviors allow animals to adapt to temporal and spatial variabilities in hazards and opportunities. Feeding behaviors are grouped roughly into decisions of what to eat (diet choice), how thoroughly to exploit a depleting feeding opportunity (patch use), and where to seek resources (habitat selection). There is a hierarchy of temporal and spatial scales in going from choosing a food item in the food patch up to habitat selection where an organism seeks a place to live and forage. Habitats may vary in the quality of climate, physical structure, predation risk, and resource productivity. Habitat quality may also vary with the abundance of competitors (the density-dependent aspect of habitat selection).

## 10.1 Habitat selection

In habitat selection, the heritable phenotype often involves a behavior of biasing effort towards an advantageous subset or mix of temporal or spatial habitats. The behavior requires the ability to assess and respond to heterogeneity. The adaptation, subject to costs and constraints, requires assessing accurately and responding appropriately.

**Density-dependent habitat selection** is a study of how an organism should allocate its time and effort among habitats. We view density-dependent habitat selection as an evolutionary game (Brown, 1998) through a progression of habitat selection scenarios starting with the classic ideal free distribution, then moving on to the ideal free distribution under resource matching, and ending with habitat selection involving resource discovery, exploitative competition and, non-linear isodars. Excellent reviews of theories and empirical studies of habitat selection are found in Rosenzweig (1985, 1987a, 1991), Kacelnik et al. (1992), Kennedy and Gray (1993), and Morris (1994).

### 10.1.1 Ideal free distribution

The concept of an **ideal free distribution** assumes that an organism has complete information. That is, it knows the quality of habitats, the distribution and abundance of other individuals among habitats, and the fitness consequences of residing within these habitats. It has free access to each habitat in the sense that, when switching habitats, there are no costs in terms of time, energy, or risk. In addition, all individuals in the patch experience the same fitness consequences of residing within a patch (all individuals are equal). Under these conditions, individuals should distribute themselves among habitats so that each habitat offers the same fitness (Fretwell and Lucas, 1970; Fretwell, 1972).

We construct a $G$-function for the two-habitat version of the ideal free distribution by letting $u \in [0, 1]$ be the population strategy describing the probability of an individual residing in habitat 1, with $(1 - u)$ the probability of residing in habitat 2. Let the fitness to an individual within a habitat be a function of the habitat's intrinsic quality and the abundance of foragers within the patch. Let $N$ be a fixed total population number, then the numbers of individuals in habitats 1 and 2 are given by

$$x_1 = uN$$
$$x_2 = (1 - u)N.$$
(10.1)

We form the $G$-function

$$G(v, u, N) = vF_1(x_1) + (1 - v)F_2(x_2)$$

where $F_i(x_i)$ gives the fitness reward from spending time in habitat $i$ as a function of the population size in habitat $i$. The quantities $x_i$ and $F_i(x_i)$ give a measure of **habitat quality**. When the presence of other individuals depresses the value of a habitat (the simplest assumption in models of density-dependent habitat selection), increasing $x_i$ will cause $F_i(x_i)$ to decline, so that $\partial F_i/\partial x_i < 0$.

This model of habitat selection has a linear structure similar to that of a matrix game. The payoff to the individual is a weighted averaging of two situations that in a $2 \times 2$ matrix game would emerge from the individual encountering two types of players. Also, as in the matrix game with mixed strategies, the strategy of the focal individual does not appear in the fitness gradient

$$\frac{\partial G}{\partial v} = F_1(x_1) - F_2(x_2).$$

Thus the slope of the adaptive landscape is independent of $v$ and is influenced by only the organism's biotic environment: $u$ and $N$. If for a given $N$, $F_1(x_1) > F_2(x_2)$ the individuals should select $u = 1$ and only occupy habitat 1. If $F_1(x_1) < F_2(x_2)$ the individuals should select $u = 0$ and only occupy habitat 2. If $F_1(x_1) = F_2(x_2)$ the two habitats are of equal value and individuals may select either habitat. Thus, as in the case of mixed-strategy ESS for the $2 \times 2$ matrix game, there are three possible ESS solutions that result from the original assumption of an ideal free distribution

Habitat selection for a given population size $N$

| ESS | Condition | |
|---|---|---|
| $u = 1$ | $F_1(x_1)\|_{x_1=N} \geq F_2(x_2)\|_{x_2=0}$ | |
| $u \in (0, 1)$ | $F_1(x_1)\|_{x_1=x_1^*} = F_2(x_2)\|_{x_2=x_2^*}$ | $x_i^* > 0$ |
| $u = 0$ | $F_1(x_1)\|_{x_1=0} \leq F_2(x_2)\|_{x_2=N}$ | |

The first and third ESSs have all of the individuals occupying just a single habitat. Fitness in the crowded, higher-quality habitat is still higher than fitness in the empty, lower-quality habitat. Habitat 1, then, is of higher quality than habitat 2 if, when both habitats are empty, it offers higher potential fitness: $F_1(x_1)|_{x_1=0} > F_2(x_2)|_{x_2=0}$; and vice versa when habitat 2 is of higher quality. As illustrated in Figure 10.1, the second situation results when, for example, habitat 1 is of higher quality at very low population densities, but declines in quality with an increase in population until a critical population size $\hat{N}$ at which $F_1(x_1)|_{x_1=\hat{N}} = F_2(x_2)|_{x_2=0}$. For population sizes $N < \hat{N}$ the ESS will be $u = 1$, but for population sizes $N \geq \hat{N}$ the ESS will have individuals in both habitats, $u \in (0, 1)$, distributed in a manner that equalizes fitness among habitats.

It follows that under density-dependent habitat selection the ESS solution obtained depends upon population size. It is possible to make an $x_2$ versus $x_1$ plot of all combinations of $x_1$ and $x_2$ such that $F_1(x_1) = F_2(x_2)$. This plot is called an **isodar** (Morris, 1988, 1992). Additional results associated with the

## 10.1 Habitat selection

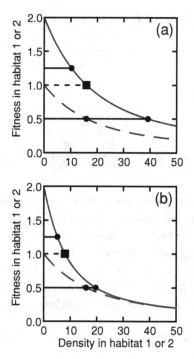

**Figure 10.1** The solid line represents the fitness in habitat 1 and the curved dashed line the fitness in habitat 2. When the density reaches a level such that the two fitnesses are equal (designated by the square), any further increase in density is divided between the two habitats.

ideal free distribution can be seen by considering specific models and their associated isodars.

**Example 10.1.1 (resource matching)** *The concept of resource matching comes from the idea that two habitats may offer different inputs of resources. If the individuals in a habitat divide equally the habitat's resource input, then the ESS distribution of individuals among habitats may match those resource inputs. Let*

$$G(v, u, N) = v\frac{R_1}{x_1} + (1-v)\frac{R_2}{x_2}$$

*where $R_1$ and $R_2$ are the resource inputs into habitats 1 and 2, respectively. In this case the condition $F_1(x_1)|_{x_1=0} > F_2(x_2)|_{x_2=0}$ cannot be satisfied and hence there will be populations in both habitats for all values of N. The isodar is found by setting*

$$\frac{R_1}{x_1} = \frac{R_2}{x_2}.$$

Using this expression with (10.1) and solving for u yields

$$u = \frac{1}{1 + \frac{R_2}{R_1}}.$$

so that

$$x_1 = \frac{N}{1 + \frac{R_2}{R_1}}$$

$$x_2 = \frac{N}{1 + \frac{R_1}{R_2}} = \frac{R_2}{R_1} x_1.$$

**Example 10.1.2 (resource discovery)** *In this model, individuals within a habitat compete in an exploitative way for a stream of resources that flow by at rate $R_i$. Organisms must find the resources and only some fraction of resources are actually discovered. Let $a_i$ represent an individual's encounter probability on a given resource item in habitat i. Let $(1 - e^{-a_i x_i})$ be the probability that an item is found, and let the total resource harvest from a habitat be divided equally among the residents of the habitat, resulting in a G-function of the form*

$$G(v, u, x) = v R_1 \frac{(1 - e^{-a_1 x_1})}{x_1} + (1 - v) R_2 \frac{(1 - e^{-a_2 x_2})}{x_2}.$$

*Increasing the number of individuals within a habitat has two effects. The first is to increase the fraction of resources harvested, the numerator of the above equation. The second is to reduce the rate of harvest by each individual, the denominators of the equation. Because the model assumes that each individual within the habitat shares equally in the harvest, the addition of another individual increases somewhat the collective harvest, at the expense of each individual's harvest. At very large and very small population sizes the model has the following properties: as $x_i \to \infty$, $F_i \to R_i/x_i$, and as $x_i \to 0$, $F_i \to a_i R_i$. At low numbers of foragers, the individual's harvest rate is limited by its ability to encounter the available resource pool or stream. At high numbers of foragers, essentially all of the resources are harvested and divided among the foragers (same as the resource matching model (Parker, 1978; Morris, 1994)). Figure 10.1 was drawn based on this model. Case a corresponds to $a_1 = a_2$ and $R_1 = 2 R_2$ and case b corresponds to $a_1 = 2 a_2$ and $R_1 = R_2$. We see that, in both cases, there is a critical population size that must be obtained before both habitats are occupied. Case b has the property that as N gets large, $x_1 = x_2$. The isodar for this situation cannot be solved analytically for $x_2$ since it is of the form*

$$\frac{x_2}{x_1} = \frac{R_2 (1 - e^{-a_2 x_2})}{R_1 (1 - e^{-a_1 x_1})}.$$

In the above model, the isodar is non-linear (Morris, 1994), and the distribution of individuals among habitats that equalizes fitness does not conform to the resource matching rule of $R_1/x_1 = R_2/x_2$ (Parker, 1978; Kennedy and Gray, 1993). Relative to the resource matching rule, individuals appear to overutilize the more productive habitat, or the habitat with the higher encounter probability on resources. If $a_1 R_1 > a_2 R_2$ then habitat 1 is the higher-quality habitat in the absence of individuals.

**Example 10.1.3 (resource renewal)** *Under resource renewal, the foragers's depletion of resources from each habitat eventually becomes balanced by resource growth. We have a consumer-resource model*

$$G(v, u, x) = a_1 v y_1 + a_2(1 - v) y_2$$

*where resources in the two habitats, $y_1$ and $y_2$, renew according to a Monod equation*

$$\dot{y}_1 = r_1(K_1 - y_1) - a_1 x_1 y_1$$
$$\dot{y}_2 = r_2(K_2 - y_2) - a_2 x_2 y_2$$

*with $r_i$ as growth rate constants and $K_i$ carrying capacities. For the forager species, $a_i$ is its encounter probability on resource $i$. We can calculate the equilibrium abundance of resources as influenced by the abundance and distribution of foragers*

$$y_i^* = \frac{r_i K_i}{r_i + a_i x_i}.$$

*At the ESS, all habitats must yield equal rewards. We can determine the isodar by setting the harvest rate from habitat 1 equal to that from habitat 2: $a_1 y_1^* = a_2 y_2^*$. This yields*

$$\frac{a_1 r_1 K_1}{r_1 + a_1 x_1} = \frac{a_2 r_2 K_2}{r_2 + a_2 x_2}.$$

*Solving for the isodar gives a straight line*

$$x_2 = \frac{r_2(a_2 K_2 - a_1 K_1)}{a_1 a_2 K_1} + \frac{r_2 K_2}{r_1 K_1} x_1.$$

*The intercept is determined by a forager's harvest rate in the pristine habitat. If $a_1 K_1 > a_2 K_2$ then habitat 1 is preferred and there is a positive x-intercept.*

## 10.2 Consumer-resource games

We introduce two models that have appeared in the literature. The first one is a plant model in which competition occurs only through the utilization of resources. The second one is a cell model where there are both competition for resources and competition between cells. In both cases we examine conditions for coexistence.

### 10.2.1 Competition between plants

Biomass allocation between roots, stems, leaves, and seeds are strategies that affect plant growth and that influence competition for sunlight and soil nutrients (see for example Cohen (1971), Grime (1977), Vincent (1979), Chapin (1980), Vincent and Brown (1984a), Bloom et al. (1985), Tilman (1988)). As in Vincent and Vincent (1996), we use a consumer-resource model from Reynolds and Pacala (1993) that only deals with biomass allocation between roots and leaves according to

$$\dot{x}_i = x_i \left[ \min \left( \begin{array}{c} \frac{ry_1 u_i}{y_1 + k_{y_1}} - R \\ \frac{ry_2(1-u_i)}{y_2 + k_{y_2}} - R \end{array} \right) - d \right] \tag{10.2}$$

where

$x_i$ = biomass of species $i$
$u_i$ = fraction of biomass allocated to root by species $i$
$y_1$ = available soil nutrient
$y_2$ = light availability
$r$ = constant per capita maximum plant growth rate
$R$ = constant per capita respiration rate
$d$ = constant per capita loss rate
$k_{y_1}$ = saturation constant for nutrient
$k_{y_2}$ = saturation constant for light.

Species $i$ is identified by the fraction of biomass allocated to root with $0 \leq u_i \leq 1$. The min function in (10.2) expresses the fact that a plant may be either nutrient limited (upper) or light limited (lower). Competition between different species of plants (using different strategies $u_i$) occurs only through the nutrient resource $y_1$ and the light resource $y_2$. These resources are not constant, but depend on $u_i$ according to

$$\dot{y}_1 = a \left( Y_1 - y_1 - \sum_{i=1}^{n_s} p x_i \right) - \sum_{i=1}^{n_s} p x_i \min \left( \begin{array}{c} \frac{ry_1 u_i}{y_1 + k_{y_1}} - R \\ \frac{ry_2(1-u_i)}{y_2 + k_{y_2}} - R \end{array} \right)$$

$$y_2 = \frac{Y_2}{1 + \sum_{i=1}^{n_s} \alpha x_i (1 - u_i)}$$

## 10.2 Consumer-resource games

where

$Y_1$ = total soil nutrient
$a$ = mineralization rate
$p$ = plant tissue nutrient concentration
$Y_2$ = solar constant
$\alpha$ = light decay rate per unit leaf biomass.

The only variables in the min function are $u_i$, $y_1$ and $y_2$ so that for given values of $y_1$ and $y_2$ the upper term increases linearly from $-R$ with an increase in $u_i$ and the lower term decreases linearly to $-R$ with an increase in $u_i$. Thus there exists a crossover value of $u_i$ obtained by setting the two terms in the min function equal to each other and solving for $u_i$ that is then labeled

$$u_m = \frac{y_2\left(y_1 + k_{y_1}\right)}{2y_1 y_2 + y_1 k_{y_2} + y_2 k_{y_1}}.$$

In other words for

$$0 \leq u_i \leq u_m$$

the min function will choose the "nutrient limiting growth curve" and for

$$u_m \leq u_i \leq 1$$

the min function will choose the "light limiting growth curve." Note that the strategy $u_m$ maximizes the growth rate of the plant for any given value of $y_1$ and $y_2$.

The $G$-function for this system is given by

$$G(v, \mathbf{u}, \mathbf{x}, \mathbf{y}) = \min \left( \begin{array}{c} \frac{r y_1 v}{y_1 + k_{y_1}} - R \\ \frac{r y_2 (1-v)}{y_2 + k_{y_2}} - R \end{array} \right) - d \qquad (10.3)$$

where the dependence on $\mathbf{u}$ and $\mathbf{x}$ is implicit through the resource equations. We can easily apply the ESS maximum principle to this case. Assuming an ESS coalition of one and given equilibrium values for $y_1^*$ and $y_2^*$ it follows that $v$ will maximize $G(v, \mathbf{u}, \mathbf{x}^*, \mathbf{y}^*)$ when

$$v = u_m^* = \frac{y_2^*\left(y_1^* + k_{y_1}\right)}{2 y_1^* y_2^* + y_1^* k_{y_2} + y_2^* k_{y_1}}. \qquad (10.4)$$

**Example 10.2.1 (root–shoot ratio from ESS maximum principle)** *We can find the root–shoot allocation for an ESS coalition of one by setting*

$$u_1 = v$$

where $v$ is given by (10.4), and simultaneously solving the following equations

$$a\left(Y_1 - y_1^* - px_1^*\right) - px_1^* \min\left(\begin{array}{c} \frac{ry_1^*u_1}{y_1^*+k_{y_1}} - R \\ \frac{ry_2^*(1-u_1)}{y_2^*+k_{y_2}} - R \end{array}\right) = 0 \qquad (10.5)$$

$$y_2^* = \frac{Y_2}{1 + \alpha x_1^*(1 - u_1)} \qquad (10.6)$$

$$\min\left(\begin{array}{c} \frac{ry_1^*u_1}{y_1^*+k_{y_1}} - R \\ \frac{ry_2^*(1-u_1)}{y_2^*+k_{y_2}} - R \end{array}\right) = d. \qquad (10.7)$$

*Equation (10.5) is the equilibrium solution for the nutrient, equation (10.6) is the algebraic (always in equilibrium) equation for sunlight, and equation (10.7) is the ESS maximum principle requirement that $G(\mathbf{v}, \mathbf{u}, \mathbf{x}^*, \mathbf{y}^*)|_{v=u_1} = 0$. Substituting (10.7) into (10.5) and using the fact that the upper and lower terms in the min function are equal at the ESS leaves us with the following four equations to solve for the four unknowns $u_1$, $x_1^*$, $y_1^*$, and $y_2^*$*

$$u_1 = \frac{y_2^*\left(y_1^* + k_{y_1}\right)}{2y_1^* y_2^* + y_1^* k_{y_2} + y_2^* k_{y_1}}$$

$$0 = a\left(Y_1 - y_1^* - px_1^*\right) - px_1^* d$$

$$y_2^* = \frac{Y_2}{1 + \alpha x_1^*(1 - u_1)}$$

$$d = \frac{ry_1^* u_1}{y_1^* + k_{y_1}} - R.$$

As a specific example, using the same parameters as in Vincent and Vincent (1996)

Parameter values used in calculations

| | |
|---|---|
| $r = 5$ | $a = 0.3$ |
| $R = 0.5$ | $p = 0.1$ |
| $d = 0.5$ | $\alpha = 0.001$ |
| $k_{y_1} = 1$ | $Y_1 = 5$ |
| $k_{y_2} = 1$ | $Y_2 = 2$ |

(10.8)

and solving the simultaneous equations[1] yields only one solution with all values positive

$$u_1 = 0.6995, \; x_1^* = 17.248, \; y_1^* = 0.4004, \; y_2^* = 1.9897.$$

[1] The software program *Maple* was used to obtain this solution.

## 10.2 Consumer-resource games

Strictly speaking, the root–shoot solution obtained in the above example represents only an ESS candidate. Because of (10.4) the $G$-function clearly takes on a maximum when $u_1 = 0.699\,48$. However, we have not demonstrated that this solution is convergent stable. We do so in the following example.

**Example 10.2.2 (root-shoot ratio from Darwinian dynamics)** *The Darwinian dynamics for the plant model is given by*

$$\dot{x}_i = x_i \, G\,(v, \mathbf{u}, \mathbf{x}, \mathbf{y})|_{v=u_i}$$

$$\dot{u}_i = \sigma_i \left.\frac{\partial G\,(v, \mathbf{u}, \mathbf{x}, \mathbf{y})}{\partial v}\right|_{v=u_i}$$

$$\dot{y}_1 = a\left(Y_1 - y_1 - \sum_{i=1}^{n_s} p x_i\right) - \sum_{i=1}^{n_s} p x_i \min\left(\frac{\frac{r y_1 u_i}{y_1 + k_{y_1}} - R}{\frac{r y_2 (1-u_i)}{y_2 + k_{y_2}} - R}\right) \quad (10.9)$$

$$y_2 = \frac{Y_2}{1 + \sum_{i=1}^{n_s} \alpha x_i (1 - u_i)}$$

*where $G\,(v, \mathbf{u}, \mathbf{x}, \mathbf{y})$ is given by (10.3) and the plant parameters are given by (10.8). We first check for an ESS coalition of one. If we obtain the same solution as in the previous example, we have demonstrated convergent stability and we need look no further since by definition such a solution must be an ESS. Using the following initial conditions*

$$u_1\,(0) = 0.5,\ x_1\,(0) = 12,\ y_1\,(0) = 3$$

*with $\sigma_1 = 0.01$ and integrating for 25 time units we obtain*

$$u_1\,(25) = 0.6995,\ x_1\,(25) = 17.248,\ y_1\,(25) = 0.4004,\ y_2(25) = 1.9897$$

*which is in total agreement with the analytical solution found in the previous example. The integration results are illustrated in Figure 10.2 plotted for only the first 15 time units. At this time, the solution is very near equilibrium. Before equilibrium is obtained, $G\,(v, \mathbf{u}, \mathbf{x})$ will have the same general shape of $G\,(v, \mathbf{u}, \mathbf{x}^*)$ as shown in Figure 10.3 but the peak of $G\,(v, \mathbf{u}, \mathbf{x})$ will not occur at zero fitness. For example, examine the strategy curve in Figure 10.2. The strategy actually decreases at first because the peak of the $G$-function is to the left of this strategy. However, the strategy very quickly climbs almost to the peak, and remains very close to the peak thereafter, "riding" the peak as the adaptive landscape continues to change until equilibrium is obtained. Figure 10.2 illustrates the convergent stability properties of the ESS coalition of one solution and Figure 10.3 illustrates that the equilibrium solution satisfies the ESS maximum principle.*

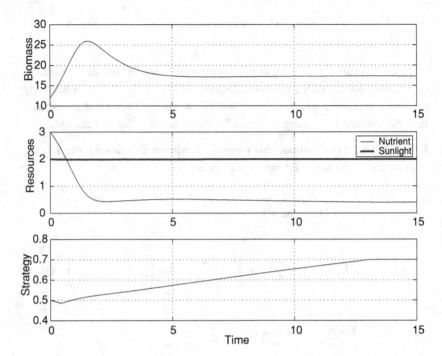

**Figure 10.2** The solution obtained in the previous example is found to be convergent stable.

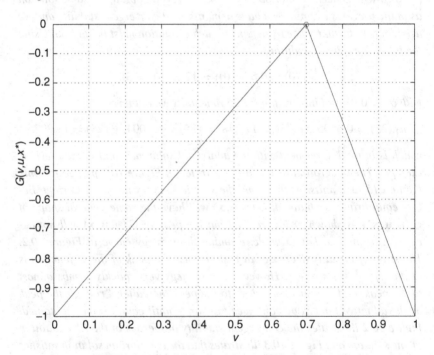

**Figure 10.3** The solution obtained satisfies the ESS maximum principle.

## 10.2 Consumer-resource games

It should be noted that any attempt to reproduce the figures in the previous example will require some special attention (when using the min function). Because the $G$-function does not have continuous slope, a variable step integrator will have difficulties in the neighborhood of the discontinuity, and the integration slows considerably. One way to avoid this difficulty (as was done here) is to simply reduce the maximum allowable step size.

It should be clear from the above examples that under the piecewise linear $G$-function given by (10.3) with no further constraints on $u$, there will never be an ESS coalition greater than one. For coexistence to occur between two or more species, the $G$-function must take on a maximum at two or more strategies. One way to obtain this with a linear $G$-function is for it to be independent of $v$ (at zero slope). In this case, any number of strategies may coexist because any particular strategy may be invaded, but not displaced, by another (Vincent et al., 1993). The $G$-function for this model can take on a zero slope only for unrealistic conditions where $r$, $y_1$, or $y_2 = 0$ and the trivial condition where only two strategies, all root and all shoot, exist. Thus, in order for the ESS to have the co-existence of two or more plant species, a non-linear dependence of the $G$-function on $v$ must be introduced (Vincent and Vincent, 1996). We use the same $G$-function as given in that reference

$$G = \min \left[ \begin{array}{c} \frac{ry_1(v-v^2)+0.2}{y_1+k_{y_1}} - R \\ \frac{ry_2(-1+v+(1-v)^2+0.5)}{y_2+k_{y_2}} - R \end{array} \right] - d. \qquad (10.10)$$

The form of this $G$-function is similar to (10.3) in concept except now the upper "nutrient limiting term" is a parabola with a positive maximum at $v = 0.5$ and the lower "light limiting term" is a parabola with a negative minimum at $v = 0.5$. The upper term implies that, if not light limited, the growth rate is greatest at an intermediate allocation to roots while the lower term implies that, if not nutrient limited, the growth rate is greatest at either a low or a high allocation to roots. Non-linear relationships between biomass allocation and growth, as given by the upper and lower functions, could occur either if the efficiency in converting or capturing a resource changes with biomass allocation or if loss rates (e.g. respiration, disturbance, herbivory, etc.) change with biomass allocation.

**Example 10.2.3 (coalition of two)** *With a G-function that now has two maxima, we suspect an ESS coalition of two strategies. We test this hypothesis by integrating the Darwinian dynamics of the previous example, (10.9), with the G-function given by (10.10) with all parameters the same as in that example. For initial conditions we choose*

$$x_1(0) = 4, x_2(0) = 10, u_1(0) = 0.2, u_2(0) = 0.8$$

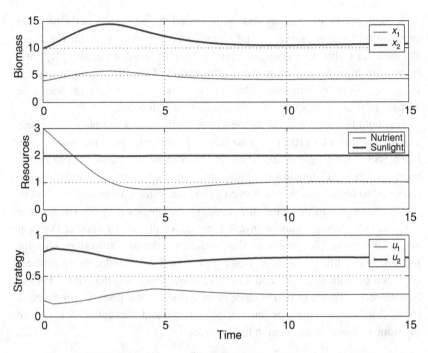

**Figure 10.4** Strategy dynamics results in an ESS coalition of two strategies.

and $\sigma_1^2 = \sigma_2^2 = 0.05$. *Integrating for 15 time units we obtain the final conditions*

$$x_1(15) = 4.2863, x_2(15) = 10.713, u_1(15) = 0.2741,$$
$$u_2(15) = 0.7258, y_1(15) = 1.0074, y_2(15) = 1.9880.$$

*The solutions as a function of time are shown in Figure 10.4. Once again, from the strategy curve, the two strategies achieve the peaks on the adaptive landscape very quickly (as shown by the initial short lines to the discontinuity in slope) and then "ride" their peaks to equilibrium. During the ride, the value of G at the peaks is above zero as the peaks shift in position until the final equilibrium configuration shown in Figure 10.5. It follows from this figure that the ESS maximum principle is satisfied.*

When the min function is linear in $v$, two species will converge towards an ESS coalition of one resulting in the extinction of one or the other species (or merging of the two into one – a kind of extinction). However, when the min function is non-linear in $v$ two species will converge to an ESS coalition of two. In this case, the two resources (light and nutrients) allow for species coexistence that is both promoted and maintained by the ESS.

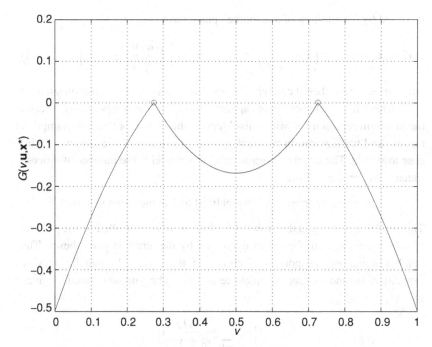

**Figure 10.5** The solution obtained satisfies the ESS maximum principle.

## 10.2.2 Carcinogenesis

Cancer appears to be an evolutionary game that represents many of the topics of this book in microcosm. It involves macroevolution (the creation of new $G$-functions), microevolution (strategy dynamics within $G$-functions), speciation, and coevolution. Cancer is a potentially fatal game of resource competition, adaptive radiation, and habitat destruction.

Carcinogenesis involves transitions from normal tissue to premalignant lesions to invasive cancer. These transitions have been modeled as an evolutionary game by Gatenby and Vincent (2003b) and we will use a simplified version of this model here (Gatenby and Vincent, 2003a). We begin by modifying the mechanistic normal–mutant cell model of Gatenby (1991). This model includes the dynamics of glucose uptake. The presence of glucose as a resource results in a consumer-resource type of model for the $G$-function. Hence, population dynamics take the form of

$$\dot{x}_i = x_i \, G\left(v, \mathbf{u}, \mathbf{x}, y\right)\big|_{v=u_i}$$

were $x_i$ is the number of cells per cubic centimeter and $y$ is the resource density. The cell dynamics is written in terms of the familiar Lotka–Volterra model modified by the addition of a resource uptake term

$$G(v, \mathbf{u}, \mathbf{x}, y) = r \left(1 - \frac{1}{K(v)} \sum_{j=1}^{n_s} a_j(v, \mathbf{u}) x_j \right) \left( \frac{E(v) y^2}{y_h^2 + y^2} - m \right) \quad (10.11)$$

where $n_s$ is the number of cell types (normal and tumor), $r$ is a common intrinsic growth rate, $K(v)$ is the carrying capacity, $a_j(v, \mathbf{u})$ is the competition term, and the last term represents a Michaelis–Menten uptake with a fixed consumption rate $m$, and half-saturation constant $y_h$. The resource $y$ is the amount of glucose available. The adaptive parameter in this model is the number of glucose transporters on the cell surface

$$u_i = \text{transporters/cm}^2 \text{ (normalized to 1 at the mean value)}.$$

This model assumes that the growth of cells in vivo is controlled by the product of two general growth factors as expressed by the terms in parentheses. The number of glucose transporters affects each of these terms through the carrying capacity $K(v)$ and the glucose uptake term $E(v)$. The glucose dynamic is given by

$$\dot{y} = R - \sum_{i=1}^{n_s} \frac{E(u_i) y^2}{y_0^2 + y^2} x_i \quad (10.12)$$

where $R$ is glucose delivery rate.

Assuming that $a$, $K$, and $E$ are normally distributed functions of $v$ with the other parameters constant, we use

$$a_j(v, \mathbf{u}) = \exp\left(-\frac{(v - u_j)^2}{2\sigma_a^2}\right)$$

$$K(v) = K_m \exp\left(-\frac{(v - u_K)^2}{2\sigma_K^2}\right)$$

$$E(v) = E_m \exp\left(-\frac{(v - u_E)^2}{2\sigma_E^2}\right),$$

where

$K_m$ = maximum tissue carrying capacity
$E_m$ = maximum glucose uptake
$u_K$ = number of glucose transporters at $K = K_m$
$u_E$ = number of glucose transporters at $E = E_m$
$\sigma_K^2$ = variance in $K$ distribution
$\sigma_E^2$ = variance in $E$ distribution
$\sigma_a^2$ = variance in $a$ distribution.

## 10.2 Consumer-resource games

### 10.2.2.1 Conditions promoting carcinogenesis

We first examine the conditions in normal tissue by seeking equilibrium solutions that satisfy an ESS maximum principle, for the cell dynamics equations, resource dynamics equations, and the strategy dynamics equations

$$\dot{u}_i = \sigma_i^2 \left. \frac{\partial G(v, \mathbf{u}, \mathbf{v}, y)}{\partial v} \right|_{v=u_i}. \tag{10.13}$$

The nature of the equilibrium solutions to (10.11), (10.12), and (10.13) depends on $n_s$. Consider first the case of $n_s = 1$. In this case equilibrium requires

$$G(v, \mathbf{u}, \mathbf{x}, y)|_{v=u} = 0 \tag{10.14}$$

$$R - \frac{Ey^2}{y_0^2 + y^2} x = 0 \tag{10.15}$$

$$\left. \frac{\partial G(v, \mathbf{u}, \mathbf{v}, y)}{\partial v} \right|_{v=u} = 0. \tag{10.16}$$

Generally we must solve all three equations simultaneously for the equilibrium values of $x^*$, $y^*$, and $u$. However, we may think of using (10.16) as determining $u$, and we will examine the equilibrium possibility as determined from (10.14) and (10.15). For a given $u$ there are two equilibrium solutions possible depending on how $G(v, \mathbf{u}, \mathbf{x}, y)|_{v=u} = 0$. If the second term in (10.11) is zero, $x^*$ is obtained from this expression and $y^*$ is obtained from (10.15), yielding the equilibrium conditions (equilibrium B)

$$x^* = K(u^*), \quad y^* = \sqrt{\frac{Ry_0^2}{E(u^*)x^* - R}}. \tag{10.17}$$

On the other hand, if the third term in (10.11) is zero, $y^*$ is obtained from this expression and $x^*$ is obtained from (10.11) and (10.15), yielding the equilibrium conditions (equilibrium C)

$$y^* = \sqrt{\frac{my_0^2}{E(u^*) - m}}, \quad x^* = \frac{R}{m}. \tag{10.18}$$

Note that the equilibrium number of cells is independent of $u$ under equilibrium C.

We can now examine both the ecological and the evolutionary stability of these two equilibrium solutions. Using an eigenvalue analysis to examine the ecological stability and the ESS maximum principle to examine evolutionary stability, we obtain the following results for equilibria B and C

| Equilibrium | $R < mK_m$ | $R > mK_m$ |
|---|---|---|
| C | ecologically stable, evolutionarily stable | ecologically unstable |
| B | ecologically unstable | ecologically stable, evolutionarily unstable |

Evolutionary stability requires ecological stability otherwise the system can never arrive at this equilibrium solution. Therefore, the two off-diagonal results are of no interest. If the critical condition $R < mK_m$ is met, Figure 10.6 illustrates that equilibrium C is at a maximum point on the adaptive landscape and cannot be invaded by mutant phenotypes. If $R > mK_m$, Figure 10.6 illustrates that equilibrium B is at a local minimum point on the adaptive landscape. This solution can be invaded by mutant phenotypes unless some cooperative mechanism is in place (e.g., the interplay between oncogenes and tumor suppressor genes) that does not allow a cell to enter a cell cycle unless

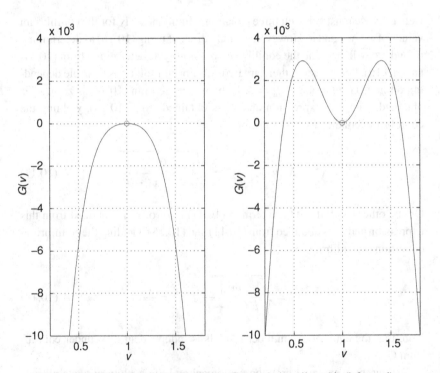

**Figure 10.6** When $R < mK_m$ equilibrium C is evolutionarily stable (left panel). When $R > mK_m$ equilibrium B is evolutionarily unstable (right panel).

## 10.2 Consumer-resource games

conditions favor producing a perfect clone. Which equilibrium state corresponds to normal tissue? We work from the following observations (Gatenby and Vincent, 2003a):

1. Under normal physiological conditions, each cell population remains at $K(u_1)$ for that particular tissue (implies equilibrium B for normal cells).
2. Substrate concentrations remain stable in normal tissue requiring regulatory mechanisms to adjust blood flow so that $R$ is slightly greater than $mK_m$ (implies equilibrium B for normal cells).
3. Mutations in either the oncogene or the tumor suppressor gene will give false information to the nucleus, overcoming normal tissue controls on cell proliferation and allowing clonal expansion of a mutant population (implies that $n_s$ goes from 1 to 2 or more during proliferation).
4. Once a population of proliferating cells develops, these cells take up more substrate at existing conditions. This will increase the number of glucose transporters on the cell membrane (implies that the value of $u$ for normal cells is smaller than the value of $u$ for tumor cells).
5. Tissue that is developing cancer becomes crowded.
6. Some tumor cell populations form small polyps that remain small after many years. Other tumor cell populations form large polyps that go on to form cancers.

Observations 1 and 2 imply that normal cells lie at a minimum point on the adaptive landscape corresponding to equilibrium B with $R > mK_m$. This explains why clonal expansion is possible in observation 3. By introducing observation 4 into our model below, we obtain observations 5 and 6.

### 10.2.2.2 A route to carcinogenesis

To make the model as realistic as possible, for the following parameters we use values available in the literature (Hatanaka, 1974)

$$R = 0.05 \times 10^6 \text{ cells per micromole glucose}$$
$$K_m = 10^9 \text{ cells per cm}^3$$
$$E_m = 14.668 \text{ micromoles glucose per } 10^6 \text{ cells per day}$$
$$m = 2.544 \text{ micromoles glucose per } 10^6 \text{ cells per day}$$
$$y_0 = 3330 \text{ micromoles glucose per cm}^3.$$

For use in the model, these values must be converted to consistent units. In addition, parameters defining the distribution functions are needed. All values used for the normal cell case are given in the following table. Units are not repeated here for brevity

| |
|---|
| $R = 0.05$ |
| $K_m = 1000$, $\sigma_K = 1$, $u_K = 1$ |
| $E_m = 14.668$, $\sigma_E = 1$, $u_E = 1$ |
| $m = 2.544$ |
| $y_0 = 3330$ |
| $\sigma_a = 0.5$ |

If normal cells could evolve, they would arrive at equilibrium B or C depending on the value used for $r$. For example, choosing $R = 0.9mN_{\text{mean}}$ we obtain equilibrium C with

$$x^* = 900, u^* = 1, N^* = 1525$$

and using $r = 1.2mN_{\text{mean}}$ we obtain equilibrium B with

$$x^* = 1000, u^* = 1, N^* = 1707.$$

These are the two cases illustrated in Figure 10.6. As previously noted, the evolutionarily unstable equilibrium B is evidently the normal cell case. Such an equilibrium presents opportunities for speciation to take place (Cohen *et al.*, 1999) (or invasion by mutant phenotypes) and can be maintained only through the "cooperative" efforts of the oncogene and tumor suppressor genes.

The situation changes somewhat when rare mutant cells are able to establish themselves, but the oncogene and tumor suppressor genes have not been sufficiently damaged that the mutant cells can evolve. With $n_s = 2$, we use the initial conditions

$$u_1(0) = 1, u_2(0) = 0.9, x_1(0) = 998, x_2(0) = 2, y(0) = 1707$$

where the subscript 1 refers to the normal cells and the subscript 2 refers to the mutant cells. With $\sigma_1^2 = \sigma_2^2 = 0$ (no evolution allowed for either the normal or the mutant cells) and keeping $R = 1.2mN_{\text{mean}}$ the system moves only slowly toward an equilibrium solution. For example, after a 10-year simulation run $t_f = 3650$ days, we obtain

$$u_1(t_f) = 1, u_2(t_f) = 0.9, x_1(t_f) = 992.7, x_2(t_f) = 7.5, y(t_f) = 1707.$$

Equilibrium under these conditions requires about 80 years ($x_1(t_f) = 640$, $x_2(t_f) = 367$, $y(t_f) = 1701$), at which time the mutant cells would have turned into a relatively large tumor.

Carcinogenesis starts when the constraints on the mutant cells that prevent them from evolving are removed. We simulate this by using as initial conditions

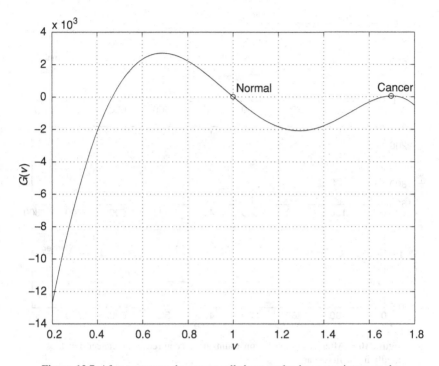

**Figure 10.7** After two years the cancer cells have evolved to a maximum on the adaptive landscape.

the final conditions obtained from the 10-year run with $\sigma_1^2 = 0$, $\sigma_2^2 = 0.1$, and set

$$u_E = 1.1, u_N = 2, N_{mean} = 1100$$

to reflect changing environmental conditions due to the presence of an increasing number of mutant cells (see observations noted above). Running the simulation for two years, $t_f = 730$, results in

$$u_1(t_f) = 1, u_2(t_f) = 1.7, x_1(t_f) = 323, x_2(t_f) = 907, y(t_f) = 1727.$$

The adaptive landscape illustrated in Figure 10.7 shows that during this period of time the mutant tumor cells have evolved into a cancer by arriving at a maximum on the adaptive landscape. Crowding is now apparent $(x_1(t_f) + x_2(t_f) > 1000)$ as the cancer has made substantial inroads.

Figure 10.8 illustrates the speed at which cancer develops once the constraints preventing evolution of mutant cells have been removed.

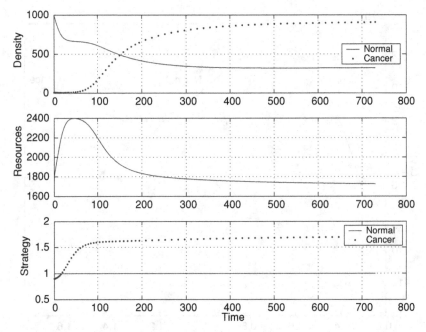

**Figure 10.8** After evolutionary constraints have been removed, cancer develops rapidly in the first year.

## 10.3 Plant ecology

### 10.3.1 Flowering time for annual plants

Cohen (1971, p. 10) developed a model for modeling the flowering time for a single annual flowering plant. During the growing season of length $T$, the plant can devote energy either to vegetative growth or to seed production. The basic question concerns how the plant should allocate growth to seed production so as to maximize the number of seeds at the end of the growing season. If we let

$x_l$ = leaf mass
$x_s$ = reproductive mass (seeds)
$R$ = net photosynthetic production per unit leaf mass
$L$ = ratio of leaf mass to remaining vegetative mass
$z(t)$ = fraction of total plant growth allocated to reproduction

then Cohen's model for leaf mass is given by

$$\dot{x}_l = RLx_l[1-z] \qquad (10.19)$$

## 10.3 Plant ecology

with the total number of seeds produced given by

$$x_s(T) = \int_0^T Rx_l z \, dt. \quad (10.20)$$

Cohen demonstrated that the optimal strategy for maximizing seed production is given by

$$u(t) = \begin{cases} 0 & \text{if } 0 \leq t \leq u \\ 1 & \text{if } u < t \leq T \end{cases}$$

where

$$u = T - \frac{1}{RL}. \quad (10.21)$$

In other words, the solution is **bang-bang** with the plant switching from growing only vegetatively up to time $u$ and then spending the remainder of the growing season producing only seeds. Instead of having to program a time-dependent strategy, the plant needs to know only the length of the growing season as well as the two other parameters in the model.

The bang-bang nature of the solution and the result given by (10.21) is also easily obtained using methods of optimal control theory (see exercises 8–10 in Vincent and Grantham, 1997). However, an easier way to obtain (10.21) is to assume the optimal solution is bang-bang, set $z = 0$ in (10.19) and integrate to $t = u$ to obtain the amount of leaf biomass at the switch time

$$x_l(u) = x_l(0) \exp(RLu).$$

Substituting this result into (10.20), setting $z = 1$, and integrating from $t = u$ until the final time $T$ gives the number of seeds at the final time

$$x_s(T) = (T - u) Rx_l(0) \exp(RLu).$$

Taking seed production to be the measure of fitness, we write

$$H(u) = (T - u) Rx_l(0) \exp(RLu) \quad (10.22)$$

and then maximize $H(u)$ with respect to $u$ to yield (10.21).

In general, plants do not grow in isolation from each other and hence, in general, there will be competition between plants for resources. Vincent and Brown (1984b) examined Cohen's problem from this perspective, assuming that, the more leaves and roots a plant has, the greater the competitive effect it will have on its neighbors. Its larger root system will draw resources more rapidly and its greater size will exert greater shading. Thus an individual plant's seed production is not only a function of its own size, as is the case of Cohen's model, but is also a function of the size and number of neighboring plants. The remainder of this section summarizes the method and results from this reference.

Assume that competition with neighboring plants does not change the general nature of the optimal seed production solution. That is, it remains bang-bang. However, competition changes the switching time strategy $u$ and its determination as an ESS becomes the focus of the reformulated problem.

Since the total number of plants (of all species present) per unit area is important, it is convenient to formulate the model in terms of strategy frequencies. The following $G$-function

$$G(v, \mathbf{u}, \mathbf{p}, N) = \frac{(T-v)f(v)}{1 + W(v, \mathbf{u}, \mathbf{p}, N)}$$

represents a generalization of plant fitness as suggested by (10.22). The variable $\mathbf{p}$ refers to the frequency of each species, $N$ is the number of plants per unit area, and $W(v, \mathbf{u}, \mathbf{p}, N)$ is a function that determines the competitive effect of neighboring plants. For the Cohen model

$$f(v) = Rx_l(0)\exp(RLv) \tag{10.23}$$

and

$$W(v, \mathbf{u}, \mathbf{p}, N) = 0.$$

Otherwise $f(v)$ is a function representing a measure of the plant's size and photosynthetic capability. We assume that

$$\left.\frac{\partial f(v)}{\partial v}\right|_{v=u_i} > 0.$$

When there is just one plant, there can be no neighbors and no competitive effect, so that

$$W(v, u_1, 1, 1) = 0. \tag{10.24}$$

When a plant has one or more neighbors ($N > 1$), we assume that a plant reduces the competitive effect of these neighbors by flowering later (growing bigger) and that the competitive effect of neighbors increases with a collective delay in flowering by all plants (a world of bigger plants experiences more severe competition than one of smaller plants). These assumptions imply

$$\left.\frac{\partial W(v, u_1, 1, N)}{\partial v}\right|_{v=u_1} < 0 \tag{10.25}$$

$$\frac{\partial W(u_1, u_1, 1, N)}{\partial u_1} > 0 \tag{10.26}$$

## 10.3 Plant ecology

and guarantee that, if there is just one species present, an invading species is able to decrease the competitive effect of other plants by flowering later, and, if there is just one species present, then the competitive effect increases if all plants increase their flowering time simultaneously.

**Example 10.3.1 (flowering time, $N = 1$)** *The maximum principle requires that*

$$\left.\frac{\partial G(v, \mathbf{u}, 1, 1)}{\partial v}\right|_{v=u_1} = \left[-f(v) + (T-v)\frac{\partial f(v)}{\partial v}\right]_{v=u_1} = 0$$

*or, alternatively*

$$\left.\frac{\partial f(v)}{\partial v}\right|_{v=u_1} = \frac{f(u_1)}{T - u_1}. \tag{10.27}$$

*This result yields (10.21) when $f(v)$ is given by (10.23). However, this maximizing condition expresses an important well-known result from economics: producing more of a product (RHS) reduces the average cost of production, but the marginal cost of production (LHS) goes up. Cost is minimized by producing at a point where the marginal cost equals the average cost. For plants, seed production is maximized when the marginal rate of increase in seed production equals the average rate of seed production taken with respect to time remaining in the growing season.*

**Example 10.3.2 (flowering time, $N > 1$)** *Assuming an ESS coalition of one strategy, the maximum principle requires that (arguments have been removed for brevity)*

$$\left.\frac{\partial G(v, u_1, 1, N)}{\partial v}\right|_{v=u_1} = \left\{\frac{(1+W)\left[-f + (T-v)\frac{\partial f}{\partial v}\right] - (T-v)f\frac{\partial W}{\partial v}}{(1+W)^2}\right\}_{v=u_1}$$

$$= 0$$

*evaluating (and putting the arguments back in) yields*

$$-f(u_1) + (T - u_1)\left.\frac{\partial f(v)}{\partial v}\right|_{v=u_1} = \frac{(T - u_1)f(u_1)}{(1 + W(u_1, u_1, 1, N))} \left.\frac{\partial W(v, u_1, 1, N)}{\partial v}\right|_{v=u_1}$$

$$= G(u_1, u_1, 1, N) \left.\frac{\partial W(v, u_1, 1, N)}{\partial v}\right|_{v=u_1}.$$

*Since $G(u_1, u_1, 1, N) > 0$ it follows from (10.25) that*

$$\left.\frac{\partial f(v)}{\partial v}\right|_{v=u_1} < \frac{f(u_1)}{(T - u_1)}. \tag{10.28}$$

Because $G(u_1, u_1, 1, N)$ will always be maximized when (10.27) is satisfied, it follows that the number of seeds produced per plant under competition will be less than the number of seeds produced without competition. In addition, for $f(v)$ of the same form as given by (10.23) the flowering time will be later than for the single plant case. This can be easily verified by plotting both $\partial f(v)/\partial v$ and $f(v)/(T-v)$ versus $v$ and noting that (10.28) is satisfied after the two functions are equal.

Delayed flowering time becomes an adaptation for mitigating the competitive effects of others. Because the other plants are also under selection to respond in kind, the resulting ESS is not one that maximizes the collective yield of the population of plants. As the plants evolve delayed flowering to reduce the competitive effects of others, the plants are consequently increasing the competitive effects imposed by others.

**Example 10.3.3 (flowering time – cooperative solution)** *It is of interest to compare the ESS coalition of one candidate solution obtained in the previous example with a cooperative solution among the N plants. In this case we set $v = u_1$ before taking the partial derivative of the G-function. In this case we get*

$$\frac{\partial G(u_1, u_1, 1, N)}{\partial v} = \frac{(1+W)\left[-f + (T-u_1)\frac{\partial f}{\partial u_1}\right] - (T-u_1)f\frac{\partial W}{\partial u_1}}{(1+W)^2} = 0$$

*evaluating (and putting the arguments back in) yields*

$$-f(u_1) + (T-u_1)\frac{\partial f(u_1)}{\partial u_1} = \frac{(T-u_1)f(u_1)}{(1+W(u_1, u_1, 1, N))}\frac{\partial W(u_1, u_1, 1, N)}{\partial v}$$

$$G(u_1, u_1, 1, N)\frac{\partial W(u_1, u_1, 1, N)}{\partial v}.$$

*Since $G(u_1, u_1, 1, N) > 0$ it follows from (10.26) that*

$$\frac{\partial f(u_1)}{\partial u_1} > \frac{f(u_1)}{(T-u_1)}. \tag{10.29}$$

*Again, for $f(v)$ of the form given by (10.23) the flowering time will be earlier than for the single plant case. However, the solution for $u_1$ obtained from (10.29) will maximize $G(u_1, u_1, 1, N)$ so that this solution will yield more seeds than the ESS solution. While desirable in agriculture, a cultivar plant that had this property would have to be artificially maintained, as this cooperative solution is not evolutionarily stable. It can be invaded by plants with a later flowering time. This helps to explain why some cultivar agricultural grain crops are competitively inferior to wild forms.*

## 10.3 Plant ecology

The results presented here are qualitatively similar to those of Cohen (1971), Gadgil and Gadgil (1975), and Schaffer (1977). However, the results apply to a large class of functions. The model predicts that plants experiencing competition from neighbors should evolve to a flowering time later than expected without competition. Furthermore the flowering time that maximizes seed production should be less than both the competitive and the non-competitive ESS. These conclusions are based on the assumptions given by (10.24)–(10.26) and that $f(v)$ is qualitatively the same as (10.23).

### 10.3.2 Root competition

We can generate a companion **root competition** model to the above flowering time model (Gersani *et al.*, 2001). In this model, the resource is below-ground space shared by the roots of a number of neighboring plants. We use a simple model of root proliferation by annual plants to consider how inter-plant root competition influences a plant's ESS level of root production and its subsequent reproductive yield. Like competition for light, the game results in a **tragedy of the commons** where the ESS represents a level of root investment that compromises the collective seed yield of the group.

Imagine $N$ individual plants sharing the same soil space. The total nutrient uptake, $y$, by all of the plants is a monotone increasing function of the total root production, $R$, by all of the plants. An individual's share of this total nutrient uptake will be in proportion to its own root biomass relative to the root biomass of others. Under these assumptions, nutrient competition is exploitative, and total nutrient harvest increases with total root mass but at a diminishing rate that eventually levels off at the total amount of available nutrients in the soil. We assume that fitness (measured as nutrients available for seed production) is the difference between an individual's nutrient uptake and the cost of growing, maintaining, and servicing its roots (this cost subsumes in units of nutrients the associated above-ground plant parts needed for photosynthesis, seed production, and root maintenance). We also assume that above-ground competition is negligible or constant within the range of root production strategies under consideration.

By combining the above assumptions, the fitness generating function determines the net nutrient profit of an individual plant as a function of its own root production and that of others. Here we consider only a single species, and we are not interested in the overall dynamics of the population. As with the competition model above, we are interested in how the number of plants competing for a shared space influences the rooting strategies used by each plant.

The $G$-function is frequency and density dependent according to

$$G(v, \mathbf{u}, \mathbf{p}, N) = \frac{v}{R(v, \mathbf{u}, \mathbf{p}, N)} y\left[R(v, \mathbf{u}, \mathbf{p}, N)\right] - C(v)$$

where $v$ is the root production of the focal individual, $R(v, \mathbf{u}, \mathbf{p}, N)$ is a the total root production of all plants, $y[R(v, \mathbf{u}, \mathbf{p}, N)]$ is total nutrient uptake, and $C(v)$ is the cost to the individual of supporting its roots and associated above-ground parts.

Although we are considering root mass as the measure of nutrient foraging effort in the experiments to follow, the rooting strategies represented by $\mathbf{u}$ at frequency $\mathbf{p}$ could also represent any plant character that increases nutrient uptake at a cost. In addition to root mass, the rooting strategy may include increasing fine root density, total root surface area (Fitter et al., 1991), and root kinetics (Drew and Saker, 1975; Lee, 1982; Jackson et al., 1990).

Assuming an ESS coalition of one strategy, the maximum principle requires that (some arguments removed for clarity)

$$\left.\frac{\partial G(v, u_1, 1, N)}{\partial v}\right|_{v=u_1} = \left[\frac{v}{R}\frac{\partial y}{\partial R}\frac{\partial R}{\partial v} + y\frac{R - v\frac{\partial R}{\partial v}}{R^2} - \frac{\partial C}{\partial v}\right]_{v=u_1} = 0 \quad (10.30)$$

where $N$ is the number of plants, including the focal individual, that share the space. Since all plants are the same species, the root production for the focal plus other individuals is given simply as

$$R(v, u_1, 1, N) = v + (N-1)u_1 \quad (10.31)$$

where $u_1$ is the root production found among the other $N-1$ individuals. It follows from (10.31) that

$$R(v, u_1, 1, N)|_{v=u_1} = Nu_1$$

$$\left.\frac{\partial R(v, u_1, 1, N)}{\partial v}\right|_{v=u_1} = 1.$$

Using these to evaluate (10.30) we obtain the necessary condition

$$\left.\frac{\partial C}{\partial v}\right|_{v=u_1} = \left[\frac{v}{R}\frac{\partial y}{\partial R}\right]_{v=u_1} + \left[\frac{y}{R}\left(\frac{R-v}{R}\right)\right]_{v=u_1}$$

$$= \left.\frac{1}{N}\frac{\partial y}{\partial R}\right|_{v=u_1} + \left(\frac{N-1}{N}\right)\left[\frac{y}{R}\right]_{v=u_1}. \quad (10.32)$$

Thus an ESS candidate $u_1$, for individual plants, has the property that sets its marginal cost for supporting roots and above-ground parts equal to a weighted sum of the average value of nutrient uptake, $y/R$, and the marginal value, $\partial y/\partial R$, of nutrient uptake. The weighting used by the plant depends upon the

## 10.3 Plant ecology

number of competitors, $N$. As $N$ increases, the individual plant weights its decision more heavily towards its average return per unit root and less on its marginal return per unit root. At $N = 1$ (no inter-plant competition), the plant bases growth entirely on the marginal value of roots. As $N \to \infty$, the plant's decision is based entirely on the average value.

We create a commons of fixed size and assume that resources (nutrients and space) per individual remain constant. Let $N_T$ be the total number of plants and let $N_D$ be the number of divisions made when the commons is subdivided into equal spaces by placing dividers into the ground. The number of competitors $N = N_T/N_D$. For example, suppose that $N_T = 10$ and we start with a commons with no subdivisions, $N_D = 1$, so that $N = 10$ where all 10 individuals share the same space. We could subdivide the commons into 5 equally sized compartments, $N_D = 5$, so that individuals are in pairs, $N = 2$. We could also create individual "owners," $N = 1$, by subdividing the commons into 10 equally sized compartments, each with one individual. If we assume that each individual within a compartment has equal access to the resources of the compartment (we are ignoring distance effects created by the exact spatial arrangement of plants); then, within the commons, it follows that values for the rates of total nutrient uptake, marginal nutrient uptake, and average nutrient uptake are independent of $N$

$$y[R(v, \mathbf{u}, \mathbf{p}, N)] = N_D y\left[\frac{R\left(v, \mathbf{u}, 1, \frac{N_T}{N_D}\right)}{N_D}\right]$$

$$\frac{\partial y[R(v, \mathbf{u}, \mathbf{p}, N)]}{\partial R} = N_D \frac{y\left[\frac{R\left(v, \mathbf{u}, 1, \frac{N_T}{N_D}\right)}{N_D}\right]}{\partial R}$$

$$\frac{y[R(v, \mathbf{u}, \mathbf{p}, N)]}{R} = N_D \frac{y\left[\frac{R\left(v, \mathbf{u}, 1, \frac{N_T}{N_D}\right)}{N_D}\right]}{R}.$$

We determine the ESS root production per individual for any subdivision by substituting the appropriate terms into equation (10.32)

$$\left.\frac{\partial C}{\partial v}\right|_{v=u_1} = \frac{N_D}{N_T}\left.\frac{\partial y}{\partial R}\right|_{v=u_1} + \left(\frac{N_T - N_D}{N_T}\right)\left[\frac{y}{R}\right]_{v=u_1}.$$

For example, when $N_T = 10$ and $N_D = 1$ (10 individuals share the entire, undivided, space)

$$\left.\frac{\partial C}{\partial v}\right|_{v=u_1} = 0.1 \left.\frac{\partial y}{\partial R}\right|_{v=u_1} + 0.9 \left.\frac{y}{R}\right|_{v=u_1}$$

and when $N_T = 10$ and $N_D = 5$ (10 individuals in 5 subdivided spaces)

$$\left.\frac{\partial C}{\partial v}\right|_{v=u_1} = 0.5 \left.\frac{\partial y}{\partial R}\right|_{v=u_1} + 0.5 \left.\frac{y}{R}\right|_{v=u_1}$$

and when $N_T = 10$ and $N_D = 10$ (10 individuals in 10 subdivided spaces)

$$\left.\frac{\partial C}{\partial v}\right|_{v=u_1} = \left.\frac{\partial y}{\partial R}\right|_{v=u_1}.$$

These necessary conditions for an ESS show how several plants sharing the same space results in a tragedy of the commons. When $N = 1$ so that the individual "owns" its space, the individual produces roots until the marginal reward from additional roots $\partial y / \partial R$ no longer exceeds the marginal cost $\partial C / \partial v$. This maximizes both individual and collective fitness. As $N$ increases, ESS root production represents a weighted averaging of the marginal reward and the average reward of root production relative to the marginal cost. In fact, as $N$ goes to infinity the plant produces roots until the average benefit no longer exceeds the marginal cost. With $y$ a convex monotone increasing function of $R$, the average of the curve will always be greater than the margin: $y/R > \partial y/\partial R$. This means that the individual's perceived benefit of producing roots will be higher when 10 plants share the commons ($N = 10$) than when 5 pairs of plants ($N = 2$) occupy 5 subdivisions of the commons. And 5 pairs of individuals will perceive greater individual benefits to producing roots than ten individuals each with its own subdivided space of the commons ($N = 1$). For all three scenarios, the total root production across the entire space is $R = 10u_1$. But, because of the individual perceived benefits of producing roots, the ESS root production per individual plant, $u_1$, will be greatest for $N = 10$ and least for $N = 1$. Furthermore, because combined reproductive yield is maximized when $N = 1$, the "owners" will produce more reproductive yield per individual than the 5 pairs of individuals ($N = 2$), which will produce more reproductive yield per individual than the 10 individuals sharing the commons ($N = 10$).

That the ESS necessary condition maximizes reproductive yield, for the owners ($N = 1$), can be shown by simply maximizing total nutrient profits, $NG(v, u_1, N)$, with respect to total root production. The condition for maximizing collective reproductive yield reduces to $\partial y(R)/\partial R = \partial C/\partial v$.

Why do individuals sharing space overproduce roots and sacrifice collective reproduction at the ESS? The presence of competitors and the opportunity to "steal" nutrients from others encourage the plant to produce more roots over and above those which would maximize collective gains. By producing more roots, the individual enhances its own reproductive yield at the expense of others. However, the other individuals respond in kind. If the plants do this, it indicates that a plant has a sense of "self" vs. "others" in modifying root production in response to intra-plant versus inter-plant competition. It also suggests that reproductive yield may be lost in the roots of crop plants if we do not recognize and breed this propensity out of cultivars.

This model has been tested successfully with soybeans (Gersani *et al.*, 2001) and a variety of Kenyan bean (Maina *et al.*, 2002) grown in greenhouse experiments. By using dividers, two plants either shared a complete space, or "owned" half of the space. While the numbers of plants and the total growth space were held constant the absence of dividers permitted inter-plant root competition (shared space) and the presence of dividers permitted only intra-plant root competition (owned space). As predicted, the individuals sharing space produced 20–80% more roots per individual plant and they suffered a 30–40% reduction in seed yield. To test this under field conditions a small plot was planted with soybeans using standard cultivation techniques. Metal dividers were placed across rows (three rows) in a manner that left some individuals separated from either row mate (owners), some individuals as isolated pairs, and others as groups of 20 individuals along the row (commons). Plant spacing along rows did not vary with treatment. In accord with the root foraging game, owners produced 20% more yield than paired individuals who produced 30% more yield than individuals in the commons [results not yet published].

## 10.4 Foraging games

### 10.4.1 Gerbil–owl fear game

Imagine a game between gerbils and owls played on a nightly basis in the Negev Desert of Israel. Each night a gerbil needs to decide when to come out of its burrow and forage for seeds. The benefits to foraging accrue from harvesting seeds, the costs of foraging include an additional energy expenditure and the risk of being captured by the owls. Similarly, each owl needs to decide when to leave its roost and hunt for gerbils. The benefits to hunting accrue through the capture of gerbils, the costs of hunting include an additional energy expenditure and a risk of injury. What sets the tone for this game is a nightly pulsing and depletion of seeds. The gerbils and owls inhabit desert sand dunes. Each afternoon before sunset, winds blow across the sand dunes uncovering some seeds, and redistributing sand and seeds alike. This wind provides the gerbils with a fresh dose of seeds at the start of the night, and this pulse of seeds depletes during the night through the harvesting activities of the gerbils. The game modeled here is based on an actual system in the Negev Desert of

Israel (Abramsky *et al.*, 1990; Ziv *et al.*, 1993; Ben-Natan *et al.*, 2004), and it is adapted from Brown *et al.* (2001).

There is some similarity in this game to the flowering time model in that the strategy depends on time. Recall that this complication was avoided in the flowering time model by using the fact that an optimal strategy for flowering time is bang-bang. This allowed us to use switching time as the strategy rather than seeking a continuous time solution for the fraction of total plant growth allocated to reproduction. Unfortunately this simplification cannot be used to formulate this foraging game. In the context of game theory, a differential equation model with time-dependent strategies is a **differential game** (Issacs, 1965; Vincent and Grantham, 1997) requiring a different method for solution. We have not presented any tools for solving evolutionary games with time-dependent strategies. A differential game approach requires laborious, computer-intensive simulations or numerical analyses. McNamara *et al.* (2001) provide examples for finding an ESS for some systems with time-dependent strategies. As yet there does not exist a formal theory for defining, finding, or verifying the ESS for evolutionary differential games. Fortunately our foraging game models are such that it is still possible to formulate the problem in terms of $G$-functions and then use a cost–benefit feature to obtain a solution that has the characteristics of an ESS.

The sand dunes are pulsed with new seeds every 24 hours. This pulse resets seed abundance, $y$, to some initial value $y_0$. Over the course of a night ($T < 24$ hours), a gerbil can spend its time either foraging or resting within its burrow. A gerbil's strategy, $u_1(t)$, specifies the probability of being active at any time during the night. Hence: $0 \leq u_1(t) \leq 1$. Let $x_1$ be the population size of gerbils. While foraging, a given gerbil harvests resources in proportion to their abundance (a type I functional response). If strategy $u_1(t)$ is the common strategy among the gerbils than seed availability declines according to

$$\dot{y} = -a_1 x_1 u_1(t) y$$

where $a_1$ is the gerbil's encounter probability on seeds, and $y$ is the current abundance of seeds at time $t$. The abundance of seeds at any given time $t$ ($0 \leq t \leq T$) is given by

$$y(t) = y_0 e^{-a_1 x_1 \int u_1(t) dt}.$$

Assume that the gerbil expends energy at rate $c_1$ while foraging and at rate $k_1 < c_1$ while resting (foraging is more costly than resting – both costs are in units of seeds). The net seed profit to a given gerbil during the night is

$$E_1 = a_1 \int y(t) u_1(t) dt - c_1 \int u_1(t) dt - k_1 \int [1 - u_1(t)] dt$$

where the first integral gives the amount of seeds harvested, the second integral gives the amount of energy expended while foraging, and the third integral gives the amount of energy expended while resting.

If we let $u_2(t)$ be the common strategy among the owls, where $u_2(t)$ specifies the probability that an owl is active at time $t$ ($0 \leq t \leq T$) then we can write the probability that a gerbil survives the night as

$$p_1 = e^{-a_2 x_2 \int u_1(t) u_2(t) dt}$$

where $x_2$ is the population size of owls, and $a_2$ is the lethal encounter probability of an owl while hunting (we give the owls a type I functional response on the gerbils).

In this formulation, it is assumed that there is no direct competition between gerbils (except through seed abundance) and no direct competition between owls (except through abundance of gerbils). The direct competition is strictly between the gerbils and owls. Hence in terms of a virtual variable for the focal gerbil

$$E_1(v_1) = a_1 \int y(t) v_1(t) dt - c_1 \int v_1(t) dt - k_1 \int (1 - v_1(t)) dt \quad (10.33)$$
$$p_1(v_1) = e^{-a_2 x_2 \int v_1(t) u_2(t) dt}.$$

We construct a $G$-function for the gerbils by letting their finite growth rate be proportional to their net seed profit multiplied by the probability that they survive the night to enjoy this profit

$$G_1(v_1(t), u_1(t), u_2(t), x_1, x_2) = p_1(v_1)[1 + b_1 E_1(v_1)]$$

where $b_1$ is the conversion factor of seed profits into new gerbils. From a focal gerbil's perspective, the strategy of the owls directly influences its fitness via survivorship, and the strategy of the other gerbils influences its fitness indirectly via their effect on seed abundances. A difference equation dynamics on the gerbil population is implied that occurs from night to night according to

$$x_1(\theta + 1) = x_1(\theta) G_1$$

where $\theta$ represents time in nights (rather than $t$ which gives time within nights).

A similar approach is used to construct the owl's $G$-function with one realistic but convenient assumption. While the gerbils may harvest a considerable fraction of the available seeds during the night (demonstrable nightly depletion of $y$), it is assumed that the owls have only a negligible nightly impact on the gerbils' population size (2% or less) and so we can approximate $x_1$ as remaining constant during any given night.

This means an owl's net profit (in units of gerbils) during the night is approximately

$$E_2(u_2) = a_2 x_1 \int u_2(t) u_2(t) dt - c_2 \int u_2(t) dt - k_2 \int (1 - u_2(t)) dt \quad (10.34)$$

where the subscripts now refer to owl harvests (first integral), foraging costs (second integral), and roosting costs (third integral).

The probability that an owl survives the night (avoids fatal injury) is given by

$$p_2(v_2) = e^{-\gamma \int u_2(t) dt}$$

where $\gamma$ is instantaneous risk of fatal injury while hunting. The owl $G$-function is written in a fashion analogous to the gerbil's (using $v_2$ in place of $u_2$ in the equations for $E_2$ and $p_2$)

$$G_2(v_2(t), u_1(t), x_1) = p_2(v_2)[1 + b_2 E_2(v_2)]$$

where $b_2$ is the conversion factor of gerbil profits into new owls, $p_2(v_2)$ gives the probability of the focal owl individual surviving injury, and $E_2(v_2)$ gives the net gerbil harvest of a owl using strategy $v_2$ in an environment whose gerbil abundance is shaped by $u_1(t)$. From a focal owl's perspective, the strategy of the gerbils directly influences its fitness via hunting success, whereas the strategy of the owl's does not influence an owls fitness. All an owl cares about on a given night is the abundance of gerbils, $x_1$, and their activity pattern, $u_1(t)$. This $G$-function for the owls results in a population dynamic of owls similar to the form given for the gerbils

$$x_2(\theta + 1) = x_2(\theta) G_2.$$

We use a cost-benefit analysis at each time $t$ to determine whether a gerbil or an owl would enhance its fitness by foraging. To a gerbil the benefit of foraging relative to remaining safely in its burrow is simply its harvest rate $f_1(t)$. The cost of foraging relative to remaining in its burrow is the additional energy expenditure incurred while foraging, $c_1 - k_1$, and the foraging cost of predation (Brown, 1992, 1998). To calculate the cost of predation one needs to multiply predation risk while foraging by the conversion rate of safety for seeds. Fortunately this conversion rate is known for both static and dynamic optimization models and we apply it here (Brown, 1992; Houston et al., 1993). Similar reasoning applies to the cost–benefit ratio for owls.

## 10.4 Foraging games

A gerbil and owl should want to forage or hunt whenever

$$f_1 = a_1 y(t) \geq (c_1 - k_1) + a_2 x_2 u_2(t) \frac{1 + E_1}{b_1} \quad (10.35)$$

$$f_2 = a_2 u_1(t) x_1 \geq (c_2 - k_2) + \frac{\gamma(1 + E_2)}{b_2} \quad (10.36)$$

where the right-most term in (10.35) is the foraging cost of predation to a gerbil and the right-most term in (10.36) is the foraging cost of injury to an owl where $1/b_i$, $i = 1, 2$ is the exchange rate of safety for food.

A solution, $u_1$ and $u_2$, is sought by using the above cost–benefit relationships to evaluate the efficacy of each time-dependent strategy, and the $G$-functions are used to provide the population dynamics towards $x_1^*$ and $x_2^*$. The solution is obtained through an iterative process. First, Equations (10.35) and (10.36) are used to find $u_i$ by assigning values for $E_i$ and $x_i$. These solutions along with $x_i$ are used to determine $E_i$ from (10.33) and (10.34). One then iterates this process until the $u_i(t)$ used to determine $E_i$ is also the $u_i(t)$ obtained by inputting $E_i$. This represents a solution to the predator–prey foraging game for the specified values of $x_i$. Now, the values for $u_i(t)$ and $E_i$ can be substituted into the $G$-functions to determine changes in $x_i$. This sets off a new round of iterations to solve for the new combination of $u_i(t)$ and $E_i$. Eventually, the system is driven to a solution for $u_i(t)$ and $x_i$.

The solution obtained has two phases, similar to density-dependent habitat selection. In the first phase, seed abundance is so high (early in the night) that it benefits all gerbils to be active even though all of the owls are active: $u_1(t) = u_2(t) = 1$. During this phase, both equations (10.35) and (10.36) are satisfied as strict inequalities. During the second phase, only a fraction of prey and predator are actually active, $u_1(t) \in (0, 1)$ and $u_2(t) \in (0, 1)$, and equations (10.35) and (10.36) are satisfied as strict equalities.

A solution always involves a subset of both phases: (1) activity by all prey and predator individuals, (2) activity by a portion of prey and predator. During the second phase the activity level of prey and predators conforms to a temporal **ideal free distribution**. The prey solution during this phase has the prey sufficiently active to exactly balance the predator's costs and benefits of hunting. Hence the prey shift at some critical time period $\hat{t}$ from the constant strategy $u_1(t) = 1$ for $t < \hat{t}$ to the constant strategy $u_1(t) = u'$ for $\hat{t} > 0$ where $0 < u' < 1$. The predator solution during the second phase has the predators restrain their activity to exactly balance the prey's costs and benefits of foraging. Following $\hat{t}$ the level of predator activity declines monotonically with the decline in seed abundance. Prior to $\hat{t}$ the system is seed driven in the sense that the high level of seeds drives complete gerbil activity which drives complete owl activity. Following $\hat{t}$, the system is driven by the predators. Throughout the rest of the night, the predators pace their activity so as to keep a constant and just profitable level of gerbil activity. To the gerbil, the fitness benefits of foraging remain constant, although earlier in the period harvest rate and risks are high while later in the night harvest rates and risks are low.

### 10.4.2 Patch-use model of fierce predators seeking wary prey

Consider a mountain lion and mule deer system. Mountain lions (*Puma concolor*) and mule deer (*Odocoileus hemionus*) in the mountains of southern Idaho approximate closely a single-prey single-predator system (Altendor, 1997; Hornocker, 1970). The mountain lions capture deer on the boundaries and interiors of forest patches, and the deer move from these forest patches to open shrub habitat as bedding and feeding areas. Mountain lions move frequently among forest patches in a manner reminiscent of patch-use models from foraging theory (Brown, 1988; Charnov, 1976). However, unlike in standard models of patch use, a mountain lion rarely harvests more than one food item per patch. Patch depletion is not the result of prey removal by the predator, but the result of resource depression as the deer either become warier and harder to catch or the deer vacate the woods for another patch. The mountain lion–deer system is a game of stealth and fear (van Balaan and Sabelis, 1993).

For simplicity, consider an environment in which prey occur as isolated individuals within patches of suitable habitat. Let the feeding rate of prey be density dependent and decline with prey number. Consider a predator that is obliged to move from patch to patch in hopes of capturing a prey. The questions of interest are: "how long should a predator remain in a prey patch before giving up and moving onto the next patch?," and "how vigilant should the prey be when increased vigilance reduces both feeding rates and predation risk?"

The following model for the prey population dynamics is used

$$\dot{x}_1 = r\left[\frac{(1-w)K}{x_1 + \chi} - c\right] - \mu x_2$$

where $x_1$ is the population size of prey, $x_2$ is the population size of predators, the first term on the right-hand side represents prey fecundity as a result of resource harvest, $w$ is the prey's vigilance level, and $\mu$ is the prey's average mortality rate from a single predator individual. Fecundity can be thought of as scramble competition in which $\frac{K}{x_1+\chi}$ represents the forager's feeding rate while not vigilant and $\frac{(1-w)K}{x_1+\chi}$ is the forager's net feeding rate when vigilance is considered, and $c$ is the forager's subsistence cost measured in the same units as the feeding rate. Vigilance can be thought of as the proportion of time that the prey spends scanning or looking for predators, $0 < w < 1$. The term $r$ scales the conversion of net energy gain into fecundity.

Let the predation risk be influenced by the prey's encounter rate with a predator, $m$, the predator's lethality in the absence of vigilance, $1/k$, the effectiveness of vigilance in reducing predator lethality, $b$, and the prey's level of vigilance, $w$

$$\mu = \frac{m}{k+bw}.$$

In this model (Brown, 1999), the prey uses vigilance to balance conflicting demands of food and safety. Increasing vigilance will increase the prey's safety (reduce $\mu$) but reduce its fecundity. Under these assumptions, Brown (1999) shows that the optimal level of vigilance that maximizes the prey's per capita growth rate is

$$w^* = \sqrt{\frac{mr(x_1 + \chi)}{bK}} - \frac{k}{b}. \qquad (10.37)$$

The prey's optimal level of vigilance increases with its encounter rate with predators, $m$, number of prey, $x_1$, the saturation constant, $\chi$, the conversion efficiency of energy for offspring, $r$, and predator lethality, $1/k$. Vigilance declines with resource abundance, $K$. The relationship between vigilance and effectiveness of vigilance is hump shaped. When vigilance is ineffective ($b$ very small), vigilance is useless, and, when vigilance is very effective ($b$ very large), little vigilance is required. When the equation for vigilance yields a value $>1$, the forager should spend all of its time vigilant, $w^* = 1$. If (10.37) yields a value $< 0$, the forager should spend no time being vigilant, $w^* = 0$.

Thus far we have an ecological model of prey population dynamics with the incorporation of fear responses, but it is not yet a game. It becomes a game by building into the model the way prey detect predators that have entered their area, and by modeling the predator's population dynamics and patch residence times.

### 10.4.2.1 Prey with imperfect information

Imagine prey that are uncertain as to the actual whereabouts of the predators but are able to make an estimate of their encounter rate with predators, $m$, based on cues emitted by the predator when it occupies a prey's patch (such cues may be auditory, olfactory, or visual). Such a prey should have some background level of apprehension even in the absence of any nearby predators. This level of apprehension is determined by its baseline expectation of encountering a predator in the absence of any cues of predatory risk. Refer to this background level of apprehension as $u_1$. The prey's background level of apprehension will be the prey's strategy in this predator–prey foraging game. When a predator actually enters a patch, the forager acquires information regarding its possible presence and, on average, adjusts its apprehension higher the longer the predator remains in the patch. For example, following the arrival of a predator, the prey's expected level of apprehension may be approximated as a learning curve

$$m(t) = u_1 + (M - u_1)(1 - e^{-at})$$

where $a$ is the rate at which a prey perceives the presence of a predator and $t$ is the time since a predator has actually been in the patch (the prey never actually knows this). As $t$ goes from $0 \longrightarrow \infty$ the prey's expectation of encountering a predator rises from $u_1 \longrightarrow M$. The level of apprehension changes with the time prior to and after the arrival of the predator in the patch. Under this model of imperfect information the prey's level of vigilance when a predator is absent from the patch, $w_{ab}^*$, is found by substituting $m = u_1$ into the equation for $w^*$. In the presence of a predator, the prey's average level of vigilance, $w_{pr}^*(t)$, can be found by setting $m = m(t)$.

The prey's challenge is to select the optimal level of background apprehension, $u_1$. Increasing apprehension reduces feeding rates and reduces predation risk. If the level of apprehension is set too high, the forager misses out on valuable feeding during periods when there are no predators in the patch. If set too low, prey experience an unacceptably high predation risk in the presence of a predator. The background level of apprehension must strike a balance between feeding rate in the absence of predators and safety in the presence of a predator. The average risk of predation experienced by a prey over time is influenced by two aspects of the predator: (1) the probability of there actually being a predator within a prey's patch, $p$, and (2) the average time that a predator remains within a patch before moving on, $u_2$ (referred to as the predator's giving-up time). The predator's giving-up time represents its strategy in this predator–prey foraging game. The average risk of predation is given by

$$\bar{\mu} = p \int_0^{u_2} \frac{M}{k + bw_{pr}^*(t)} dt$$

where $p$ is the probability that a predator is actually in the prey's patch, and the risk is integrated from 0 to $u_2$ (the giving-up time of the predator).

The forager's average level of vigilance over time is given by

$$\bar{w} = (1 - p)w_{ab}^* + p \int_0^{u_2} w_{pr}^*(t) dt.$$

These mean values for predation risk, $\bar{\mu}$, and vigilance, $\bar{w}$, can be substituted

## 10.4 Foraging games

into the prey's population growth model to produce the prey's $G$-function where an individual's prey's fitness is now a function of its baseline level of apprehension, the population size of prey, the predator's giving-up time and the predator's population size

$$G_1(v, u_2, x_1, x_2) = r\left[\frac{(1-\bar{w})K}{x_1 + \chi} - c\right] - \bar{\mu}x_2.$$

Given the probability that a predator is actually within the prey's patch, $p$, and given the predator's patch residence time, $u_2$, it is possible to find the ESS value of $u_1$ (background level of apprehension) that maximizes the prey's finite rate of growth (as we shall see below, $p$ is influenced by $x_1, x_2, u_2$).

The optimal baseline level of apprehension has the following properties. Baseline apprehension increases with the predator's population size, $x_2$, the predator's encounter rate with prey, $M$, and the prey's net feeding rate, $\bar{f}$. Baseline apprehension decreases with the prey's population size, $x_1$, effectiveness of vigilance, $b$, the prey's detection rate of predators, $a$, the prey's intrinsic growth rate, $r$, and the predator's patch residence time, $u_2$.

This characterizes the prey's fear responses. But what of the predator's behavior? How long should a predator remain in a patch before moving to another?

### 10.4.2.2 Predator's response to prey with imperfect information

Consider a predator that is aware of the prey's level of apprehension. Let the predator's population growth rate increase linearly with the predator's average harvest rate of prey, $H$. The predator's average harvest rate will be a function of prey abundance, $x_1$, the prey's baseline level of apprehension, $u_1$, and the predator's patch residence time, $u_2$. This produces the predator's $G$-function

$$G_2(v, u_1, u_2, x_1, x_2) = r_P H - d$$

where $r_P$ is the conversion rate of harvested prey into new predators and $d$ is the predator's death rate.

The predator's expected harvest rate within the patch, $\mu(t)$, declines with time spent in the patch. The longer the predator pursues the prey the less catchable it becomes. There comes a point, $u_2$, at which the prey is no longer worth pursuing and the predator is better off abandoning the prey and seeking another. The predator's optimal stay time within the patch before abandoning the hunt, $u_2^*$, will satisfy a marginal value theorem (Charnov, 1976). The predator should abandon the hunt for a given prey at the point where its expected capture rate, $\mu(t)$, drops to its average capture rate from seeking and pursuing a new prey.

The ESS satisfies

$$\mu(u_2) = H(u_2)$$

where the predator's average harvest rate, $H(u_2)$, is given by the probability of making a kill within a given patch divided by the time required to find the patch

$$H(u_2) = \frac{1 - e^{\left(\int_0^{u_2} \mu(t)dt\right)}}{T + u_2}$$

where the integral is evaluated from $t = 0$ to $t = u_2$, and $T = 1/sN$ is the predator's travel time between one prey patch and the next ($s$ being the predator's efficiency at finding potential prey patches).

For any given prey and predator population size, $x_1, x_2$, and prey background level of apprehension, $u_1$, it is possible to find a given predator's ESS patch residence time $u_2$. All else being equal, increasing $x_1$ increases a predator's harvest rate without altering the likelihood of capturing a prey within its patch. Hence, $u_2$ declines. Increasing $x_2$ increases the prey's baseline level of apprehension, $u_1$, reduces the predator's harvest rate, and increases $u_2$.

To a fierce predator, the number, $x_1$, and catchability, $w(t)$, of prey determine patch quality. Increasing patch quality increases the amount of time that a predator should hunt within a patch before giving up if it has not yet captured any prey. Increasing patch quality also increases the probability that a predator will have captured a prey before it gives up on the patch. All else being equal, increasing the background apprehension level of the prey or increasing the energy state of the prey will reduce patch quality. All else being equal, increasing the number of prey, increasing the prey marginal value of energy, and increasing the prey's feeding rate will increase patch quality.

This foraging game between predator and prey does not produce equations that can be solved analytically for the ESS. The ESS values for the prey's baseline level of apprehension, $u_1$, and the predator's giving-up time, $u_2$, must be found iteratively.

# 11
# Managing evolving systems

Nature is not an art gallery with a static display of species. Rather, Nature is a rich dynamical system involving both ecological changes in population sizes and evolutionary changes in species' number and strategies. This is never so apparent as when humans modify the environment or attempt the daunting task of managing species. Habitat destruction, competing land use objectives, harvesting, and even relatively benign changes to environments can threaten the viability of a species' population. At the other end of the management spectrum, many exotic invasive species, diseases, and agricultural pest species defy control efforts. Understandably, most management and conservation of species focus on ecological dimensions and principles. Yet, all species have evolutionary dynamics as well. Rapid evolutionary responses are commonplace (Ashley et al., 2003). Antibiotic resistance in bacteria can evolve within months or years. Pesticide resistance and herbicide resistance in insects and weedy plants can occur within years or decades. Within the span of a few decades (and less), the hand weeding of plants from Asian rice paddies has selected for weed seedlings that mimic the look of rice plants. The mechanical seed-sorting technologies of mechanized agriculture have produced weed ecotypes that mimic the flowering phenology and seed characteristics of grain crops. Newly established island populations of rodents show rapid evolutionary changes (usually increases) in body size and concomitant changes in skull morphology. Red squirrels have expanded their range south a few hundred miles into central Indiana, USA. These squirrels already show cranial adaptations to a diet that includes more thick-shelled nuts (such as black walnuts) and fewer pine nuts. Intensive fish harvesting can cause a variety of life-history changes as witnessed by the North Sea cod and Pacific salmon fisheries.

It is not unexpected, then, to see human-induced evolutionary changes. This is particularly true as species face novel circumstances such as human-modified habitats, changed resource availabilities, novel resources or absences of mortality, and even drastically altered climatic regimes. The term pristine describes environments free from human perturbations that remain, more or less, in some perceived original state. We describe a **pristine environment** as one that offers a particular set of niche archetypes and evolutionarily stable strategies that influence and determine the current ecological and evolutionary mix of species. Any habitat alteration of a pristine environment that is consequential to a species' ecology will cause shifts in the positions, number, and diversity of niche archetypes and the make up of an ESS. After such an alternation, a species' strategy may no longer be ESS in an environment with changed niche archetypes. Species will evolve according to the changed adaptive landscape. Without knowledge of species' $G$-functions and the salient variables influencing the resultant adaptive landscapes, we humans can be only mere observers of the evolutionary changes that we have wrought. Ongoing strategy dynamics will remain obscure and any knowledge of change will occur only after the fact.

We are reminded of an amusing anecdote on the selective influence of human harvesting of prey (related to us by P. Smallwood). Researchers bred laboratory mice for feeding experiments with raptors. As is frequently the case, the docile mice were handled by gently lifting them by their tails. Soon, to the amusement of the researchers, the colony of mice consisted of individuals that held their tails flat to the ground!

The objective of this chapter is to provide tools for understanding the evolutionary implications of management practices. The novel selective pressures created by management programs will often be less humorous and more profound than in the mice example.

## 11.1 Evolutionary response to harvesting

While most literature on environmental management considers ecological rather than evolutionary impacts (Goh, 1980; Vincent and Skowronski, 1981; see also several papers in Vincent *et al.*, 1987), the need for including evolutionary considerations in ecological models is recognized (Ricker, 1981). However, only a few attempts have been made to incorporate evolutionary effects into management models (Vincent and Brown, 1987*b*; Rijnsdorp, 1993; Vincent, 1994; Lynch, 1996; Law, 2000). Fortunately, evolutionary effects are easily incorporated into management models by using $G$-functions. As in previous chapters, one begins with a $G$-function for the threatened species, the undesirable pest, etc. Next the $G$-function is modified by introducing the influence of humans on the models' parameters and structure.

For example, adding a harvesting term to the Lotka–Volterra model (see Example 4.1.1) may be thought of as subjecting a prey population to harvesting.

## 11.1 Evolutionary response to harvesting

The predator (e.g., humans) doing the harvesting is assumed to be external to the system, so that the harvesting strategy of the predator is not influenced by the prey. The $G$-function in this case is obtained by simply adding the harvesting term

$$G(v, \mathbf{u}, \mathbf{x}) = \frac{r}{K(v)} \left[ K(v) - \sum_{j=1}^{n_s} a(v, u_j) x_j \right] - h(v)$$

where the function $h(v)$ describes a harvest-induced mortality rate.[1] The model assumes that the strategy of the focal prey, $v$, influences the likelihood that it is harvested. The prey strategy also affects its fitness in other ways. For example, $v$ could represent body size. A small body size might reduce harvesting pressure, but it could also decrease the carrying capacity, $K$, and increase the negative effects of competition with others through the function $a$.

In previous examples (e.g., Example 4.3.1) we have used a competition function that has the following symmetry properties

$$\left. a(v, u_j) \right|_{v=u_j} = 1$$
$$\left. \frac{\partial a(v, u_j)}{\partial v} \right|_{v=u_j} = 0. \quad (11.1)$$

In order to make the subsequent results more general, we assume that $a$ satisfies only (11.1) before considering specific examples.

### 11.1.1 Necessary conditions for an ESS coalition of one

It follows that for a coalition of one

$$G\left(v, u_1, x_1^*\right) = \frac{r}{K(v)} \left[ K(v) - a(v, u_1) x_1^* \right] - h(v).$$

From the ESS maximum principle we have the necessary condition that both

$$\left. G\left(v, u_1, x_1^*\right) \right|_{v=u_1} = \frac{r}{K(u_1)} \left[ K(u_1) - a(u_1, u_1) x_1^* \right] - h(u_1)$$

and

$$\left. \frac{\partial G\left(v, u_1, x_1^*\right)}{\partial v} \right|_{v=u_1} = \frac{r}{K(u_1)} \left[ \frac{\partial K(v)}{\partial v} - x_1^* \frac{\partial a(v, u_1)}{\partial v} \right]_{v=u_1}$$

$$- \frac{r \left[ K(u_1) - a(u_1, u_1) x_1^* \right]}{\left[ K(u_1) \right]^2} \left. \frac{\partial K(v)}{\partial v} \right|_{v=u_1} - \left. \frac{\partial h(v)}{\partial v} \right|_{v=u_1}$$

---

[1] Since the rate at which prey $x_i$ are removed is given by $h(v) x_i$, the term $h(v)$ corresponds to a level of harvest effort rather than a level of harvest rate.

are equal to zero. Using (11.1) these expressions reduce to

$$\frac{r}{K(u_1)}\left[K(u_1) - x_1^*\right] = h(u_1) \tag{11.2}$$

$$\frac{r}{K(u_1)}\frac{\partial K(v)}{\partial v}\bigg|_{v=u_1} - \frac{r\left[K(u_1) - x_1^*\right]}{[K(u_1)]^2}\frac{\partial K(v)}{\partial v}\bigg|_{v=u_1} = \frac{\partial h(v)}{\partial v}\bigg|_{v=u_1}.$$

The second expression is simplified by using (11.2) in the second term and factoring to yield

$$\frac{r - h(u_1)}{K(u_1)}\frac{\partial K(v)}{\partial v}\bigg|_{v=u_1} = \frac{\partial h(v)}{\partial v}\bigg|_{v=u_1}. \tag{11.3}$$

Since $x_1^*$ does not appear in (11.3) it may be used to solve for $u_1$ and then $x_1^*$ may be determined from (11.2).

### 11.1.2 Necessary conditions for an ESS coalition of two

For a coalition of two we have

$$G(v, \mathbf{u}, \mathbf{x}^*) = \frac{r}{K(v)}\left[K(v) - a(v, u_1)x_1^* - a(v, u_2)x_2^*\right] - h(v)$$

with the four necessary conditions

$$G(v, \mathbf{u}, \mathbf{x}^*)\big|_{v=u_1} = 0$$

$$G(v, \mathbf{u}, \mathbf{x}^*)\big|_{v=u_2} = 0$$

$$\frac{\partial G(v, \mathbf{u}, \mathbf{x}^*)}{\partial v}\bigg|_{v=u_1} = 0$$

$$\frac{\partial G(v, \mathbf{u}, \mathbf{x}^*)}{\partial v}\bigg|_{v=u_2} = 0$$

to solve for $x_1^*$, $x_2^*$, $u_1^*$, and $u_2^*$. Following the same procedure for the coalition of one, using (11.1) we obtain

$$x_1^* + x_2^* a(u_1, u_2) = K(u_1)\left[1 - \frac{h(u_1)}{r}\right]$$

$$x_1^* a(u_2, u_1) + x_2^* = K(u_2)\left[1 - \frac{h(u_2)}{r}\right]$$

$$\frac{r - h(u_1)}{K(u_1)}\frac{\partial K(v)}{\partial v}\bigg|_{v=u_1} = \frac{\partial h(v)}{\partial v}\bigg|_{v=u_1} + \frac{r}{K(u_1)}x_2^*\frac{\partial a(v, u_2)}{\partial v}\bigg|_{v=u_1}$$

$$\frac{r - h(u_2)}{K(u_2)}\frac{\partial K(v)}{\partial v}\bigg|_{v=u_2} = \frac{\partial h(v)}{\partial v}\bigg|_{v=u_2} + \frac{r}{K(u_2)}x_1^*\frac{\partial a(v, u_1)}{\partial v}\bigg|_{v=u_2}.$$

## 11.1 Evolutionary response to harvesting

In this case all four equations must be solved simultaneously in order to obtain an ESS candidate.

### 11.1.3 Specific examples

For the specific examples below we use the following functions for $r$, $K$, and $a$.

$$r = 0.25$$
$$K(v) = 100 \exp\left[-\frac{(v-1)^2}{2}\right]$$
$$a(v, u_j) = 1 - \frac{(v-u_j)^2}{16}.$$

**Example 11.1.1 (an ESS under no harvesting)** *Under no harvesting*

$$h(v) = 0.$$

*The necessary conditions for a coalition of one reduce to*

$$x_1^* = K(u_1)$$
$$\left.\frac{\partial K(v)}{\partial v}\right|_{v=u_1} = 0$$

*yielding*

$$u_1 = 1$$
$$x_1^* = 100.$$

*Plotting $G(v, u_1, x_1^*)$ vs. $v$ demonstrates that the ESS maximum principle is satisfied. Simulation demonstrates that this solution is convergent stable, so that this solution is, indeed, an ESS coalition of one.*

**Example 11.1.2 (an ESS under linear harvesting)** *Under linear harvesting we use*

$$h(v) = 0.1v.$$

*If $v$ is body size, then a linear harvest corresponds to a situation where a premium is paid for a catch of larger body size or where nets or traps select linearly for larger body size. In this case the necessary conditions for a coalition of one reduce to*

$$\frac{r}{K(u_1)}\left[K(u_1) - x_1^*\right] = 0.1u_1$$
$$\left.\frac{r - h(u_1)}{K(u_1)} \frac{\partial K(v)}{\partial v}\right|_{v=u_1} = 0.1$$

yielding the solutions

$$u_1 = 0.5$$
$$x_1^* = 70.6$$

and

$$u_1 = 3$$
$$x_1^* = -2.7.$$

Clearly the second solution is not valid. Plotting $G\left(v, u_1, x_1^*\right)$ vs. $v$ demonstrates that the ESS maximum principle is satisfied with the first solution. Simulation demonstrates that this solution is convergent stable, so that it is, indeed, an ESS coalition of one. Thus the effect of a linear harvest is to cause the species to evolve to a body size of half its non-harvesting ESS. We can now calculate the steady-state yield before and after evolution takes place. Before evolution ($u_1 = 1$) we have

$$\frac{r}{K(1)}\left[K(1) - x_1^*\right] = 0.1$$

resulting in $x_1^* = 60$ with a steady-state yield of

$$\text{yield} = x_1^* h(u_1) = 6.$$

After evolution has taken place ($u_1 = 0.5$) we have

$$\frac{r}{K(0.5)}\left[K(0.5) - x_1^*\right] = 0.05$$

resulting in $x_1^* = 70.6$ with a steady-state yield of

$$\text{yield} = x_1^* h(u_1) = 3.53.$$

We have the interesting before-and-after evolutionary effect that the yield will decrease while the equilibrium population increases with individuals at one-half the body size of their original ESS.

**Example 11.1.3 (an ESS under hump-shaped harvesting)** Under a hump-shaped harvest we use

$$h(v) = 0.15 \exp\left[-\frac{(v-1)^2}{2}\right].$$

This could correspond to a situation where body size is again the adaptive parameter, but there is a premium for a body size of 1. Using the methods of the previous example we obtain the following ESS candidates

$$u_1 = 1.604$$
$$x_1^* = 41.67$$

## 11.1 Evolutionary response to harvesting

and

$$u_1 = 0.396$$
$$x_1^* = 41.67.$$

However, a plot of $G\left(v, u_1, x_1^*\right)$ vs. $v$ demonstrates that neither solution satisfies the ESS maximum principle. Using the necessary conditions for an ESS coalition of two, we obtain only one ESS candidate

$$\mathbf{u} = \begin{bmatrix} 0.275 & 1.725 \end{bmatrix}$$
$$\mathbf{x}^* = \begin{bmatrix} 22.16 & 22.16 \end{bmatrix}.$$

A plot of $G(v, \mathbf{u}, \mathbf{x}^*)$ vs. $v$ shows that this solution does satisfy the ESS maximum principle and simulation demonstrates that the solution is convergent stable. The effect of harvesting in this case is to produce the coevolution of two phenotypes. The effect on yield is similar to the previous example. Before the population evolves, the equilibrium population under harvesting is $x_1^* = 40$ with a corresponding

$$\text{yield} = x_1^* h(u_1) = 6.$$

After evolution the equilibrium population is $x_1^* + x_2^* = 44.32$ with a corresponding

$$\text{yield} = x_1^* h(u_1) + x_2^* h(u_2) = 5.11.$$

Again, the before-and-after evolution effect is to decrease yield at a higher population number (of the combined two phenotypes).

The results obtained from the Lotka–Volterra model are generic. Whether intentional or accidental, harvesting and cropping of prey species by humans introduces new selective pressures on the population. Body size, for example, is one such evolutionary trait which may be brought under selection. The inclination of humans to harvest the largest individuals (it is often illegal to harvest crabs, shellfish, or fish below a threshold size) or the use of traps and nets that selectively collect larger individuals corresponds to a harvest function qualitatively similar to the linear harvest function. This results in directional selection that modifies the ESS resulting in a body size that leads to a decline in yield.

Often, harvest techniques are designed to be most efficient at collecting individuals using the strategy which predominates in the population. Insecticides are an example of this. They are designed to be effective against insect traits that actually occur, not those that may occur. The hump-shaped harvest function is most effective at collecting individuals using the existing ESS. For this example, harvesting results in disruptive selection that drives the system to a new ESS coalition of two strategies.

## 11.2 Resource management and conservation

Taking a proactive, predictive modeling approach to the joint ecological and evolutionary consequences of human management is still in its infancy. The imperative is clear. In the face of global climate change, urbanization, agriculture, forestry, animal husbandry, and oceanic fisheries there may no longer be any pristine ecosystems. Humans have likely altered the form of most, if not all, species' $G$-functions. While evolutionary concerns may seem distant, slow, or secondary to more immediate ecological challenges, evolution does and will matter. Whether evolution will occur is the wrong question. How fast and how much of an evolutionary change due to human impacts are the correct questions to ask. By way of several examples we will explore how the evolutionary game theory of this book can be brought to bear on these questions. These examples do not provide a treatise on managing evolving systems, rather they suggest how one can expand the ecological models used in the conservation of threatened species and communities, or in the management of pest species to include evolutionary consequences. When management is viewed as a bio-economic evolutionary game, managers are co-players when their utility functions depend on the strategies used by the species they are managing.

### 11.2.1 Evolutionarily stable harvest strategies

The harvesting of commercially valuable species represents a new and often large source of mortality for a population. The harvested species experiences selection to evolve adaptations that either mitigate or accommodate the actions of the new predator. Several fish stocks have shown striking changes in their size, age, and fertility patterns (Murphy, 1967; Borisov, 1978; Garrod, 1988). Such changes can be ascribed to phenotypic plasticity in response to changing environmental conditions or to evolved changes in life history strategies (MacDonald and Chinnappa, 1989; Trexler and Travis, 1990; Trexler *et al.*, 1990). However, whenever heritable variation exists, evolution will or has occurred in exploited populations. Worse yet, fisheries are not only prone to an **ecological tragedy of the commons** (overharvesting beyond that which would maximize sustainable yield) but are equally prone to an **evolutionary tragedy of the commons** that results when a harvested species evolves to less desirable strategies.

## 11.2 Resource management and conservation

Harvesting is an evolutionary game that involves humans. Generally a harvested species possesses a strategy that influences interactions within the species, the desirability of the species to humans, and the ease of human harvesting. The rate of harvesting can be thought of as a predator strategy that influences the prey and the value of the harvest. With this in mind, Law and Grey (1989) define and apply the concept of an **evolutionarily stable optimal harvesting strategy** (ESOHS). At the ESOHS, the manager considers not the just the ecological but the evolutionary consequences of harvesting. The ESOHS is an example of evolutionarily enlightened management. For example, knowing that the harvesting of a fish species will drive them to a new ESS, the ESOHS of the fishery is the harvest pattern that maximizes yield at the new evolutionarily stable equilibrium (Law and Grey, 1989).

As an illustration, we will draw on a fisheries model used by Blythe and Stokes (1990), and Brown and Parman (1993). The exploited species consists of juveniles and adults. Juveniles grow linearly at rate $g$ and achieve adult size $v$ at age $T$ where $v = gT$. Upon reaching adulthood, growth ceases and reproduction commences. Fecundity increases linearly with adult size, $v$. Let $s$ be the maximum fecundity rate per unit of adult size. We assume that $s$ is the product of neonate production (newborn per unit of adult size) and neonate survivorship. Once a neonate has survived it becomes a juvenile of effectively zero size and zero age. For simplicity it is assumed that juveniles and adults suffer no mortality other than human harvest. Let $z$ be the threshold harvesting size and let $H$ be the instantaneous harvest rate on all individuals (juveniles and adults) sized greater than or equal to $z$. Given a threshold size $z$ and a harvest rate $H$, the probability $P$ that a juvenile with strategy $u$ survives to adulthood is given by

$$P(v) = e^{-\frac{H(v-z)}{g}} \tag{11.4}$$

where $(v - z)/g$ gives the amount of time that a juvenile is exposed to harvesting and

$$0 \leq P \leq 1 \text{ for } v \geq z$$
$$P = 1 \text{ for } v < z.$$

It follows that when $P = 1$ then $v = z$. The probability of surviving to adulthood declines with increasing harvest, increasing size at maturity, and a decreasing harvest threshold.

To formulate a $G$-function, we introduce density-dependent effects by assuming that competition for food among adults reduces neonate production below its maximal level. Blythe and Stokes use a Ricker stock recruitment relationship by letting actual neonate production scale as a negative exponential of adult biomass

$$\text{actual neonate production} = sve^{-\frac{vx}{R}}$$

where $x$ is adult population size, and $R$ scales the intensity of food competition.

The fitness generating function for the harvested species is written as

$$G(v, \mathbf{u}, \mathbf{x}) = svP(v)e^{-\frac{\sum_{i=1}^{n_s} u_i x_i}{R}} - H. \tag{11.5}$$

The first term of the $G$-function gives the expected rate of adult recruitment from the production of neonates by an adult. It is the product of the maximal rate of neonate production per unit adult size, parent size, probability of surviving the juvenile period of growth, and the density-dependent effect of other adult fish. The second term is simply the rate of adult mortality due to harvesting. In this formulation, the strategy of the individual, $v$, influences its maximal fecundity rate and influences its probability of surviving to adulthood, $P(v)$.

A candidate ESS solution is obtained by using the ESS maximum principle. We seek a coalition of one ESS for adult size. Because we have not imposed any constraints on $u_1$, the necessary conditions for an ESS are given by

$$G(v, u_1, x_1^*)\big|_{v=u_1} = 0 \tag{11.6}$$

$$\frac{\partial G(v, u_1, x_1^*)}{\partial v}\bigg|_{v=u_1} = 0. \tag{11.7}$$

Setting $n_s = 1$ it follows from (11.5) and (11.6) that

$$x_1^* = \frac{R}{u_1} \ln \frac{su_1 P(u_1)}{H}. \tag{11.8}$$

Evaluating the partial derivative of (11.5) yields

$$\frac{\partial G(v, u_1, x_1)}{\partial v} = \left[se^{-\frac{u_1 x_1}{R}}\right]\left[u_1 \frac{\partial P}{\partial v} + P(v)\right].$$

Imposing the requirement (11.7) it follows that

$$\left[se^{-\frac{u_1 x_1^*}{R}}\right]\left[u_1 \frac{\partial P}{\partial v}\bigg|_{v=u_1} + P(u_1)\right] = 0. \tag{11.9}$$

From the definition of $P$ given by (11.4), it follows that

$$\frac{\partial P}{\partial v}\bigg|_{v=u_1} = -\frac{H}{g}P(u_1)$$

and because only the second term of (11.9) can equal zero we obtain

$$-u_1\frac{H}{g}P(u_1) + P(u_1) = 0. \tag{11.10}$$

Thus from (11.10) we have $u_1 = \frac{g}{H}$. Thus

$$u_1 = \frac{g}{H} \quad \text{if} \quad \frac{g}{H} \geq z$$
$$u_1 = z \quad \text{if} \quad \frac{g}{H} < z.$$

The two necessary conditions result from the discontinuity of $P$ at $u = z$. When harvest rates are sufficiently severe the fish evolve up to but not beyond the threshold harvest size, $u_1 = z$. Adult size increases with the growth rate of juveniles and declines with the harvest rate. Fishing decreases size at maturity.

#### 11.2.1.1 Yield

The yield function has two components. The first represents the harvest rate of juveniles between sizes $z$ and $u_1$ weighted by their size, and the second represents the harvest of adults weighted by their size $u_1$. The first component requires integrating yield across the time period during which juveniles are exposed to harvesting weighted by their size and size-specific population sizes (there are successively fewer juveniles as juvenile size increases). The sum of these two components gives the rate at which biomass is harvested

$$Y(H, z, u_1, x_1^*) = H \int_{\frac{z}{g}}^{\frac{u_1}{g}} g \left[ \frac{H x_1^*}{P(u_1)} \right] e^{-(\frac{u_1}{g} - \frac{z}{g})} q \, dq + H u_1 x_1^*$$

where $x_1^*/P(u_1)$ gives the equilibrium number of juvenile fish alive at age $z/g$ required to maintain the equilibrium adult population size of $x_1^*$ given the harvest rate $H$ and the adult size $u_1$. By substituting $x_1^*$ as given by (11.8) into the yield function and solving the integral gives

$$Y(H, z, u_1) = \frac{R}{u_1} \left[ \frac{Hz + g}{P(u_1)} - g \right] \ln \left[ \frac{s u_1 P(u_1)}{H} \right].$$

Yield increases with $R$ (a decrease in the density-dependent effect of others), an increase in the maximum rate of neonate production, $s$, and an increase in juvenile growth rate, $g$.

If $Y$ comprises the objective function of the fishery manager, there are two ways the manager can maximize yield. The manager can look at the problem from an ecological point of view (ecologically enlightened manager), who considers the consequences of harvesting on the population size of fish, but does not assume that the fish will evolve to a new adult size. Or the manager can look at the problem from an evolutionary point of view (evolutionarily enlightened manager), who considers both the ecological and evolutionary consequences of harvesting. Such a management strategy will conform to the ESOHS of Law and Grey (1989).

### 11.2.1.2 Ecologically enlightened manager

The **ecologically enlightened manager** maximizes $Y$ with respect to $H$ by considering the consequences of the harvest rate $H$ on $x_1^*$. But, the ecologically enlightened manager does not consider the consequences of harvesting on the ESS corresponding to the adult size of fish. To find $H_{\text{ECOL}}$ the partial derivative of $Y$ with respect to $H$ is set equal to zero. In taking this derivative (11.4) must be used to account for the dependence of $P$ on $H$. We obtain

$$\left.\frac{\partial Y}{\partial H}\right|_{H=H_{\text{ECOL}}}$$
$$= \{u_1 z H_{\text{ECOL}} + g u_1 - z^2 H_{\text{ECOL}}\} \left\{\ln\left[\frac{s u_1 P(u_1, H_{\text{ECOL}})}{H_{\text{ECOL}}}\right] - 1\right\} = 0.$$

The first term in brackets is always positive and so $H_{\text{ECOL}}$ must satisfy

$$\left\{\ln\left[\frac{s u_1 P(u_1, H_{\text{ECOL}})}{H_{\text{ECOL}}}\right] - 1\right\} = 0. \qquad (11.11)$$

This necessary condition cannot be solved analytically. But it can be solved numerically. And the consequences of the fish evolving can be considered by substituting $u_1 = \frac{g}{H_{\text{ECOL}}}$ into the necessary condition. With these substitutions, analysis determines the yield, the manager's harvest rate, the size of the fish, and the population size of the fish.

### 11.2.1.3 Evolutionarily enlightened manager

The **evolutionarily enlightened manager** maximizes $Y$ with respect to $H$ by including the consequences of $H$ on both $x_1^*$ and $u_1$. To find $H_{\text{EVOL}}$, $u_1 = \frac{g}{H_{\text{EVOL}}}$ is first substituted into $Y$ and then the derivative of $Y$ with respect to $H$ is set equal to zero

$$\left.\frac{\partial Y}{\partial H}\right|_{H=H_{\text{EVOL}}} = \{g z H_{\text{EVOL}} + g^2 [1 - P(u_1, H_{\text{EVOL}})] u_1 - z^2 H_{\text{EVOL}}\}$$
$$\times \left\{\ln\left(\frac{sg P(u_1, H_{\text{EVOL}})}{(H_{\text{EVOL}})^2}\right) - 1\right\}$$
$$- g z P(u_1, H_{\text{EVOL}}) H_{\text{EVOL}} - g^2 [1 - P(u_1, H_{\text{EVOL}})]$$
$$= 0.$$

This necessary condition has terms similar to those for the ecologically enlightened manager with two additional, negative terms. Substituting (11.11) into this necessary condition shows that with the harvest rate used by the ecologically enlightened manager, $H_{\text{ECOL}}$, the evolutionarily enlightened manager will

## 11.2 Resource management and conservation

perceive a decline in yield with harvesting. That is

$$-gzP(u_1, H_{ECOL})H_{ECOL} - g^2[1 - P(u_1, H_{ECOL})] < 0.$$

This result shows that the ecologically enlightened manager will always select a higher optimal harvest rate than the evolutionarily enlightened manager. The reason for this is that the ecologically enlightened manager selects for a smaller adult fish size. Using the harvest rate $H_{ECOL}$ on a pristine stock of fish will achieve a higher initial yield than $H_{EVOL}$, but this situation is not evolutionarily sustainable. As the fish begin to evolve a smaller size, the evolutionarily enlightened manager will achieve a higher evolutionarily sustainable yield with a lower harvest rate on adults with a larger body size. The manager sacrifices harvest rate in exchange for maintaining a larger ESS adult body size in the fish.

### 11.2.2 Sustainable yield

A fisheries management policy resulting in **sustainable yield** is currently in vogue. Such a policy is one that seeks to obtain a stable harvest indefinitely. Traditional analyses (by modeling or not) of fish communities address stability on an ecological time scale, ignoring the evolutionary consequences of management policies. However, given enough time, fish species will evolve in response to management policy. For example, a fish population managed using catch-release regulations will have fish of a certain size range removed from the population in larger proportion than others. Given that fish growth and size are controlled genetically, this selective pressure on the fish population will result in a continual decrease in that size range. Thus harvesting changes the parameters (e.g., fecundity, growth rate) of the system to which the sustainable yield analysis was originally applied. Furthermore, sustainable yield does not address problems that may arise from invasion by exotic species or mutants.

These issues are addressed next from an evolutionary point of view using the Schaeffer **stock recruitment** model. Management that results in an ESS protects the fish community from invasion by exotic species (at least species from within the same $G$-function) and mutants because such an evolutionarily stable community will be simultaneously ecologically and evolutionarily stable.

### 11.2.3 The Schaeffer model in an evolutionary context

The Schaeffer model (discrete logistic model subject to harvesting) is given by

$$x(t+1) = x\left[1 + \frac{r}{K}(K - x)\right] - qEx$$

where $x$ is the number of fish at time $t$, $r$ is the growth rate, $K$ is the equilibrium size of the population in the absence of exploitation (carrying capacity), $q$ is the catchability coefficient, and $E$ is the fishing effort. We put this model into an evolutionary framework, by assuming that fish length, $u_i$, is the phenotypic trait of interest and modify the model so that an individual's fitness is affected by its own length and by the length of the other species in the population to obtain

$$x_i(t+1) = x_i \left[ 1 + \frac{r}{K(u_i)} \left( K(u_i) - \sum_{j=1}^{n_s} a(u_i, u_j) x_j(t) \right) - q(u_i) E \right].$$

Given a fixed environment (food source, temperature, etc.), assume that there is a length, $\beta$, at which $K$ is maximized. Fish longer or shorter than $\beta$ will not achieve the maximum unexploited population density $K_m$. The parameter $\sigma_k$ determines by how much the population will decrease when individuals vary from $\beta$, as given by

$$K(u_i) = K_m \exp\left[ -\left( \frac{u_i - \beta}{\sigma_k} \right)^2 \right].$$

Fish of various lengths compete. The strength of the competition is a function of the individual's length and the length of all other individuals in the population. Furthermore, the strongest competition, $a_m$, occurs for fish of similar length. As fish differ in length, the strength of the competition between them decreases, and the amount by which it decreases is determined by the parameter $\sigma_a$. Thus, for competition between individuals of length $i$ and length $j$ we have

$$a(u_i, u_j) = a_m \exp\left[ -\left( \frac{u_i - u_j}{\sigma_a} \right)^2 \right].$$

In terms of a fitness generating function we have

$$G(v, \mathbf{u}, \mathbf{x}) = \frac{r}{K(v)} \left( K(v) - \sum_{j=1}^{n_s} a(v, u_j) x_j(t) \right) - q(v) E$$

with

$$K(v) = K_m \exp\left[ -\left( \frac{v - \beta_K}{\sigma_K} \right)^2 \right]$$

and

$$a(v, u_j) = a_m \exp\left[ -\left( \frac{v - u_j}{\sigma_a} \right)^2 \right].$$

## 11.2 Resource management and conservation

**Example 11.2.1** (**no harvest** ($E = 0$)) *We start by examining the ESS of the fish when there is no harvesting. To flavor the no-harvest analysis with reality, the following parameter values applicable to the Namibian hake fishery are used (Hilborn and Mangel, 1997)*

| $K_m = 3000$ | $a_m = 1$ |
|---|---|
| $\beta_K = 10$ | $\sigma_a = 4$ |
| $\sigma_K = 2$ | $r = 0.39$ |

(11.12)

*Assuming an ESS coalition of one, we determine from the ESS maximum principle that $u_1 = 10$, and $x_1^* = 3000$. At these values, $G\left(v, u_1, x_1^*\right)$ as a function of $v$ (the adaptive landscape) is maximized with a zero maximum value as illustrated in the first panel of Figure 11.1. Simulation demonstrates convergent stability and we conclude that $u_1 = 10$ is an ESS with $x_1^* = 3000$.*

We now examine the effects of size-restricted harvesting.

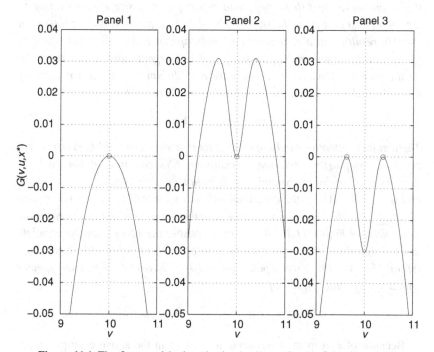

**Figure 11.1** The first panel is the adaptive landscape for the Schaeffer model with no harvest ($E = 0$). The second and third panels illustrate how the adaptive landscape changes with size-restricted harvesting both before and after speciation.

**Example 11.2.2 (size-restricted harvest)** *This policy removes a "window" of fish lengths from the population according to*

$$q(v) = q_m \exp\left[-\left(\frac{v - \beta_q}{\sigma_q}\right)^2\right]. \quad (11.13)$$

*By removing fish of size $\beta_q$ catchability is maximized. The variation in removing fish of a different size is determined by the parameter $\sigma_q$. We maintain the parameter values given by (11.12) and add the following additional harvesting parameters*

| $E = 100$ | $\beta_q = 10$ |
|---|---|
| $q_m = 0.00045$ | $\sigma_q = 0.25$ |

(11.14)

*Here the catchability $q_m$ is obtained from Hilborn and Mangel (1997). Using these parameters with a single species, Darwinian dynamics produce the results shown in the second panel of Figure 11.1. The initial conditions used to obtain this result were the fish strategy and fish population size corresponding to the no-harvest example. The size-restricted harvest function is most intense at $v = 10$, resulting in a change in the landscape such that the species is now located at an evolutionarily unstable, but convergent stable local minimum on the adaptive landscape. In this case equilibrium conditions are given by $u_1 = 10$, and $x_1^* = 2654$, giving a steady-state harvest of*

$$q(u_1) E x_1^* = q_m E x_1^*.$$

*The resultant adaptive landscape suggests that size-restricted harvesting will result in sympatric speciation. Seeking an ESS coalition of two, we obtain, using Darwinian dynamics with initial conditions $\mathbf{x} = \begin{bmatrix} 1327 & 1327 \end{bmatrix}$ and $\mathbf{u} = \begin{bmatrix} 9.9 & 10.1 \end{bmatrix}$, the candidate solution $u_1 = 9.621$, $u_2 = 10.379$, with $x_1^* = x_2^* = 1456$. The corresponding adaptive landscape is illustrated in the third panel of Figure 11.1. This solution satisfies the ESS maximum principle and through simulation can be shown to be convergent stable and we conclude it is an ESS. With this harvest policy ($u_1 = 10$) we have created two new species with a steady-state harvest of*

$$q(u_1) E x_1^* + q(u_2) E x_2^* = 0.201 q_m E x_1^*.$$

Because of a drop in steady-state harvest as in the above example, management policy will usually change as population characteristics change. Thus, one can expect that the effort $E$ or size-restricted harvesting policy may be changed in order to try to track the evolving fish population. This could introduce instability in the population. This (evolutionary, not necessarily ecological) instability can facilitate invasion by exotic species. Thus, even if a fisheries management policy based on some mathematical non-evolutionary model seems to produce a stable "sustainable yield" this need not be the case. This situation speaks to the need for evolutionarily enlightened management rather than just ecologically enlightened management.

## 11.3 Chemotherapy-driven evolution

Application of chemotherapy for the control of cancer is somewhat analogous to harvesting. As an illustration of the likely outcome of applying chemotherapy to cancer, we work with the familiar Lotka–Volterra model introduced in Example 4.3.1. The main idea is that cancer cells can and do evolve (see Subsection 10.2.2) and that normal cells do not. Hence, unless all cancer cells are destroyed during a chemotherapy session, they will ultimately recover to possibly a more virulent form by the process that could be called chemotherapy-driven evolution (Gatenby and Vincent, 2003a).

Assume differential equation dynamics for the cell population $x_i$

$$\dot{x}_i = x_i H_i(\mathbf{u}, \mathbf{x}) = x_i \, G(v, \mathbf{u}, \mathbf{x})|_{v=u_i}. \tag{11.15}$$

The normal cells are designated by $x_1$ and mutant or cancer cells by $x_2, \ldots, x_{n_s}$ with the $G$-function given by the Lotka–Volterra model

$$G(v, \mathbf{u}, \mathbf{x}) = r - \frac{r}{K(v)} \sum_{j=1}^{n_s} \alpha(v, u_j) x_j$$

where $r$ is intrinsic growth rate common to all cells, $K(v)$ is carrying capacity, and $\alpha(v, u_j)$ determines the competitive effect from using different strategies. It is assumed that cancer cells evolve according to

$$\dot{u}_i = \sigma_i^2 \left. \frac{\partial G(v, \mathbf{u}, \mathbf{x})}{\partial v} \right|_{v=u_i} \quad i = 2, \ldots, n_s. \tag{11.16}$$

As in Example 4.3.1, we assume the following distribution functions for $K$ and $\alpha$

$$K(v) = k_m \exp\left[-\frac{v^2}{2\sigma_k^2}\right]$$

$$\alpha(v, u_j) = 1 + \exp\left[-\frac{(v - u_j + \beta)^2}{2\sigma_\alpha^2}\right] - \exp\left[-\frac{\beta^2}{2\sigma_\alpha^2}\right]$$

where the $\sigma_k^2$ and $\sigma_\alpha^2$ variables are variances in the distribution functions. By varying the environmental parameter $\sigma_k^2$ associated with the carrying capacity of a given cell type, the dynamical system can have equilibrium solutions composed of one or more cell types. That is, given a constant strategy vector **u** there exists at least one equilibrium solution for $x^*$ such that not every component of $x^*$ is zero. The corresponding strategies are those that can co-exist in the population of cells.

Using the parameter values of Example 6.2.1

$$r = 0.25 \quad k_m = 100 \quad \sigma_k^2 = 4 \quad \sigma_\alpha^2 = 4 \quad \beta = 2, \tag{11.17}$$

there is only one non-zero equilibrium solution to the system (11.15). Choosing different strategies will result in different equilibrium values. For example, $u_1 = 0$ results in $x_1^* = 100.0$. However, this solution is not evolutionarily stable, and as we showed in Example 6.2.1 an ESS coalition of one exists with $u_1^* = 1.213$ and $x_1^* = 83.2$. Viewing this as the strategy employed by normal cells, the normal cell populations will remain evolutionarily stable so long as environmental conditions do not change parameter values.

As was demonstrated in Example 6.2.1, if $\sigma_k^2$ is increased to $\sigma_k^2 = 12.5$ (e.g., due to damage or changes in surrounding tissue), two equilibrium solutions to (11.15) exist. As a consequence, if we introduce a mutant cell at some strategy other than $u_2 = 1.213$, it can coexist. In fact, if we allow both the normal cells and mutant cells to evolve according to (11.15) and (11.16) we arrive at an ESS composed of two strategies (the normal cells are no longer at $u_1 = 1.213$). However, normal cells, because of their low basal mutation rate, are limited in their ability to evolve significantly within the lifetime of the host, whereas tumor cells typically possess an increased mutation rate (due to alterations in DNA repair, chromosomal stability, or a mutagenic environment) and have no such limitation and can evolve. This is incorporated into the integration of (11.15) and (11.16) by setting $\sigma_1^2 = 0$ (no evolution of normal cells) and $\sigma_2^2 = 0.25$ (allowing evolution of the tumor cells).

Introducing mutant (tumor) cells in small numbers at a strategy different from the normal cell strategy with the normal cells at their carrying capacity (in the absence of tumor cells) with the initial conditions

$$x_1(0) = 83.2, \ x_2(0) = 5, \ u_1(0) = 1.213, \ u_2(0) = 5$$

we find that, over time, the tumor cells evolve into an invasive cancer. For example after 500 days ($t_f = 500$)

$$x_1(t_f) = 31.4, \ x_2(t_f) = 45.5, \ u_1(t_f) = 1.213, \ u_2(t_f) = 3.516.$$

Figure 11.2 illustrates that, at this time, the cancer cells are near equilibrium (zero fitness) and that they are at a local maximum on an adaptive fitness landscape. This maximum represents a point of local evolutionary stability for the cancer cells. At this point it is impossible for other cells with similar strategy

## 11.3 Chemotherapy-driven evolution

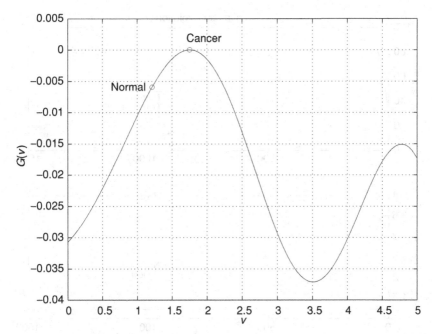

**Figure 11.2** Before treatment, the cancer cells are at a local maximum on the adaptive landscape.

values to invade. The normal cells have a slight negative fitness and with time they would completely die out.

We model the treatment of a cancer population, at this point in time, by using cell-specific drugs to eliminate the cancer by adding an appropriate "harvesting" term so that the $G$-function becomes

$$G(v, u, \mathbf{x}) = R - \frac{R}{k(u)} \sum_{j=1}^{n} \alpha(v, u_j) x_j - k_h \exp\left[-0.5 \left(\frac{v - \bar{u}}{\sigma_h}\right)^2\right] \quad (11.18)$$

where $k_h$ is a term expressing the level of drug dosage, $\bar{u}$ is the cancer cell strategy at which the drug is most effective and $\sigma_h^2$ is the variance in effectiveness. The following values are used

$$k_h = 0.1, \ \bar{u} = 3.516, \ \sigma_h^2 = 1.$$

Starting with the final conditions listed above and integrating (11.15) and (11.16) with the $G$-function defined by (11.18), it is found that cytotoxic chemotherapy is effective initially as shown in the first panel of Figure 11.3. At first, there is an apparent recovery of the normal cells. However, the cancer cells ultimately take over because they can evolve to a new equilibrium state

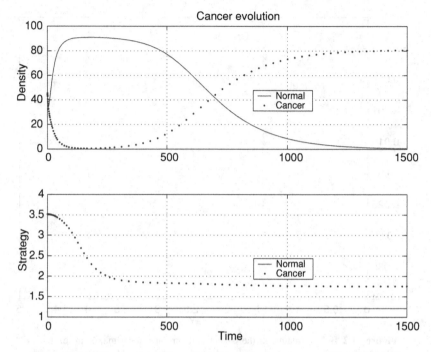

**Figure 11.3** During treatment the cancer cells evolve to a new, more deadly strategy.

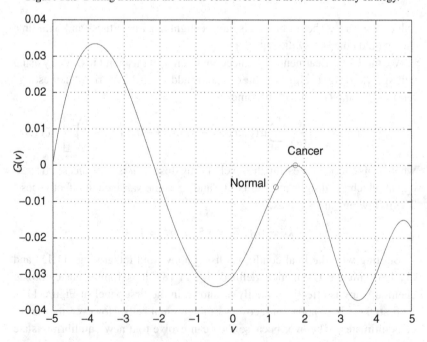

**Figure 11.4** After treatment, the cancer cells are again at a local maximum on the adaptive landscape.

## 11.3 Chemotherapy-driven evolution

illustrated in the second panel of Figure 11.3. Rather than curing the cancer, the cell-specific drug causes the cancer to evolve to a new form that is now highly resistant to the current and any similar therapeutic procedure. This is illustrated in Figure 11.4 by the fact that the cancer cells are again sitting at a local maximum. Note that the normal cells are at a fitness less than zero resulting in a rapid decrease to a zero equilibrium population. These results are essentially identical to evolution of multi-drug resistance observed in treated human tumors (Matsumoto *et al.*, 1997; Ichihashi and Kitajima, 2001).

Cancer cells, with their ability to evolve, adapt to chemotherapy and a resistant population readily emerges. Ultimately, the cancer regrows as the resistant population proliferates, rendering the therapy ineffective. In other words, any therapy relying solely on tumor cytotoxic effects will be curative only if it is sufficiently effective to kill all the cancer cells in a time period sufficiently short to prevent evolution of resistance. Thus, while a therapeutic strategy that relies solely on killing cancer cells may be effective in reducing the number of cancer cells, this model suggests that cancer will inevitably rebound unless all such cells within the population are eliminated.

# References

Abrams, P. A. (1987), "Alternative models of character displacement: I. Displacement when there is competition for nutritionally essential resources", *Evolution* **41**, 651–661.

(2001a), "Adaptive dynamics: neither F nor G", *Evolutionary Ecology Research* **3**, 369–373.

(2001b), "Modelling the adaptive dynamics of traits involved in inter- and intraspecific interactions: an assessment of three methods", *Ecology Letters* **4**, 166–175.

(2003), "Can adaptive evolution or behavior lead to diversification of traits determining a trade-off between foraging gain and predation risk?", *Evolutionary Ecology Research* **5**, 653–670.

Abrams, P. A. and Harada, H. (1996), "Fitness minimization and dynamic instability as a consequence of predator prey coevolution", *Evolutionary Ecology* **10**, 167–186.

Abrams, P. A., Harada, Y. and Matsuda, H. (1993a), "On the relationships between quantitative genetics and ESS", *Evolution* **47**, 982–985.

Abrams, P. A., Matsuda, H. and Harada, Y. (1993b), "Evolutionary unstable fitness maxima and stable fitness minima of continuous traits", *Evolutionary Ecology* **7**, 465–487.

Abramsky, Z., Rosenzweig, M. L., Pinshow, B. P., Brown, J. S., Kotler, B. P. and Mitchell, W. A. (1990), "Habitat selection: an experimental field test with two gerbil species", *Ecology* **71**, 2357–2358.

Altendor, K. B. (1997), F assessing the impact to predation risk by mountain lions (*Puma concolor*) on the foraging behavior of mule deer (*Odocoileus hemionus*) M.Sc. thesis. Pocatello: Idaho State University.

Anderson, C., Boomsma, J. J. and Bartholdi, J. (2002), "Task partitioning in insect societies: bucket brigades", *Insectes Sociaux* **49**, 171–180.

Apaloo, J. (1997), "Revisiting strategic models of evolution: the concept of neighborhood invader strategies", *Theoretical Population Biology* **52**, 71–77.

Ashley, M. V., Willson, M. F., Pergams, O. R. W., O'Dowd, D. J., Gende, S. M. and Brown, J. S. (2003), "Evolutionarily enlightened management", *Biological Conservation* **111**, 115–123.

Auslander, D. J., Guckenheimer, J. M. and Oster, G. (1978), "Random evolutionary stable strategies", *Theoretical Population Biology* **13**, 276–293.

Balaan, M. van and Sabelis, M. W. (1993), "Coevolution of patch strategies of predator and prey and the consequences for ecological stability", *American Naturalist* **142**, 646–670.
Ben-Natan, G., Abramsky, Z., Kotler, B. P. and Brown, J. S. (2004), "Seed redistribution in sand dunes: a basis for coexistence of two rodent species", *Oikos* **105**, 325–335.
Bernstein, H., Byerly, H. C., Hopf, F. A., Michod, R. E. and Vemulapalli, G. K. (1983), "The Darwinian dynamic", *Quarterly Review of Biology* **58**, 185–207.
Bishop, D. T. and Cannings, C. (1976), "Models of animal conflict", *Advanced Applied Probability* **8**, 616–621.
Bloom, A. J., Chapin, F. S. and Mooney, H. A. (1985), "Resource limitation in plants – an economic analogy", *Annual Review of Ecology and Systematics* **16**, 363–392.
Blythe, S. P. and Stokes, T. K. (1990), "Some consequences of size-selective harvesting on fitness and on yield", *IMA Journal of Mathematics Applied in Medicine and Biology* **7**, 41–53.
Borisov, V. M. (1978), "The selective effect of fishing on the population structure of species with a long life cycle", *Journal of Ichthyology* **18**, 896–904.
Brooks, D. R. and McLennan, D. A. (1991), *Phylogeny, Ecology, and Behavior: A Research Program in Comparative Biology*. Chicago, Ill.: University of Chicago.
Brown, J. S. (1988), "Patch use as an indicator of habitat preference, predation risk, and competition", *Behavioral Ecology and Sociobiology* **22**, 37–47.
  (1990), "Habitat selection as an evolutionary game", *Evolution* **44**, 732–746.
  (1992), "Patch use under predation risk: I. Models and predictions", *Annales Zoologici Fennici* **29**, 301–309.
  (1998), "Game theory and habitat selection", in L. A. Dugatkin and H. K. Reeve (eds.) *Game Theory and Animal Behavior*. Oxford, England: Oxford University Press, pp. 188–220.
  (1999), "Vigilance, patch use and habitat selection: foraging under predation risk", *Evolutionary Ecology Research* **1**, 49–71.
Brown, J. S. and Parman, A. O. (1993), "Consequences of size-selective harvesting as an evolutionary game", in T. K. Stokes, J. M. MacGlade and R. Law (eds.) The exploitation of evolving resources. Vol. 99 of *Lecture Notes in Biomathematics*. Berlin: Springer-Verlag, pp. 248–261.
Brown, J. S. and Pavlovic, N. B. (1992), "Evolution in heterogeneous environments: effects of migration on habitat specialization", *Evolutionary Ecology* **6**, 360–382.
Brown, J. S. and Vincent, T. L. (1987a), "Coevolution as an evolutionary game", *Evolution* **41**, 66–79.
Brown, J. S. and Vincent, T. L. (1987b), *Predator-Prey Coevolution as an Evolutionary Game*, Vol. 73 of *Lecture Notes in Biomathematics*. Heidelberg: Springer-Verlag.
Brown, J. S. and Vincent, T. L. (1987c), "A theory for the evolutionary game", *Theoretical Population Biology* **31**, 140–166.
Brown, J. S. and Vincent, T. L. (1992), "Organization of predator-prey communities as an evolutionary game", *Evolution* **46**, 1269–1283.
Brown, J. S., Kotler, B. P. and Bouskila, A. (2001), "Ecology of fear: foraging games between predators and prey with pulsed resources", *Annales Zoologici Fennici* **38**, 71–87.
Brown, J. S., Cohen, Y. and Vincent, T. L. (2005), "Adaptive dynamics with vector-valued strategies", Submitted to *Evolutionary Ecological Research*.

Caswell, H. (1989), *Matrix Population Models: Construction, Analysis and Interpretation*. Sunderland, Mass.: Sinauer.

Chapin, F. S. (1980), "The mineral nutrition of wild plants", *Annual Review of Ecology and Systematics* **11**, 233–260.

Charlesworth, B. (1990), "Optimization models, quantitative genetics, and mutation", *Evolution* **44**, 520–538.

Charnov, E. L. (1976), "Optimal foraging, the marginal value theorem", *Theoretical Population Biology* **9**, 129–136.

Christiansen, F. B. (1991), "On conditions for evolutionary stability for a continuously varying character", *American Naturalist* **138**, 37–50.

Cohen, D. (1971), "Maximizing final yield when growth is limited by time or by limiting resources", *Journal of Theoretical Biology* **33**, 299–307.

Cohen, Y., Vincent, T. L. and Brown, J. S. (1999), "A G-function approach to fitness minima, fitness maxima, evolutionary stable strategies and adaptive landscapes", *Evolutionary Ecology Research* **1**, 923–942.

(2001), "Does the G-function deserve an f?", *Evolutionary Ecology Research* **3**, 375–377.

Cressman, R. (1992), *The Stability Concept of Evolutionary Game Theory (a Dynamical Approach)*, Vol. 94 of *Lecture Notes in Biomathematics*. Berlin: Springer-Verlag.

(2003), *Evolutionary Dynamics and Extensive Form Games*. Cambridge, Mass.: MIT Press.

Crow, J. F. and Kimura, M. (1970), *An Introduction to Population Genetics*. New York: Harper and Row.

Darwin, C. (1859), *On the Origin of Species by Means of Natural Selection, or The Preservation of Favoured Races in the Struggle for Life*. London: Murray.

Dawkins, R. (1976), *The Selfish Gene*. Oxford: Oxford University Press.

(1986), *The Blind Watchmaker*. Essex: Longman Scientific.

Deevey, E. S. (1947), "Life tables for natural populations of animals", *Quarterly Review of Biology* **22**, 283–314.

del Solar, E. (1966), "Sexual isolation caused by selection for positive and negative phototaxis and geotaxis in *Drosophila pseudobscura*", *Proceedings of the National Academy of Science, USA* **56**, 484–487.

Dennett, D. C. (1995), *Darwin's Dangerous Idea: Evolution and the Meanings of Life*. New York: Simon & Schuster.

Dieckmann, U. and Doebeli, M. (1999), "On the origin of species by sympatric speciation", *Nature* **400**, 354–357.

Dieckmann, U. and Law, R. (1996), "The dynamical theory of coevolution: a derivation from stochastic ecological processes", *Journal of Mathematical Biology* **34**, 579–613.

Dieckmann, U., Marrow, P. and Law, R. (1995), "Evolutionary cycling in predator-prey interactions: populations dynamics and the red queen", *Journal of Theoretical Biology* **176**, 91–102.

Dobzhansky, T. G. (1937), *Genetics and the Origin of Species*. New York: Columbia University Press.

Doebeli, M. and Dieckmann, U. (2000), "Evolutionary branching and sympatric speciation caused by different types of ecological interactions", *American Naturalist* **103**, 99–111.

Drew, M. C. and Saker, L. R. (1975), "Nutrient supply and the growth of seminal root system in barley: II. Localized compensatory increase in lateral root growth and rates of nitrate uptake when nitrate supply is restricted to only one part of the root system", *Journal of Experimental Botany* **26**, 79–90.

Eiseley, L. (1958), *Darwin's Century*. New York: Anchor Books.

Eldridge, N. and Gould, S. J. (1972), "Punctuated equilibria: an alternative view to phyletic gradualism", in T. J. Schopf (ed.), *Models in Paleobiology*. San Francisco, Calif.: Freeman, Cooper, pp. 82–115.

Endler, J. A. (1986), *Natural Selection in the Wild*. Princeton, N. J.: Princeton University Press.

Eshel, I. (1983), "Evolutionary and continuous stability", *Journal of Theoretical Biology* **103**, 99–111.

(1996), "On the changing concept of population stability as a reflection of a changing point of view in the quantitative theory of evolution", *Journal of Mathematical Biology* **34**, 485–510.

Eshel, I. and Motro, U. (1981), "Kin selection and strong evolutionary stability of mutual help", *Theoretical Population Biology* **19**, 420–433.

Falconer, D. S. (1960), *Introduction to Quantitative Genetics*. New York: Ronald Press.

Feder, J. L., Chilcote, C. A. and Bush, G. L. (1988), "Genetic differentiation and sympatric host races of the apple maggot fly, *Rhagoletis pomonella*", *Nature* **336**, 61–64.

Fisher, R. A. (1930), *The Genetical Theory of Natural Selection*. Oxford: Clarendon Press.

Fitter, A. H., Strickland, T. R., Harvey, M. L. and Wilson, G. W. (1991), "Architectural analysis of plant root systems: I. Architectural correlates of exploitation efficiency", *New Phytologist* **119**, 383–389.

Freeman, S. and Herron, J. C. (2004), *Evolutionary Analysis*, 3rd edition. Upper Saddle River, N.J.: Pearson Education, Inc.

Fretwell, S. D. (1972), *Populations in a Seasonal Environment*. Princeton, N.J.: Princeton University Press.

Fretwell, S. D. and Lucas, H. L. (1970), "On territorial behavior and other factors influencing habitat distributions in birds: I. theoretical developments", *Acta Biotheoretica* **19**, 16–36.

Frieden, B. R. (1998), *Physics from Fisher Information*. Cambridge: Cambridge University Press.

Gadgil, S. M. and Gadgil, M. (1975), "Can a single resource support many consumer species?", *Journal of Genetics* **62**, 33–47.

Garrod, G. J. (1988), *North Atlantic Cod: Fisheries and Management to 1986*, 2nd edition. London: Wiley, pp. 185–218.

Gasue, G. F. (1934), *The Struggle for Existence*. Baltimore, Md.: Williams and Wilkins.

Gatenby, R. A. (1991), "Population ecology issues in tumor growth", *Cancer Research* **51**, 2542–2547.

Gatenby, R. A. and Vincent, T. L. (2003a), "Application of quantitative models from population biology and evolutionary game theory to tumor therapeutic strategies", *Molecular Cancer Therapeutics* **2**, 919–927.

(2003b), "An evolutionary model of carcinogenesis", *Cancer Research* **63**, 6212–6220.

Geritz, S. A. H. (1998), "Co-evolution of seed size and seed predation", *Evolutionary Ecology* **12**, 891–911.
Geritz, S. A. H., Metz, J. A. J., Kisdi, E. and Meszena, G. (1997), "The dynamics of adaptation and evolutionary branching", *Physical Review Letters* **78**, 2024–2027.
Geritz, S. A. H., Kisdi, E., Meszena, G. and Metz, J. A. J. (1998), "Evolutionarily singular strategies and the adaptive growth and branching of the evolutionary tree", *Evolutionary Ecology* **12**, 35–57.
Gersani, M., Brown, J. S., O'Brien, E., Maina, G. and Abramsky, Z. (2001), "Tragedy of the commons as a result of root competition", *Journal of Ecology* **89**, 660–669.
Getz, W. M. (1999), *Population and Evolutionary Dynamics of Consumer–Resource Systems*. Oxford: Blackwell.
Goh, B. S. (1980), *Management and Analysis of Biological Populations*. New York: Elsevier.
Goldschmidt, R. B. (1940), *The Material Basis of Evolution*. New Haven, Connt.: Yale University Press.
Gomulkiewicz, R. and Holt, R. D. (1995), "When does evolution by natural selection prevent extinction?", *Evolution* **49**, 201–207.
Gould, S. J. (1998), *Wonderful Life: The Burgess Shale and the Nature of History*. New York: Replica Books.
Gould, S. J. and Lewontin, R. C. (1979), "The spandrels of San Marco and the panglossian paradigm: a critique of the adaptationist programme", *Proceedings of the Royal Society of London, Series B* **205**, 581–598.
Grafen, A. (1979), "The hawk–dove game played between relatives", *Animal Behavior* **27**, 905–907.
Grant, P. R. and Grant, B. R. (2002), "Unpredictable evolution in a 30-year study of Darwin's finches", *Science* **296**, 707–711.
Grantham, W. J. and Vincent, T. L. (1993), *Modern Control Systems Analysis and Design*. New York: Wiley.
Grime, J. P. (1977), "Evidence for the existence of three primary strategies in plants and its relevance to ecological and evolutionary theory", *American Naturalist* **111**, 1169–1194.
Haldane, J. B. S. (1932), *The Causes of Evolution*. London: Harper and Row.
Hamilton, W. D. (1963), "The evolution of altruistic behavior", *American Naturalist* **97**, 354–356.
  (1964), "The genetical evolution of social behviour, I and II", *Journal of Theoretical Biology* **7**, 1–52.
Hammerstein, P. (1996), "Darwinian adaptation, population genetics and the streetcar theory of evolution", *Journal of Mathematical Biology* **34**, 511–532.
Hanski, I. (1991), *Metapopulation Ecology*, Oxford: Oxford University Press.
Hardin, G. (1968), "The tragedy of the commons", *Science* **162**, 1243–1248.
Hardy, G. H. (1908), "Mendelian proportions in a mixed population", *Science* **28**, 49–50.
Hatanaka, M. (1974), "Transport of sugars in tumor cell membranes", *Biochemica et Biophysica Acta* **355**, 77–104.
Heino, M., Metz, J. A. J. and Kaitala, V. (1998), "The enigma of frequency-dependent selection", *Trends in Ecology and Evolution* **13**, 367–370.

Hilborn, R. and Mangel, M. (1997), *The Ecological Detective*. Princeton, N.J.: Princeton University Press.
Hines, W. G. S. (1987), "Evolutionary stable strategies", *Theoretical Population Biology* **31**, 195–272.
Hines, W. G. S. and Maynard Smith, J. (1979), "Games between relatives", *Journal of Theoretical Biology* **79**, 19–30.
Hofbauer, J. and Sigmund, K. (1988), *The Theory of Evolution and Dynamical Systems*. Cambridge: Cambridge University Press.
  (1998), *Evolutionary Games and Population Dynamics*. Cambridge: Cambridge University Press.
Hofer, H. and East, M. L. (2003), "Behavioral processes and costs of co-existence in female spotted hyenas: a life history perspective", *Evolutionary Ecology* **17**, 315–331.
Holling, C. S. (1965), "The functional response of predators to prey density and its role in mimicry and population regulation", *Memoirs of the Entomological Society of Canada* **45**, 1–60.
Holt, R. D. and Gaines, M. S. (1992), "Analysis of adaption in heterogenous landscapes: implication for the evolution of fundamental niches", *Evolutionary Ecology* **6**, 433–447.
Holt, R. D. and Gomulkiewicz, R. (1997), "How does immigration influence local adaptation? A re-examination of a familiar paradigm", *American Naturalist* **149**, 563–572.
Hornocker, M. G. (1970), "An analysis of mountain lion predation upon mule deer and elk in the Idaho primitive area", *Wildlife Monographs* **21**, 1–39.
Houston, A. I., McNamara, J. M. and Hutchinson, J. M. C. (1993), "General results concerning the trade-off between gaining energy and avoiding predation", *Philosophical Transactions of the Royal Society of London Series B* **341**, 375–397.
Hutchinson, G. E. (1965), *The Ecological Theater and Evolutionary Play*. New Haven, Conn.: Yale University Press.
Huxley, J. S. (1942), *Evolution: The Modern Synthesis*. London: George Allen and Unwin Ltd.
Ichihashi, M. and Kitajima, Y. (2001), "Chemotherapy induces or increases expression of multi-drug resistance-associated protein in malignant melanoma cells", *British Journal of Dermatology* **144**, 745–750.
Issacs, R. (1965), *Differential Games*. New York: Wiley.
Jackson, R. B., Manwaring, J. H. and Caldwell, M. M. (1990), "Rapid physiological adjustment of roots to localized soil enrichment", *Nature* **344**, 58–60.
Jones, C. G., Lawton, J. H. and Shachak, M. (1994), "Organisms as ecosystem engineers", *Oikos* **69**, 373–386.
Kacelnik, A., Krebs, J. R. and Bernstein, C. (1992), "The ideal free distribution and predator–prey populations", *Trends Ecology Evolution* **7**, 50–55.
Kennedy, M. and Gray, R. D. (1993), "Can ecological theory predict the distribution of foraging animals? A critical analysis of experiments on the ideal free distribution", *Oikos* **68**, 158–166.
Kisdi, E. and Geritz, S. A. H. (1999), "Adaptive dynamics in allele space: evolution of genetic polymorphism by small mutations in a heterogeneous environment", *Evolution* **53**, 993–1008.

Law, R. (2000), "Fishing, selection, and phenotypic evolution", *Journal of Marine Science* **57**, 659–668.
Law, R. and Grey, D. R. (1989), "Evolution of yields from populations with age-specific cropping", *Evolutionary Ecology* **3**, 343–359.
Lawlor, R. L. and Maynard Smith, J. (1976), "The coevolution and stability of competing species", *American Naturalist* **110**, 76–99.
Lee, R. B. (1982), "Selectivity and kinetics of ion uptake by barley plants following nutrient deficiency", *Annals of Botany* **50**, 429–449.
Lenski, R. E. and Travisano, M. (1994), "Dynamics of adaptation and diversification: A 10,000 generation experiment with bacterial populations", *Proceedings of the National Academy of Science, USA* **91**, 6808–6814.
Leslie, P. H. (1945), "On the use of matrices in certain population mathematics", *Bometrika* **33**, 183–212.
Levins, R. (1962), "Theory of fitness in a heterogeneous environment: I. The fitness set and adaptive function", *American Naturalist* **96**, 361–373.
  (1968), *Evolution in Changing Environments*. Princeton, N.J.: Princeton University Press.
Lewontin, R. C. (1961), "Evolution and the theory of games", *Journal of Theoretical Biology* **1**, 382–403.
  (1974), *The Genetic Basis of Evolutionary Change*. New York: Columbia University Press.
Lotka, A. J. (1932), *Elements of Physical Biology*. Baltimore, Md.: Williams and Wilkins.
Luce, R. D. and Raiffa, H. (1957), *Games and Decisions*. New York: J. Wiley and Sons.
Lyell, C. (1830), *Principles of Geology*, Vol. I-3, London: John Murray.
Lynch, M. (1996), *A Quantitative-Genetic Perspective on Conservation Issues*. New York: Chapman and Hall, pp. 471–501.
Lyons, S. K., Smith, F. A. and Brown, J. H. (2004), "Of mice, mastodons and men: human-mediated extinctions on for continents", *Evolutionary Ecology Research* **6**, 339–358.
MacDonald, S. E. and Chinnappa, C. C. (1989), "Population differentiation for phenotypic plasticity in the *Stellaria longipes* complex", *American Journal of Botany* **76**, 1627–1637.
Maina, G. G., Brown, J. S. and Gersani, M. (2002), "Intra-plant versus inter-plant root competition in beans: avoidance, resource matching or tragedy of the commons?", *Plant Ecology* **160**, 235–247.
Malthus, R. (1796), *Essay on Population*.
Marrow, P., Law, R. and Cannings, C. (1992), "The coevolution of population interactions: ESSs and red queen dynamics", *Proceedings of the Royal Society of London, Series B* **250**, 133–141.
Martin, P. S. (1984), *Prehistoric Overkill: The Global Model*. Tucson, Ariz.: University of Arizona Press, pp. 354–403.
Matsumoto, Y., Takano, H. and Fojo, T. (1997), "Cellular adaptation to drug exposure: evolution of the drug-resistant phenotype", *Cancer Research* **57**, 5086–5092.
May, R. M. (1973), *Stability and Complexity in Model Ecosystems*. Princeton, N.J.: Princeton University Press.
  (1976), "Simple mathematical models with very complicated dynamics", *Nature* **266**, 459–467.

Mayer, E. and Provine, W. B. (1980), *The Evolutionary Syntheses: Perspectives on the Unification of Biology.* Cambridge, Mass.: Harvard University Press.

Maynard Smith, J. (1966), "Sympatric speciation", *American Naturalist* **100**, 637–650.

  (1974), "The theory of games and the evolution of animal conflicts", *Journal of Theoretical Biology* **47**, 209–221.

  (1976), "Evolution and the theory of games", *American Scientist* **64**, 41–45.

  (1982), *Evolution and the Theory of Games.* Cambridge: Cambridge University Press.

Maynard Smith, J. and Price, G. R. (1973), "The logic of animal conflicts", *Nature* **246**, 15–18.

Mayr, E. (1970), *Populations, Species and Evolution: An Abridgment of Animal, Species and Evolution.* Cambridge, Mass.: Harvard University Press.

McNamara, J. M., Houston, A. I. and Collins, E. J. (2001), "Optimality models in behavioral biology", *SIAM Review* **43**, 413–466.

Mendel, G. J. (1866), "Experiments in plant hybridization", *Proceeedings of the Natural History Society in Brünn* **iv**, 3–47.

Meszena, G., Czibula, I. and Geritz, S. A. H. (1997), "Adaptive dynamics in a 2-patch environment: a toy model for allopatric speciation", *Journal of Biological Systems* **5**, 265–284.

Metz, J. A., Geritz, S. A. H., Meszena, G., Jacobs, F. J. A. and van Heerwaarden, J. S. (1996), *Adaptive Dynamics, a Geometrical Study of the Consequences of Near Faithful Reproduction.* Amsterdam: North-Holland, Royal Academy of Arts and Sciences.

Metz, J. A. J. and Gyllenberg, M. (2001), "How should we define fitness in structured metapopulation models? Including an application to the calculation of evolutionarily stable dispersal strategies", *Proceedings of the Royal Society of London, Series B* **268**, 499–508.

Michod, R. E. (1998), "Origin of sex for error repair – III. Selfish sex", *Theoretical Population Biology* **53**, 60–74.

  (1999), *Darwinain Dynamics: Evolutionary Transitions in Fitness and Individuals,* Princeton, N.J.: Princeton University Press.

Mirmirani, M. and Oster, G. (1978), "Competition, kin selection and evolutionary stable strategies", *Theoretical Population Biology* **13**, 304–339.

Mitchell, W. A. (2000), "Limits to species richness in a continuum of habitat heterogeneity: an ESS approach", *Evolutionary Ecology Research* **2**, 293–316.

Mitchell, W. A. and Valone, T. J. (1990), "The optimization research programme: studying adaptations by their function", *Quarterly Review of Biology* **65**, 43–52.

Morris, D. W. (1988), "Habitat-dependent population regulation and community structure", *Evolutionary Ecology* **2**, 253–269.

Morris, D. W. (1992), "Scales and costs of habitat selection in heterogeneous landscapes", *Evolutionary Ecology* **6**, 412–432.

  (1994), "Habitat matching: alternatives and implications to populations and communities", *Evolutionary Ecology* **8**, 387–406.

Murphy, G. I. (1967), "Vital statistics of the Pacific sardine (*Sardinops caerulea*) and the population consequences", *Ecology* **48**, 731–736.

Nash, J. F. (1951), "Non-cooperative games", *Annals of Mathematics* **54**, 286–295.

Nowak, M. and Sigmund, K. (2004), "Evolutionary dynamics of biological games", *Science* **303**, 793–799.

Odling-Smee, F. J., Laland, K. N. and Feldman, M. W. (2003), *Niche Construction: the Neglected Process in Evolution*. Princeton, N.J.: Princeton University Press.

Paine, R. T. (1966), "Food complexity and species diversity", *American Naturalist* **100**, 65–76.

Pareto, V. (1896), *Cours d'economie politique*. Lausanne, Switzerland: Rouge.

Parker, G. A. (1978), *Searching for Mates Behavioral Ecology: an Evolutionary Approach*, 1st edition. Oxford: Blackwell Scientific Publications.

Pearl, R. (1924), *Studies in Human Biology*. New York: Williams and Wilkins.

Pearson, K. (1904), "On a generalized theory of alternative inheritance, with special reference to Mendel's laws", *Philosophical Transactions of the Royal Society Series A* **203**, 53–86.

Pierce, G. J. and Ollason, J. G. (1987), "Eight reasons why optimal foraging theory is a complete waste of time", *Oikos* **49**, 111–118.

Portha, S., Deneubourg, J. L. and Detrain, C. (2002), "Self-organized asymmetries in ant foraging: a functional response to food type and colony needs", *Behavioral Ecology* **13**, 776–781.

Queller, D. C. (1995), "The spaniels of St. Marx and the Panglossian paradox: a critique of a rhetorical programme", *Quarterly Review of Biology* **70**, 485–489.

Rand, D. A., Wilson, H. and McGlade, J. M. (1994), "Dynamics and evolution: evolutionary stable attractors, invasion exponents and phenotype dynamics", *Philosophical Transactions of the Royal Society of London Series B – Biological Sciences* **343**(1305), 261–283.

Rees, M. and Westoby, M. (1997), "Game-theoretical evolution of seed mass in multi-species ecological models", *Oikos* **78**, 116–126.

Reeve, H. K. and Sherman, P. W. (1993), "Adaptation and goals of evolutionary research", *Quarterly Review of Biology* **68**, 1–32.

Reynolds, H. L. and Pacala, S. W. (1993), "An analytical treatment of root–shoot ratio and plant competition for soil nutrient and light", *American Naturalist* **141**, 51–70.

Rice, W. R. and Salt, G. W. (1990), "The evolution of reproductive isolation as a correlated character under sympatric conditions: experimental evidence", *Evolution* **44**, 1140–1152.

Ricker, W. E. (1981), "Changes in the average size and average age of Pacific salmon", *Canadian Journal of Fisheries and Aquatic Sciences* **38**, 1636–1656.

Riechert, S. E. and Hammerstein, P. (1983), "Game theory in the ecological context", *Annual Review of Ecology and Systematics* **14**, 377–409.

Rijnsdorp, A. D. (1993), "Fisheries as a large-scale experiment on life-history evolution – disentangling phenotypic and genetic effects in changes in maturation and reproduction of North Sea plaice", *Pleuronectes platessa L. Oecologia* **96**, 391–401.

Roberts, A. (1974), "The stablity of feasible random systems", *Nature* **250**, 607–608.

Rosenberg, A. (1985), *The Structure of Biological Science*. New York: Cambridge University Press.

Rosenzweig, M. L. (1978), "Competitive speciation", *Biological Journal of the Linnean Society* **10**, 275–289.

　(1985), *Some Theoretical Aspects of Habitat Selection*. New York: Academic Press, pp. 517–540.

　(1987a), *Community Organization from the Point of View of Habitat Selectors*. Oxford: Blackwell Scientific Publications, pp. 469–490.

(1987b), "Habitat selection as a source of biological diversity", *Evolutionary Ecology* **1**, 315–330.

(1991), "Habitat selection and population interactions", *American Naturalist* **137**, 5–28.

Rosenzweig, M. L. and MacArthur, R. H. (1963), "Graphical representation and stability of predator–prey interaction", *American Naturalist* **97**, 209–223.

Rosenzweig, M. L. and McCord, R. D. (1991), "Incumbent replacement: evidence of long-term evolutionary progress", *Paleobiology* **17**, 202–213.

Rosenzweig, M. L., Brown, J. S. and Vincent, T. L. (1987), "Red queens and ESS: the coevolution of evolutionary rates", *Evolutionary Ecology* **1**(1), 59–94.

Roughgarden, J. (1976), "Resource partitioning among competing species – a coevolutionary approach", *Theoretical Population Biology* **9**, 388–424.

(1983), *The Theory of Coevolution*. Sunderland, Mass.: Sinauer, pp. 383–403.

(1987), "Community coevolution: a comment", *Evolution* **41**, 1130–1134.

Rummel, J. D. and Roughgarden, J. (1985), "A theory of faunal buildup for competition communities", *Evolution* **39**, 1009–1033.

Ruse, M. (1979), *The Darwinian Revolution*. Chicago, Ill.: Chicago University Press.

Samuelson, L. (1997), *Evolutionary Games and Equilibrium Selection*. Cambridge, Mass.: MIT Press.

Schaffer, W. M. (1977), "Some observations on the evolution of reproductive rate and competitive ability in flowering plants", *Theoretical Population Biology* **11**, 90–104.

Schluter, D. (2000), "Ecological character displacement in adaptive radiation", *American Naturalist* **156**, S4–S16.

Schmidt, K. A., Earnhardt, J. M., Brown, J. S. and Holt, R. D. (2000), "Habitat selection under temporal heterogeneity: sinking the ghost of competition past", *Ecology* **81**, 2622–2630.

Schmidt-Nielson, K. (1979), *Desert Animals: Physiological Problems of Heat and Water*. New York: Dover Press.

Schoombie, S. W. and Getz, W. M. (1998), "Evolutionary stable strategies and tradeoffs in generalized Beverton and Holt models", *Theoretical Population Biology* **53**, 216–235.

Schroeder, G. L. (1997), *The Science of God*. New York: Broadway Books.

Simaan, M. and Cruz, J. B. (1973), "On the stackelberg strategy in nonzero-sum games", *Journal Optimization Theory and Applications* **11**, No. 5.

Slobodkin, L. B. (2001), "The good, the bad and the reified", *Evolutionary Ecology Research* **3**, 1–13.

Smith, C. C. and Fretwell, S. D. (1974), "The optimal balance between size and number of offspring", *American Naturalist* **108**, 499–506.

Sober, E. (1984), *The Nature of Selection: Evolutionary Theory in Philosophical Focus*. Cambridge, Mass.: MIT Press.

Speiss, E. B. (1977), *Genes in Populations*. New York: John Wiley and Sons.

Stanely, S. M. (1979), *Macroevolution: Pattern and Process*. San Francisco, Calif.: W. H. Freeman and Co.

Stearns, S. C. (1992), *The Evolution of Life Histories*. Oxford: Oxford University Press.

Takada, T. and Kigami, J. (1991), "The dynamical attainability of ESSs in evolutionary games", *Journal of Mathematical Biology* **29**, 513–529.

Taper, M. L. and Case, T. J. (1985), "Quantitative genetic models for the coevolution of character displacement", *Ecology* **66**, 335–371.

(1992), "Models of character displacement and the theoretical robustness to taxon cycles", *Evolution* **46**, 317–333.

Taylor, P. D. (1989), "Evolutionary stability in one-parameter models under weak selection", *Theoretical Population Biology* **36**, 125–143.

(1997), "Evolutionary stability under the replicator and the gradient dynamics", *Evolutionary Ecology* **11**, 579–590.

Taylor, P. D. and Jonker, L. B. (1978), "Evolutionarily stable strategies and game dynamics", *Mathematical Biosciences* **40**, 145–156.

Thoday, J. M. and Gibson, J. B. (1962), "Isolation by disruptive selection", *Nature* **193**, 1164–1166.

Tilman, D. (1982), *Resource Competition and Community Structure*, Princeton, N.J.: Princeton University Press.

(1988), *Plant Strategies and the Dynamics and Structure of Plant Communities*, Monographs in Population Biology. Princeton, N.J.: Princeton University Press.

Trexler, J. C. and Travis, J. (1990), "Phenotypic plasticity in the sailfin molly *Poecilia latipinna* (Pisces: Poeciliidae). I. Field experiments", *Evolution* **44**, 143–156.

Trexler, J. C., Travis, J. and Trexler, M. (1990), "Phenotypic plasticity in the sailfin molly, *poecilia latipinna* (Pisces: Poeciliidae): II. Laboratory experiment", *Evolution* **44**, 157–167.

Turchin, P. (2001), "Does population ecology have general laws?", *Oikos* **94**, 17–26.

Tuttle, M. D. and Ryan (1981), "Bat predation and the evolution of frog vocalizations in the neotropics.", *Science* **214**, 677–678.

Verhulst, P. F. (1844), "Recherches mathématiques sur la loi d'accroissement de la population", *Mémoires de l'Académie Royale de Bruxelles* **43**, 1–58.

Vermeij, G. J. (1994), "The evolutionary interaction among species: selection, escalation, and coevolution", *Annual Reviews of Ecology and Systematics* **25**, 219–236.

Vincent, T. L. (1979), "Yield for annual plants as an adaptive response", *Rocky Mountain Journal of Mathematics* **9**(1), 163–173.

(1985), "Evolutionary games", *Journal of Optimization Theory and its Applications* **46**(4), 605–612.

(1994), "An evolutionary game theory for differential equation models with reference to ecosystem management", in T. Bazar and A. Havrie (eds.) *Advances in Dynamic Games and Applications*, Boston, Mass.: Birkhaüser, pp. 356–374.

Vincent, T. L. and Brown, J. S. (1984*a*), *The Effects of Competition on Flowering Time of Annual Plants*, Vol. 54 of *Lecture Notes in Biomathematics*. Heidelberg: Springer-Verlag.

(1984*b*), "Stability in an evolutionary game", *Theoretical Population Biology* **26**, 408–427.

(1987*a*), "Evolution under nonequilibrium dynamics", *Mathematical Modelling* **8**, 766–771.

(1987*b*), *An Evolutionary Response to Harvesting*. Heidelberg: Springer-Verlag, pp. 80–97.

(1988), "The evolution of ESS theory", *Annual Review of Ecology and Systematics* **19**, 423–443.

(2001), "Evolutionary stable strategies in multistage biological systems", *Selection* **2**, 85–102.
Vincent, T. L. and Cressman, R. (2000), "An ESS maximum principle for matrix games", *Theoretical Population Biology* **58**, 173–186.
Vincent, T. L. and Fisher, M. E. (1988), "Evolutionarily stable strategies in differential and difference equation models", *Evolutionary Ecology* **2**, 321–337.
Vincent, T. L. and Grantham, W. J. (1981), *Optimality in Parametric Systems*. New York: Wiley.
 (1997), *Nonlinear and Optimal Control Systems*. New York: Wiley.
Vincent, T. L. and Skowronski, J. M., eds. (1981), *Renewable Resource Management*. Vol. 72 in *Lecture Notes in Biomathematics*. Heidelberg: Springer-Verlag.
Vincent, T. L. and Vincent, T. L. S. (2000), "Evolution and control system design", *IEEE Control System Magazine* **20**(5), 20–35.
Vincent, T. L., Cohen, Y., Grantham, W. J., Kirkwood, G. P. and Skowronski., J. M., eds. (1987), *Modeling and Management of Resources under Uncertainty*, Vol 72 of *Lecture Notes in Biomathematics*. Heidelberg: Springer-Verlag.
Vincent, T. L., Cohen, Y. and Brown, J. S. (1993), "Evolution via strategy dynamics", *Theoretical Population Biology* **44**, 149–176.
Vincent, T. L., Van, M. V. and Goh, B. S. (1996), "Ecological stability, evolutionary stability and the ESS maximum principle", *Evolutionary Ecology* **10**, 567–591.
Vincent, T. L. S. and Vincent, T. L. (1996), "Using the ESS maximum principle to explore root–shoot allocation, competition and coexistence", *Journal of Theoretical Biology* **180**, 111–120.
Volterra, V. (1926), "Variazioni e fluttuazioni del numero d'individui in specie animali conviventi", *Memorie Rendiconti Acccademia Nazionale dei Lincei, Roma Seriale VI* **2**, 31–113.
Von Neumann, J. and Morgenstern, O. (1944), *Theory of Games and Economic Behavior*. Princeton, N.J.: Princeton University Press.
von Stackelberg, H. (1952), *Theory of Games and Economic Behavior*. Oxford: Oxford University Press.
Weibull, J. W. (1997), *Evolutionary Game Theory*. Cambridge, Mass.: MIT Press.
Weiner, J. (1994), *The Beak of the Finch*. New York: Vintage Books.
Werner, E. E. and Gilliam, J. F. (1984), "The ontogenetic niche and species iterations in size-structured populations", *Annual Review of Ecology and Systematics* **15**, 393–425.
Wright, S. (1931), "Evolution in mendelian population", *Genetics* **16**, 97–159.
 (1932), "The roles of mutation, inbreeding, crossbreeding, and selection in evolution", in D. Jones, ed., *Proceedings of the Sixth International Congress on Genetics*, Brooklyn, N.Y.: Brooklyn Botanical Gardens.
 (1960), *Physiological Genetics, Ecology of Populations, and Natural Selection*, Chicago, Ill.: University of Chicago Press.
 (1969), *Evolution and Genetics of Populations*, Vol. 2. Chicago, Ill.: University of Chicago Press.
 (1977), *Evolution and Genetics of Populations*, Vol. 3, *Experimental Results and Evolutionary Deductions*. Chicago, Ill.: University of Chicago Press.

Zeeman, E. C. (1980), "Population dynamics from game theory", in Z. Nitecki and C. Robinson, eds. *Global Theory of Dynamical Systems, Proceedings Northwestern University 1979*. Berlin: Springer-Verlag, pp. 471–497.
  (1981), "Dynamics of the evolution of animal conflicts", *Journal of Theoretical Biology* **89**, 249–270.
Ziv, Y., Abramsky, Z., Kotler, B. P. and Subach, A. (1993), "Interference competition and temporal and habitat partitioning in two gerbil species", *Oikos* **66**, 237–246.

# Index

active edge, 14, 154
adaptation, 20
   research program, 10
adaptive dynamics, 21
adaptive landscape, 109, 113
   compound, 226
   flexible, 13
   rigid, 266
adaptive radiation, 260
   Lotka–Volterra example, 262
   predator–prey example, 266
adaptive speciation, 210
advertising game, 11
allometry, 98
allopatric speciation, 251
   Lotka–Volterra example, 258
altruism, 14, 86
anisogamy, 234
archetype, 232, 237
assortative mating, 129, 231, 252
asymmetric matrix game, 66
asymptotic stability
   Leslie predator–prey continuous example, 56
   Leslie predator–prey discrete example, 54
   local, 58
   logistic equation example, 51

ball, 162
bang-bang solution, 325
battle of sexes, 66
bauplan, 2, 22, 81
   multistage, 106
   two or more, 99
   unique, 92, 93, 96, 103
Bergmann's rule, 97, 141, 173, 212

bi-linear game, 69, 279
bully function, 95

carrying capacity, 43
chaotic attractor, 60
chaotic motion, 60
characteristics
   that define a species, 232
clump of strategies
   distribution becomes bimodal, 249
   following a mean, 247
clumped strategies, 243
clumping
   of individuals, 231
co-adaptation, 21
coalition, 127
coalition vector, 164
   non-equilibrium, 191
co-evolutionary stable strategy, 159
competition coefficient, 43
competitive speciation, 159, 252
compound
   adaptive landscape, 226
   fitness example, 189
   fitness function, 189
   fitness generating function, 225
consumer-resource models, 46
continuous games, 70
continuous strategy, 68
convergent
   stability, 148, 158
   stable, 152
   stable point, 165
converting density to frequency, 80
corollary
   sufficient condition for a matrix-ESS, 282

# Index

critical value, 108, 185
  gradient of, 144
  of the G-matrix, 222
  of **H**, 109

Darwin's postulates, 5
Darwinian dynamics, 22, 27, 112
definitions
  coalition vector – frequency, 181
  coalition vector – multistage, 186
  coalition vector for the scalar bauplan, 164
  critical value, 109
  dominant eigenvalue, 108
  ecological equilibrium, 162
  ecological equilibrium – multistage, 183
  ecological equilibrium for the resource bauplan, 171
  ecologically stable equilibrium for the resource bauplan, 171
  ESC, 189
  ESE – frequency, 180
  ESE – multiple, 175
  ESE for the scalar bauplan, 162
  ESS – frequency, 181
  ESS – multistage, 186
  ESS for the scalar bauplan, 164
  fitness for multistage $G$-functions, 109
  fitness generating function, 62, 77
  fitness matrix, 108
  $G$-function, 77
  $G$-function for multistage systems, 109
  $G$-function in terms of population frequency, 105
  $G$-function with resources, 97
  $G$-function with scalar strategies, 93
  $G$-function with vector strategies, 94
  G-matrix, 109
  matrix ecological equilibrium, 280
  matrix ecologically stable equilibrium, 280
  matrix-ESS, 281
  multiple $G$-functions, 102
  single-mutant-ESS, 283
  species, 242
density-dependent selection, 79, 116
difference equations, 34
  solution by iteration, 35
differential equations
  solution by integration, 37
differential games, 334
domain of attraction, 52, 163
dominant eigenvalue, 108

ecological cycle, 188, 225
  equilibrium requirement, 190
ecological equilibrium, 162
  multiple, 174
ecological theater, 15, 75, 304
ecologically enlightened manager, 354
ecologically keystone, 264
ecologically stable, 152
  cycle (ESC), 189, 225
  equilibrium (*see also* ESE), 162
eigenvalue
  dominant, 108
epistasis, 86
equilibrium point, 50, 162
  asymptotically stable, 51, 58
  globally asymptotically stable, 51
  stable, 50, 51, 58
  unstable, 58
ESC, 189
ESE, 162, 163
  global, 163, 175
  local, 163, 175
  multiple, 175
ESS, 18, 21, 22, 151, 164
  candidate, 197
  global, 164
  local, 164
  maximum principle, 197
  non-equilibrium, 191, 225
  optimal harvesting (ESOHS), 351
ESS maximum principle, 197
evolutionarily enlightened manager, 354
evolutionarily identical, 62
  individuals, 17
evolutionarily stable
  minima, 159
  optimal harvesting strategy, 351
  strategy (*see* ESS)
evolutionary
  branching, 159
  game, 16
  keystone, 264
  play, 15, 75, 304
  strategies, 61
evolutionary stability
  convergence stability, 19
  resistant to invasion, 19
existence
  struggle for, 62
expected payoff, 69, 279
extinction
  contexts for, 233

# Index

families, 82
fast tracking, 196
first-order approximation, 124
Fisher's theorem of natural selection, 125
fit of form and function, 1
fitness, 40
  density dependent, 8
  density independent, 8
  frequency dependent, 8
  frequency independent, 8
  function, 40
  generating function, 22, 62, 77
  generating function (compound), 225
  landscape, 13, 113
  matrix, 108
  multistage $G$-functions, 109
  set, 154
fixed point, 50, 162
forming a $G$-fuction, 89
frequency, 27
  of individuals, 104
  of phenotypes, 122
  space, 180, 277
  vector, 277
frequency-dependent selection, 79

$G$-function, 22
$G$-functions, 77
  categorizing, 92
  multiple, 24, 99
  multistage, 106
  in terms of population frequency, 103
  with resources, 96
  with scalar strategies, 92
  with vector strategies, 93, 106
**G**-matrix, 109
game of chicken, 67
games
  ant wars, 71
  asymmetric matrix, 66
  battle of the sexes, 66
  bi-linear, 69
  cancer chemotherapy, 359
  continuous, 70
  ESS under hump-shaped harvest, 348
  ESS under linear harvest, 347
  ESS under no harvest, 347
  flowering time – cooperative solution, 328
  flowering time, $N > 1$, 327

flowering time, $N = 1$, 327
game of chicken, 67, 290, 293
gerbil-owl fear, 333
kin selection, 294
L–V big bully, 95, 139, 169
L–V big bully – coalition of one, 206
L–V big bully – coalition of two, 208
L–V competition, 93, 127, 131, 166
L–V competition – coalition of one, 200
L–V competition – coalition of two, 201
L–V competition in terms of frequency, 105, 143, 182, 220
life cycle, 109, 144
modified game of chicken, 295
multistage tutorial, 187
non-equilibrium L–V, 145, 192, 227
non-equilibrium L–V dependent on $x$, 229
non-equilibrium with **x** dependence, 193
offspring size vs. number, 64
predator–prey coevolution, 102, 142, 177, 215
predators seeking wary prey, 338
prisoner's dilemma, 66, 288
reciprocal altruism, 294
resource discovery, 308
resource matching, 307
resource renewal, 309
rock–scissor–paper, 290
root competition, 329
root–shoot ratio, 311, 313
root–shoot with an ESS coalition of two, 315
Schaeffer model with no harvest, 357
Schaeffer model with size-restricted harvest, 357
symmetric competition, 89
symmetric matrix, 66
war of attrition, 71
zero-sum, 63
genetic drift, 266
genetic interactions
  epistatic, 7
  pleiotropic, 7
genetics
  epistasis, 86
  pleiotropy, 86
  population, 7
  quantitative, 7
group selection, 78
group-optimal strategy, 70

habitat quality, 306
habitat selection
  density dependent, 305
  ideal free distribution, 305
  identity matrix, 32, 48
Hamilton's rule, 294
heritable variability, 13
heritable variation, 62
  invasion-driven, 117
  strategy-driven, 118

ideal free distribution, 305, 337
identity matrix, 32, 48
inclusive fitness, 14
incumbent replacement, 272
individual selection, 78
inner game, 17, 75
intrinsic growth rate, 43
invariant set, 52
isodar plot, 306

kin selection, 294

landscape
  adaptive, 113
  fitness, 113
  rigid, 266
lemmas
  ecologically stable cycle, 190
  ESC, 190
  ESE, 163
  ESE – frequency, 181
  ESE – multiple, 175
  ESE – multistage, 185
  ESE for the resource bauplan, 171
  multistage eigenvalues, 184
life
  diversity, 1
  procession, 1
life cycle example, 223
life-history stages, 106
limit cycle
  stable, 60
linearization, 52
logistic equation, 35
  continuous, 38
  discrete, 35
  discrete exponential, 36
Lotka–Volterra predator–prey model, 51
Lyapunov's first method, 52

macroevolution, 9, 233, 269
map, 35
mating
  assortative, 231
matrix, 31
  fitness, 48, 108
  identity, 32, 48
  population projection, 48
  square, 32
  transpose, 48
matrix games
  as an evoutionary game, 72
  bi-linear, 275, 279
  non-linear, 275
  symmetric, 66, 278
matrix-ESE, 280
matrix-ESS, 281
  sufficient condition, 282
max-min strategy, 69
microevolution, 9, 233, 268
mixed strategy, 68, 292
Modern Synthesis, 7
monomorphic population, 276

Nash
  equilibrium, 18
  solution, 70
natural selection, 5
  density dependent, 7
  Fisher's fundamental theorem, 13
  frequency dependent, 7
niche, 260
  construction, 30
no-regret strategy, 70
nominal operating condition, 50
non-equilibrium dynamics, 58
non-negative orthant, 161
notation example
  fitness matrix, 48
  species, strategies, and resources, 30
  transpose, 48

optimization problem, 64
organism
  distribution and abundance, 2
outer game, 17, 75

pangenesis, 4, 85
Pareto-optimal
  set, 154
  solution, 15

# Index

payoff
  bi-linear, 279
  expected, 69
periodic orbits, 59
perturbation
  equations, 147
  solutions, 52
phenotype, 121
pleiotropy, 86
polymorphic population, 276
population dynamics, 93, 112
population projection function, 40
population projection matrix, 106
positively invariant set, 52
predator–prey coevolution, 102, 142
prisoner's dilemma, 66
pristine environment, 344
procession of life, 273
punctuated equilibria, 274
pure strategy, 69

quasi-periodic orbit, 60

rational reaction set, 157
reciprocal altruism, 294
reproduction
  asexual, 117
  sexual, 117
resistance to invasion, 158
resource dynamics, 96
resources, 30
rising number of species, 99
root competition, 329

scalar, 28
Schaeffer model, 355
sector stability, 162
selection
  density-dependent, 116
social systems
  despotic, 73
  eusocial, 73
speciation, 128
  adaptive, 210
  adaptive radiation, 260
  allopatric, 251
  competitive, 159, 252
  sympatric, 251, 252
species, 1, 27, 121, 242
  archetype, 237
  biological species concept, 121, 234

ecologically keystone, 264
evolutionarily keystone, 264
morphological species concept, 121, 235
phylogenetic species concept, 235
strategy species concept, 121
strategy-species definition, 28
species archetype, 237
speed, 126
stable equilibrium point, 50
stability, 50
  convergent, 148, 152
  ecological, 49, 152
  evolutionary, 49
  global, 53
  linear systems, 56
  local, 53
  Lotka–Volterra predator–prey example, 51
  periodic orbits, 59
state, 33
  perturbation equations, 52
  variables, 33
state-space notation, 33
  difference equations, 33
  differential equations, 34
stock recruitment, 355
strategies, 27, 73
  as heritable phenotypes, 8
  concatenation, 94
  continuous, 68
  evolutionarily stable, 151
  fixed, 73
  group-optimal, 70
  max-max, 70
  max-min, 69
  mean, 121
  mixed, 68, 276
  Nash solution, 70
  no-regret, 70
  pure, 69, 276
  scalar, 92, 106
  variable, 73
  vector, 93, 96, 99, 103
strategy dynamics, 21, 112, 114
strategy species concept, 121, 236, 242
struggle for existence, 62
sustainable yield, 355
symmetric competition game, 89
symmetric matrix game, 66, 278
sympatric speciation, 251, 252
  gene flow example, 253

theorems
  ESS – frequency, 219
  ESS – multiple, 213
  ESS – multistage, 222
  ESS – resource, 211
  ESS – scalar, 198
  ESS – vector, 205
  game against relatives, 303
  matrix-ESS, 282
  matrix-ESS maximum principle, 281
time scale, 126
  ecological, 119
  evolutionary, 119
tragedy of the commons, 12, 96, 329

  ecological, 350
  evolutionary, 350
transpose, 48, 108

variance, 125
variance dynamics
  in the L–V competition game, 244
vector, 28
  partitioning of, 94
virtual strategy, 78

war of attrition, 71

zero-sum game, 63

Printed in the United States
By Bookmasters